UX for Enterprise ChatGPT Solutions

A practical guide to designing enterprise-grade LLMs

Richard H. Miller, Ph.D.

‹packt›

UX for Enterprise ChatGPT Solutions

Group Product Manager: Niranjan Naikwadi

Publishing Product Manager: Tejashwini R

Book Project Manager: Aparna Ravikumar Nair

Content Development Editor: Manikandan Kurup

Technical Editor: Kavyashree K S

Copy Editor: Safis Editing

Proofreader: Manikandan Kurup

Indexer: Rekha Nair

Production Designer: Gokul Raj

Senior DevRel Marketing Executive: Vinishka Kalra

First published: September 2024

Production reference: 2270924

Published by Packt Publishing Ltd.
Grosvenor House
11 St Paul's Square
Birmingham
B3 1RB, UK

ISBN 978-1-83546-119-8

www.packtpub.com

It is no surprise to anyone who knows me that I dedicate this book to my loving wife, Jill, and my two amazing kids, Madison and Max. They make me proud every day, even though they don't know what I do for a living.

For those who know about UX, UXD, UI, experience design, or whatever flavor of the month we call our field, I dedicate this book to all the design professionals who go to great lengths to understand their customers and apply exceptional design practices to make their customers successful.

- Richard H. Miller, Ph.D.

Foreword

Anyone who has tried ChatGPT, Google's BARD, or any other **large language model** (LLM) knows that getting useful answers from them requires knowing how to feed the LLM the relevant inputs and formulate the right queries (called "*prompts*" in the LLM world). The desire to create LLM-based applications that are actually useful and not just entertaining has given rise to a new field of expertise: conversational design.

Riding the AI wave, AI/LLM experts are churning out books and courses on how to incorporate AI and LLMs into software applications, e.g., chat systems, smart speakers, and business software. However, just knowing how to focus an LLM on a specific domain and how to compose instructions for it is insufficient to create applications that people can easily learn and use productively. That requires a separate type of expertise: how to design applications to meet user and task requirements, often known as **User Experience Design (UXD)** or **User/Task Centered Design (UCD)**.

Of course, there are many training courses and books on UXD and UCD. As the author of some, I can say that perhaps too many. However, UXD and UCD books don't teach how to incorporate LLMs into focused applications.

Richard Miller has extensive experience in both LLMs and UXD/UCD, so his book is unique: it blends those two seemingly disparate disciplines, teaching *both*—how to create and incorporate into enterprise applications specialized versions of ChatGPT that focus on domains relevant to the application, *and* how to ensure that those applications are easy to learn and use, meet the requirements their intended users, and provide value for the enterprises that deploy them.

The book is structured as a tutorial on building ChatGPT-based enterprise applications, interspersed with lessons on the methods used in UXD and UCD. It starts by summarizing the histories of AI/LLMs and UXD/UCD and explains the benefits of each, but then jumps into a tutorial on creating a custom instance of ChatGPT with proprietary data. Subsequent chapters teach how to perform user research and task analysis, prioritize features and improvements, choose the most suitable type of application, how the book's recommendations fit into Agile development processes, and more.

One of the book's most useful features is that it is designed mainly as an e-book with live links to sources, examples, and other external resources. For the benefit of readers of the book's printed version, Richard created a webpage with all the links in the e-book version.

Conclusion: This book is the first of its kind and a significant and welcomed addition to the growing body of books on maximizing the value of LLMs.

- Jeff Johnson, Ph.D.

Former Professor, Department of Computer Science,

University of San Francisco

Acknowledgment

From the day I started this book-writing journey, it has been a wild ride. I appreciate the efforts of the Packt publishing team, who reached out to inquire about me writing this book. It started as a book on how to use ChatGPT to help be a good designer, but the more valuable contribution to our field is how to make ChatGPT do what we want in the enterprise. Thank you to Aparna, Tejashwini, Vandita, Shambhavi, and Manikandan for making this process easy.

The book has a lot of UX specifics, and I certainly don't want to understate the value of a good technical review team. Dan Miller and Martin Yanev brought thoughtful insight to the UX chapters, which were mainly new to them while helping me refine the more technical portions of the book. Kevin Mullet's book, which he authored with Darrell Sano, *Designing Visual Interfaces*, was so thoughtful and insightful that I never thought I could write a book myself. However, his great efforts in his technical review of this book dramatically improved how I wrote and thought about this book from the reader's perspective. Also, I thank Jeff Johnson, a pillar of the UX community, for his extraordinary effort to include his thoughts in the forward for this book. His wisdom has already been so insightful.

The feedback from the Wove.com team, especially Jay Edlin and David Xu, hugely improved our in-depth case study. In my need to reach out and get approvals for images and references for the book, I have to thank a slew of authors and AI experts for allowing me to share some of their work in the book. Thank you to Dan Miller from Opus Research for allowing me to quote his Conversational AI Survey, Chris Spalton for his amazing UX cartoon storyboards, Christian Roher from `xdstrategy.com` for his landscape of user research methods, Mathew Leverone from ScaledAgile for the various Agile material, Jakob Nielsen for being so open with his usability heuristics, Keyvan Mohajer and Fiona McEvoy for the SoundHound image, Ryan Patrick from Occamonics, Haofen Wang for his images from their paper on RAG, Kevin Dewalt from Prolingo for the image from his Lessons Learned video, Chen Qian for the ChatDev image, Jindong Wang for their figure from his article, Jim Ekanem for his insights into accessibility, Mihael Cacic for his wonderful training class and use of his graphics and fine-tuning examples, and Joe Huang for the ODA demo screenshot.

In addition, I have learned so much from my peers: linguists, writers, engineers, developers, designers, researchers, and engineering leaders; there is probably no one idea here that wasn't touched by their expertise. Thank you to Toff van Alphen, Andrew Bulloch, Juliette Fleming, Jason Fox, Miranda Glasbergen, Jason Goecke, Philip Hayne, Joe Huang, Peggy Larson, Jacob Nielsen, David Price, Ken Rehor, Grant Ronald, Dalila Rosales, Aita Salasoo, Ben Schneiderman, David Stowell, Bruce Tognazzi, and Hardeep Walla. I have learned so much from y'all on my journey.

Icons for some images were provided by flaticon.com:

- Speaker icon by Eklip Studio (`https://www.flaticon.com/free-icon/audio-input_13430774?term=voice+input&page=1&position=6&origin=search&related_id=13430774`)
- Edit icons created by Kiranshastry (`https://www.flaticon.com/free-icons/edit`)
- Food icons created by Freepik (`https://www.flaticon.com/free-icons/food`)

Contributors

About the author

Richard H. Miller, Ph.D., is a dynamic leader in user experience and conversational AI. With over 20 years of experience in UX design strategy and 7 years in conversational AI, he has founded and managed four global teams, delivering user-centered design solutions to Fortune 500 organizations. At Oracle Corp., he led a team that developed the Oracle support portal, generating over $15B of in-service support revenue.

Richard was at the forefront of Oracle's conversational AI deployments on Slack, Teams, and the web. He developed the Expense Assistant AI and designed Oracle's first conversational AI platform. After multiple start-ups, some successful and some not so much, he has grown his design expertise across many disciplines, platforms, toolkits, and technologies. Dr. Miller, as he is known in academic circles or when teaching, specializes in global team building, innovative UX design, Agile design, and growing the expertise of the next generations of UI leaders. Richard still gets to apply what he learned from his Ph.D. in UX design and his MBA. He is committed to excellence and innovation in design and conversational AI.

About the reviewers

Kevin Mullet is a software designer and UX innovator whose user-centered experience designs span a wide range of product types. From GUI platforms (OPEN LOOK) to design systems (the Macromedia User Interface, Oracle's Redwood User Experience), to multimedia authoring tools (Macromedia Director, Extreme 3D), from enterprise applications (Icarian Workforce, Edgenuity, My Oracle Support) to consumer apps and applications (eBay, Kijiji, Parker, Show Evidence, and most recently, Node), there aren't many experience design problems he hasn't run up against over three decades of practice. His latest work on AI-powered conversational design applies his unique perspective and "best of both worlds" approach to supercharging the traditional chat experience.

Martin Yanev is a highly accomplished software engineer with expertise in aerospace and medical technology. With over eight years of experience, Martin excels in developing and integrating software solutions for critical domains such as air traffic control and chromatography systems. As a computer science professor at Fitchburg State University, he has empowered over 280,000 students worldwide. Martin's proficiency in frameworks such as Flask, Django, Pytest, and TensorFlow, combined with his mastery of OpenAI APIs, highlights his instructional prowess. He holds dual master's degrees in aerospace systems and software engineering, driving innovation and advancements in software engineering.

Dan Miller is the founder of Opus Research, where he defines conversational commerce by authoring reports regarding automated speech, natural language processing, conversational AI, analytics, and customer experience.

As the Director of the New Electronic Media Program at LINK Resources (IDC) from 1980-1983, he helped define one of the first continuous advisory services in the information industry. He held management positions at Atari, Warner Communications, and Pacific Telesis Group. He also published Telemedia News & Views, a monthly newsletter regarding developments in voice processing and intelligent telephony.

Dan received a BA from Hampshire College and an MBA from Columbia University.

Table of Contents

Part 1: UX Foundation for Enterprise ChatGPT

1

2

3

Identifying Optimal Use Cases for ChatGPT   67

4

Scoring Stories   91

5

Defining the Desired Experience   121

Part 2: Designing

6

Gathering Data – Content is King   177

7

Prompt Engineering 227

8

Fine-Tuning 261

Part 3: Care and Feeding

9

Guidelines and Heuristics   297

10

Monitoring and Evaluation   347

Preface

This book combines **User Experience (UX)** expertise with ChatGPT and related **Large Language Models (LLMs)** to create enterprise-grade applications that can solve real business problems. This is done in a way that almost all of the learnings of the books will continue to apply to the latest LLMs as they evolve and improve. We focus on the integration of LLMs with business solutions. This includes creating customer chatbots for customer service, creating recommender solutions to offer suggestions for sales and service, making purchase choices, solving other business problems in any vertical, or helping create more effective behind-the-scenes solutions that contain little or no UX. We take the science and art of UI design and research methods, techniques, and recommendations to make LLM solutions functional, usable, necessary, and engaging. Tips and expert secrets on applying UX to every stage of the design of LLM solutions at scale are shared. Almost none of this material is on the Internet or shared at vendor sites, so it is a unique resource for the design and design adjacent community.

Who this book is for

This book would appeal to individuals interested in enhancing their knowledge and skills in UI/UX design and looking for a comprehensive guide incorporating the latest technologies to apply UX principles to create enterprise-grade ChatGPT-powered solutions. It is suitable for seasoned designers looking to expand their knowledge, as well as writers, linguists, product managers, and design-savvy engineers who need to know UI/UX design fundamentals as they apply to ChatGPT.

The book follows a design-centered approach to producing ChatGPT-based solutions to solve business or "enterprise" problems. It helps decide and prioritize customer use cases for generative AI, accelerates the value from an LLM, extends the platform to serve customer needs, and explains monitoring and improving the quality of that service. Learning these skills will give you conversational AI design superpowers. An enterprise or business-class ChatGPT-powered solution should focus on providing customers with a unique skill, something more intelligent and focused than they can get from generic generative AI. To imagine and create world-class LLM-powered solutions, this book is for you.

What this book covers

Chapter 1, Recognizing the Power of Design in ChatGPT, begins with a brief introduction to the relationship between design and LLMs, including the art and science of UX and the history of LLMs. The various design frameworks for deploying an LLM are discussed, including a chat UI, a hybrid UI that includes chat with graphical user interfaces, recommender UIs that are not interactive, and designs intended to work behind the scenes with backend solutions. There is a small hands-on lab for building a simple model using a no-code playground.

Chapter 2, Conducting Effective User Research, provides many tips and tricks for using some of the most critical user research tools to evaluate adding ChatGPT and LLMs to enterprise solutions. We cover methods such as surveys, needs analysis, interviews, and digging into data to create a conversational analysis.

Chapter 3, Identifying Optimal Use Cases for ChatGPT, teaches how to identify the breadth of solutions to which an LLM can add value and explains when an LLM is not suited for a use case. We briefly cover classic use case design and then spend time aligning an LLM's capabilities with user goals. We ensure you know ChatGPT's limitations and biases and how to handle inappropriate responses.

Chapter 4, Scoring Stories, helps you become an expert in prioritizing user stories. This chapter is also valuable after a product or service goes to customers. You learn to prioritize updates, patches, and bug fixes so the customer gets the most value from the team's efforts. You will be able to balance customer priorities with the cost of development and make rational decisions to help plan and deliver the most value for the least cost. It explains in simple terms how to apply some special Agile tools to prioritize all this work. No road is without some bumps, but we share some complexities so you can navigate this successfully with the entire team.

Chapter 5, Defining the Desired Experience, is the final chapter before we get serious about the inner workings of ChatGPT. You will uncover specific considerations, design issues, and solutions for the full range of contexts of use. These include chat experiences, hybrid UIs (a graphical user interface merged with chat intelligence), recommendation UIs, and backend solutions (those without a customer-facing UI). We will address overarching considerations for these desired experiences, ensuring you know how to handle accessibility and internationalization while creating trust and handling security in any of these solutions.

Chapter 6, Gathering Data – Content Is King, dives into the complex nature of enterprise data, which is fundamental to creating a ChatGPT solution based on customers' needs. Explore how data sources such as knowledge bases, databases, spreadsheets, and other systems provide a source of truth. This helps connect customers to actions and explains how product people like yourself can contribute at this stage. Hands-on activities and a case study on annotating and cleaning data help explain the key points. We will cover retrieval augmented generation to help bridge the gap between an enterprise's vast data sources and the LLM.

Chapter 7, Prompt Engineering, coaches you on creating instructions that control, adapt, and personify the communications from the LLM to the customer. You will learn the difference between prompts anyone can give to an LLM and the more refined nature of instructional prompts for enterprise solutions.

Chapter 8, Fine-Tuning, explains what happens within the fine-tuning process, provides a tutorial on how to start fine-tuning, and continues our in-depth case study. You will be shown different methods to apply when training models. This includes a hands-on exercise to fine-tune a very sarcastic chatbot.

Chapter 9, Guidelines and Heuristics, steps past the technical nature of ChatGPT design to examine how to interpret ChatGPT style, tone, and voice. Essential guidelines and heuristics adapted and applied to evaluating ChatGPT solutions are reviewed so you can learn how to use design thinking to create clarity in the output from your LLM solutions. Dozens of examples are provided, along with a case study and example prompts that tie together the suite of heuristics covered in the chapter.

Chapter 10, Monitoring and Evaluation, focuses on knowing if the solution is doing well. It covers evaluating successes and failures, defining quality, and judging whether the UX improves. Our approach is one of care and feeding, following the life cycle of learning from the product's users and feeding back any learnings to have it grow and mature. Statistical measures of model performance, user quality metrics, and heuristic evaluation methods are covered, with tips on improving quality.

Chapter 11, Process, focuses on adapting traditional Agile and modern development methods to more interactive and customer-driven needs to improve ChatGPT solutions rapidly. We cover practical strategies to integrate a care and feeding approach into traditional Agile or Agile-like development while explaining why you should advocate for a continuous improvement life cycle.

Chapter 12, Conclusion, is the final chapter and provides additional suggestions and coaching to wrap up the entire life cycle covered in this book to set you up for success.

To get the most out of this book

We make no assumptions about any existing use of ChatGPT to build business solutions. We expect everyone to use some form of LLM for personal use. We would like some basic familiarity with UI terms and techniques, even if you are not an expert or a UI professional. We talk about creating surveys, carrying out customer interviews, and giving lots of tips and tricks, but assume a basic understanding of these techniques. References are provided to get you up to speed in places where a knowledge roadblock might appear.

The entire book can be followed without coding; we rely on ChatGPT's free playground experience. We dabble in a few other free resources and LLMs. You will need an account to access our GitHub files and ChatGPT. If you code or are more technical, explore some of the more advanced topics and links provided. Although the book and tutorials are focused on ChatGPT, the learnings in the book can apply to any LLM.

For those in the digital version of the book, you can cut and paste examples directly into ChatGPT. However, no programming or code examples are needed, so missing a comma, for instance, will not impact your ability to learn and follow along. We provide files with sample data; you can use those without issue and test and experiment with the latest LLMs.

We have a solution for readers of the physical book who want quick access to the references and resources or those who notice an out-of-date link because web pages and companies come and go. We are maintaining a single-page online reference guide for all links, articles, demos, videos, and books mentioned in the book. Each chapter has a QR code that links to online references listed from every chapter. This means the links will be able to be updated on the reference website as this emerging field grows.

Online references and links: `Book References` (`https://uxdforai.com/references`)

Download the example and files

You can download the example code files for this book from GitHub at https://github.com/PacktPublishing/UX-for-Enterprise-ChatGPT-Solutions. If the files are updated, they will be updated in the GitHub repository.

This chapter's links, book recommendations, and GitHub files are posted on the reference page. Book References (https://uxdforai.com/references)

We also have other code bundles from our rich catalog of books and videos available at https://github.com/PacktPublishing/. Check them out!

Conventions used

Several text conventions are used throughout this book.

Code in text: Indicates code words in text, database table names, folder names, filenames, file extensions, pathnames, dummy URLs, user input, and Twitter handles. Here is an example: "We can compare the answer provided by the base model to the same question answered after adding the Full Thesis.pdf file, a draft of almost 200 pages. "

A block of code or examples to type into ChatGPT is set as follows:

```
You are a helpful assistant named Alli, short for the name of our
bank. Be courteous and professional. Prioritize information in any
files first. Format output using lists when appropriate.
```

When we wish to draw your attention to a particular part of an example, the relevant lines or items are set in bold:

```
(Mac operating systems only)</li></ul><p><strong>Note:</strong> Our
latest site features will not work with older unsupported browsers. </
p><p>
```

We sometimes include conversations between a user and a chat solution. You can read along by following the standard chat convention: messages sent to the chat are right-justified, and the responses are left-justified.

```
                                        Is solar power a renewable resource?
Solar power is a renewable resource.
Because solar power is an infinite
resource, it has unlimited potential.
```

Bold: Indicates a new term, an important word, or words you see onscreen. For instance, words in menus or dialog boxes appear in **bold**. An example is "Alistair Cockburn's **Writing Effective Use Cases** is the definitive guide when I teach use case design"

> Tips, secondary resources, or important notes
> Appears like this.

Get in touch

Feedback from our readers is always welcome.

General feedback: For questions about any aspect of this book, email us at `customercare@packtpub.com` and mention the book title in the subject of the message.

Errata: Although we have taken every care to ensure the accuracy of our content, mistakes do happen. If you have found an error in this book, we would be grateful if you would report this to us. Please visit `www.packtpub.com/support/errata` and fill out the form.

Piracy: If you come across any illegal copies of our works in any form on the internet, we would be grateful if you would provide us with the location address or website name. Please get in touch with us at `copyright@packt.com` with a link to the material.

If you are interested in becoming an author: If there is a topic that you have expertise in and you are interested in either writing or contributing to a book, please visit `authors.packtpub.com`.

Share Your Thoughts

Once you've read *UX for Enterprise ChatGPT Solutions*, we'd love to hear your thoughts! Scan the QR code below to go straight to the Amazon review page for this book and share your feedback.

`https://packt.link/r/1-835-46119-0`

Your review is important to us and the tech community and will help us make sure we're delivering excellent quality content.

Download a free PDF copy of this book

Thanks for purchasing this book!

Do you like to read on the go but are unable to carry your print books everywhere?

Is your eBook purchase not compatible with the device of your choice?

Don't worry, now with every Packt book you get a DRM-free PDF version of that book at no cost.

Read anywhere, any place, on any device. Search, copy, and paste code from your favorite technical books directly into your application.

The perks don't stop there, you can get exclusive access to discounts, newsletters, and great free content in your inbox daily

Follow these simple steps to get the benefits:

1. Scan the QR code or visit the link below

https://packt.link/free-ebook/978-1-83546-119-8

2. Submit your proof of purchase
3. That's it! We'll send your free PDF and other benefits to your email directly

Part 1: UX Foundation for Enterprise ChatGPT

Every good story has a beginning, a middle, and an end. In this part, we'll start our journey by exploring how traditional user experience methods and best practices can be applied to creating world-class solutions powered by ChatGPT. We will then explore essential user research methods and provide tips and secrets of the trade that work when creating conversational designs. This will lead us to explore how to define and pick the use cases that **large language models** (**LLMs**) are best suited to solve. A user experience approach is taught to prioritize use cases. When combined with a method to include development costs into the mix, this powerful method from Agile allows for prioritizing the most valuable solutions to build. Although we'll consider how these use cases play out with ChatGPT, almost all the learnings can apply to any LLM model. We'll look at LLM-powered applications, chat-only experiences, and robust chat-powered graphical user interfaces, and we'll even explain how to work with ChatGPT when there is no UI.

This part includes the following chapters:

- *Chapter 1, Recognizing the Power of Design in ChatGPT*
- *Chapter 2, Conducting Effective User Research*
- *Chapter 3, Identifying Optimal Use Cases for ChatGPT*
- *Chapter 4, Scoring Stories*
- *Chapter 5, Defining the Desired Experience*

1

Recognizing the Power of Design in ChatGPT

If you only like to play with ChatGPT, this book is not for you. Read on to make a quality user experience for customers by incorporating ChatGPT or various alternative language models that span the gamut of cost, quality, and expertise. Every new technology has too many people jumping on the bandwagon only to abandon it with failure. It is widespread. Why? Because they don't know what they don't know. But we know what it takes to make successful ChatGPT solutions for the enterprise. We will take you from zero to hero in your organization. Do you want poor-quality interactions or targeted, intelligent results that resonate with customers? Should they feel empowered and able to explore further with the confidence that they are understood? If it is the latter, this book will focus on user experience design methods, practices, and tools to help decide what to do, design the most effective solutions, and verify that they do what they should do. We can apply user interface practices to the ChatGPT life cycle, leaving you confident to create quality solutions.

This book is intended for designers or design-related professionals, such as product managers, product owners, writers, linguists, or developers, who want to understand how to apply design principles and practices to improve the generative AI experience of customers and employees. For those exposed to design methodologies, they won't be novel, but their application will be. For those with limited exposure to the science of **user experience** (**UX**) design (or UXD), we will provide enough learning to make you dangerous and help create enterprise-grade ChatGPT solutions.

You might have yet to work with generative AI products such as ChatGPT in a production way, maybe only using some of these tools at home or to supplement work. We will only get into some of the explanations of how ChatGPT works in *Chapter 6, Gathering Data – Content is King* as we have some ground to cover first. The book follows a typical design. First, we help figure out what to do, prioritize that work, then how to do it, and finally, how to interpret and improve on what was done. We have found the design skills and tips in this book to work for a wide range of design challenges, and through our experience in the last seven years with AI solutions and 30 years of UXD, we have adapted those insights to the creation of generative AI solutions. Following design processes will help one craft high-quality solutions driven by ChatGPT.

In this chapter, we're going to cover the following main topics:

- Traversing the history of conversational AI

- Appreciating the importance of UX design

- Understanding the science and art of UX design

- Setting up a customized model

Technical requirements

There are two ways to work with this book: follow along and learn the principles and practices and use one of the OpenAI playgrounds, which is typically a *no-code* approach, or use the APIs provided by ChatGPT.

We have examples in our book's GitHub repository. If this is your first time using GitHub, it is a place where we will store any materials needed to download to complete the examples in the book. It is an online folder of resources.

GitHub: `Repository for book materials (https://github.com/PacktPublishing/ UX-for-Enterprise-ChatGPT-Solutions/)`

GitHub is the repository for all files from the book. Click the download button, highlighted in *Figure 1.1*, to download the file to the desktop. There is no viewer for most of the files on the GitHub repository.

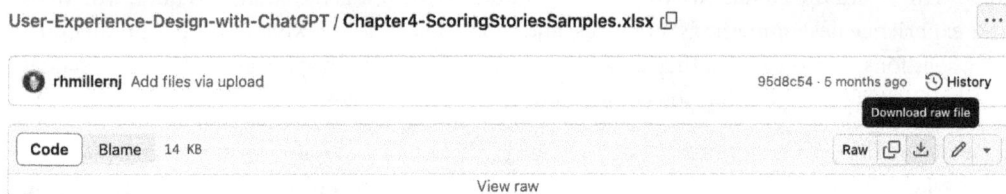

User-Experience-Design-with-ChatGPT / **Chapter4-ScoringStoriesSamples.xlsx**   ...

rhmillernj Add files via upload	95d8c54 · 5 months ago History

Download raw file

Code Blame 14 KB	Raw

View raw

Figure 1.1 – How to download a file from GitHub

Make sure you have a ChatGPT account.

Website: `OpenAI Chat (https://chat.openai.com/)`

That is easy; everyone should have that. We will try out some of the material as we go. This will also allow us to use the Playground, essential for some demos. We have a QR code at the end of each chapter, so all of the references we provide, such as the preceding links, are available online for easier access.

Approach 1 – The no-code approach

You can learn about 80% of this material by just reading, but some folks do better by doing. If you go this way, focus on the design practices and methods and learn how to apply those to any generative AI solution. We will give demos and samples to try without coding. Give the examples a try; understanding how an LLM reacts is critical.

Approach 2 – code with Node.JS, Python, or curl

If you don't already have a ChatGPT account, set one up. Then, head to the quick start guide to ensure Node.js (curl or Python) works. The URL has step-by-step instructions for getting your environment up and running.

Website: `Quickstart guide for developers (https://platform.openai.com/docs/quickstart?context=node)`

> **Note**
>
> This book does not require coding. A more technical reader can mirror some of our no-code approaches with a code version, but we won't discuss this path.

To use the APIs, follow the instructions on the link:

1. Install the essential software (Node.js, Python, or curl are all documented on the same page; choose the tab that suits you).
2. Install the OpenAI library package.
3. Set up your API key.

And give it a try! If you have used other **Large Language Models** (**LLMs**) or have never used one, try it out and ask anything; we call this input from the user a prompt. *The material in this book can be quickly learned without doing any coding.* Some models don't have the most recent data, so asking about today's weather or sports scores won't work, but if asked to give five ways to clean a clogged toilet, it has the answer. The power we want to expose here is the combination of this powerful experience and UX design practices to create a high-quality, customer-centric experience. Now we have something to discuss!

We should be on the same page concerning the basic history of conversational AI. With all the news, ChatGPT should be well known so we can cover just the basics for a few minutes.

Traversing the history of conversational AI

Interaction design intersected with AI well before the LLM revolution. This history is helpful to appreciate when applying design principles to the latest conversational experiences. Any discussion of AI at least mentions Alan Turing and the question posed by his article in Mind (a peer-reviewed academic journal).

Article: `Computing Machinery and Intelligence (1950) (https://redirect.cs.umbc.edu/courses/471/papers/turing.pdf)`

This is routinely referenced as the **Turing Test**. The ability of a machine to seem intelligent and be indistinguishable from a human.

Article: `Wikipedia on the Turing Test (https://en.wikipedia.org/wiki/Turing_test)`

When this was published in 1950, we were still far from a computer being indistinguishable from a human, at least in a text-only interaction. We must skip ahead to the mid-1960s before we see something that appears to engage in discourse.

If we try out one of ELIZA's conversational interfaces from 1964 to 1967, we quickly see its limitations based on its *natural* responses when recognizing keywords or phrases.

Article: `Wikipedia on ELIZA (https://en.wikipedia.org/wiki/ELIZA)`

The well-known version is called DOCTOR. It turns written questions asked of it back onto the patient. Give it a try to interact with the psychotherapist chatbot.

Demo: `ELIZA` – The psychotherapist chatbot `(https://web.njit.edu/~ronkowit/eliza.html)`

ELIZA was considered one of the first attempts at passing the Turing Test. With its simple psychotherapist banter to simulate a doctor ("Why do you feel this way?"), it was perceived as human-like. Without going too deep, the discussion around its design looked at the rank of essential words, and it included *transformation* rules that dictated how it treats what the user types. Maybe LLMs are paying homage to this since they are based on **transformers**. We will explain transformers and the terms common to LLMs in later chapters. ELIZA had the superficial appearance of a conversation and could not go *off-topic* or even provide an answer. The psychology of conversational interaction was fundamental to this experience. But it wasn't going to solve anyone's psychological problems. However, things did get better as chatbots; it just took a few decades. Visit Wikipedia to learn a brief history of chatbots.

Article: `Wikipedia's history of chatbots (https://en.wikipedia.org/wiki/Chatbot)`

The idea of a *natural language* experience was not lost on the research community from the 1960s to the 2000s. Still, the next step in evolution came with the conversational assistants or chatbots we have seen since around 2016. And this is where interaction design had a significant impact, even though most chatbots were not worth anyone's time. About 100,000 chatbots were created on Facebook Messenger in the first year of its support. I would suggest that 99% of them failed quickly. Very few survived for all the reasons we will explain shortly. But a few lived on when teams were willing to mature the solution. Support use cases, such as for airlines ("How much is a second bag going to cost?" or "Can

I get a refund for a canceled ticket?"), are a great place to answer specific questions with specific answers. Although it seems evident that this can save a company a lot of money in support costs versus a phone call, there is also value to the consumer. For them, time is also valuable. If a customer gets a reliable answer in seconds, they will gladly trade that for 10 minutes of holding onto the phone. It is a win-win. Additionally, this experience can be a frontend for required interactions with humans. In support cases, it can gather details reliably before engaging a human, making it more likely to connect with the right human and give them the details they need to help more directly.

Imagine a young child before going to school. If no one interacts with kids, teaches them, or plays with them, by the time they go to first grade, they lack primary language and interpersonal skills and might not be potty trained. Even the great comedian Steve Martin understood this. Please take a minute and laugh at his bit.

Video: `Steve Martin teaching a kid for the first day of school` (https://www.youtube.com/watch?v=4OK6rApRnhQ)

However, remarkable changes can be made by investing in a child's growth and care and feeding them physically and mentally. This maturity is what we can see in chatbots. They won't typically become a Ph.D., but they can coached to be smarter than a 5th grader. We can use design skills to make a chatbot (or any LLM-based solution) knowledgeable, dependable, and articulate. We will apply what we have learned to our next generation of conversational assistants built with ChatGPT. We will critically explore ChatGPT to form robust solutions, and you will learn to notice when there might be other tools out there to use in conjunction with ChatGPT.

There is one other related area worth mentioning. Everyone has experience with phone trees when calling a business. We mentioned this example earlier. Eventually, those "Press 1 for service, press 2 for sales…" gave way to experiences that listened for more than the touch tone of a key press. But how many of us have struggled with these? Probably all of us. Why? Because the experience isn't designed well, and the technology is likely lacking. These, too, will benefit from ChatGPT. So, if you come from creating voice experiences (probably using Voice XML, the de facto standard for modeling interactions for years) or from chatbots within Alexa, Siri, Google, or dozens of other vendors, the learnings and practices of making great experiences apply to ChatGPT. We will go through that extensively in this book.

Wikipedia: `Wikipedia's Voice XML background` (https://en.wikipedia.org/wiki/VoiceXML)

And yet so many of these chatbots, phone trees, or conversational experiences fail to help a primary user accomplish their task. Why? Because of a few critical reasons:

- The features or services in the chatbots don't match the user's needs
- The models don't support the complexity of the user's language
- The user's primary spoken language might not be supported, requiring them to be understood in a secondary language or not be understood at all

- The chatbots only know what they know, so they will return seemingly random results, which is discouraging

- The chatbots do not respond in a voice or tone the customer expects

- The chatbot should have been monitored and improved to address these issues

Your goal should be to set a higher quality bar than expected from a human performing the same task. Sounds crazy? It isn't. A typical support person might be able to help with a narrow topic (say, a website password reset). Another agent would be needed to resolve billing issues or find missing payments. So, the average support person will be less helpful than the future state of well-trained ChatGPT advisors with access to all the institutional data and processes.

This brings us to the founding of companies such as OpenAI. This long history of machine learning models and the increased computational capabilities allow this very large language model to work. OpenAI didn't come into the world view with the December 2022 release of ChatGPT 3.5, the company was founded seven years earlier with non-profit roots. It took over three years to go from GPT-2, which could generate human-like text, to the 3.5 version that gained worldwide attention. For those who like tech history, dive into a brief background on OpenAI.

Article: `The origins of OpenAI (https://www.britannica.com/money/OpenAI)`

Like many Silicon Valley companies, engineers from Google Brain (and DeepMind, which merged with Google), Facebook, and other AI came together at OpenAI. Then, 11 OpenAI employees left to form Anthropic around the start of 2021. None of this happened overnight, so we need to remind ourselves that it will take years for this new technology to weave its way into our everyday lives. The phone, the car, the computer, and the mobile phone have all become fundamental to today's modern society. This will impact all of us more than all these previous inventions, but it will take time to happen. There will be a lot of failures along the way.

Imagine a solution that is 60% accurate at every interaction. Does it sound high for a computer to get something right 60% of the time? Before ChatGPT, some didn't consider this too bad. I routinely used to ask this in my classes (typically from 20 to 100 people per class) on conversational AI. And many folks consider a 50–80% success rate to be "pretty good."

With some simple assumptions, we can understand why these systems fail. Every time a question is asked, the likelihood of a failure increases, as modeled in *Figure 1.2*. To keep this simple, we base this on the independent probability of each turn having the same chance of failure. The system doesn't know it has failed, and if the user trusts it, they might not even notice the failure, thus causing more failures.

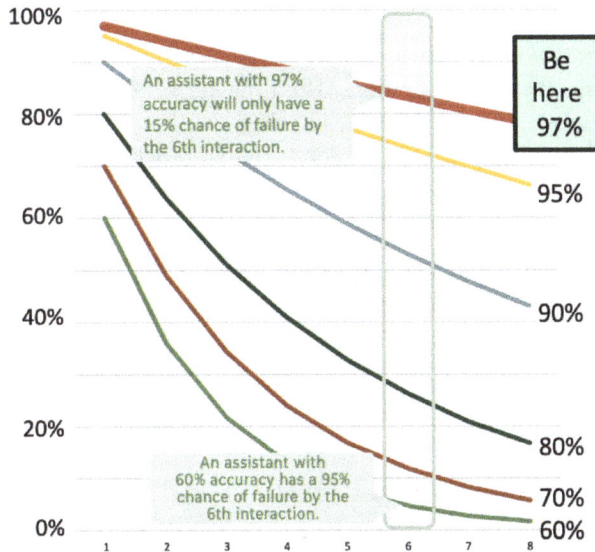

Figure 1.2 – The chance of failure increases at each turn

When asked six questions at a 60% likelihood of success rate, there is a 95% chance of one of those answers being wrong. And what if your next question is dependent on the previous answer? The interaction will go off the rails. I have seen this time and time again. If the user trusts the (wrong) answer, they make the next decision based on that (incorrect) answer. And failure can be assured. This relationship will sour. Customers will go elsewhere if they see these failures (likely a more expensive channel) to get their needs addressed, or worse, they will go to another vendor.

We can consider strategies to improve these curves so that each turn is more likely to succeed. *Chapters 6, 7*, and *8* explain strategies to use multiple generative AI components to do different forms of validation. Yes, the AI can watch over another AI. While traditional LLMs such as ChatGPT have improved and will continue to improve, we want to provide tools and measurement skills to help ensure success. But look at *Figure 1.2* again. Look at raising the bar to 97% accuracy. After the same number of turns, there is a very good chance (85%) that all interactions were successful. So, raise the bar on expectations.

It is possible to achieve these levels of quality. We will also show how to measure and scope improvements to give the most significant return on investment.

> **Chatbot failures**
>
> To get a laugh at how bad *bad* can be, read this article on chatbot fails. We aim to teach enough design methods to never fall into these black holes of disgrace.
>
> Article: `Chatbot failures (https://research.aimultiple.com/chatbot-fail/)`

With purpose-built experiences, for example, focused on filing business expenses or getting answers to common questions around internal business processes, users spend less energy trying to break the chat experience or ask off-the-wall questions. This behavior, expected in widely available public ChatGPT and chatbot experiences, is likely seen less than 1% of the time when building a custom ChatGPT tool. It will still get questions it might not be able to answer, but it is more likely that they are questions it should *eventually* answer. We will show how to prioritize that backlog to be in the business of continuous improvement.

This brings us to ChatGPT and the new class of LLMs, which are indistinguishable from humans in many ways. Google's LaMDA, Meta's Llama, Anthropic's Claude, and OpenAI's GPT models are all in the same class of software.

- Article: `Wikipedia survey of LLMs (https://en.wikipedia.org/wiki/Large_language_model)`
- Article: `Google's LaMDA (https://en.wikipedia.org/wiki/LaMDA)`
- Article: `Meta's Llama (https://llama.meta.com)`
- Article: `Anthropic's Claude (https://www.anthropic.com/claude)`
- Article: `OpenAI's GPT Models (https://platform.openai.com/docs/models)`

But even if they are like humans, we must ask which humans in the enterprise space they mimic. Does this represent my company? Does it have the knowledge it needs to solve my customers' problems? How will my customer handle a wrong answer? LLMs have a lot of potential and will evolve rapidly. We aim to give you the tools to evaluate whether an LLM solution will fit at every stage of your development.

The importance of UX design for ChatGPT

Does ChatGPT even need an introduction at this point? The innovative model developed by OpenAI is in a new class of LLMs, which are trained on billions of data elements from the internet's vast supply of articles, books, and knowledge. It achieved over 100 million users in about two months. It can generate human-like conversational interactions in text or voice in many languages and converse on vast information. And it does it pretty darn fast.

ChatGPT has undoubtedly come on like a firestorm. Unique, fun, fast, intelligent? However, when designing solutions for your business or enterprise, they should be accurate, have the most current business-related information, and communicate to customers in the voice, style, and tone expected from the business. So, how do you take such a fast-paced moving target and wrap it into a product that exceeds customer's expectations? Can one ensure that it doesn't give random answers that are off-brand? You can, but it takes design. It needs to be monitored. And it would be best if a process was in place to improve it. For that, this is the right place.

Let me define *design* because I see a lot of really horrible definitions. Design is the process and practice of clearly communicating an experience for a user. Good software UX design accounts for human behavior and limitations by applying the scientific method to solving human and machine interface issues. This means we can use what we know about the visual, auditory, and kinesthetic systems and combine them with understanding how the mind works to make decisions on how to create an experience that is functional, usable, needed, and even engaging. We see design all around us: visual design, graphic design, software design, conversational design, building architecture, and many other fields. We use the expertise of user researchers to guide our designs based on subjective and objective feedback from our customers, using formal and informal methods to better understand our users' needs. We then combine the inputs from customers, primary research, and the goals of the product and company and mix in a bit of magic to make great experiences.

If you just put icons on screens or write conversational copy without these efforts, you do production work, not design. We want everything done for a reason. The more done by creating fitness to purpose, the more our customer experience will improve. We don't always get it right. We will get a higher quality product if we know how to fulfill the user's needs. That is where the iterative design concept plays a role. We learn from and improve our designs even if we don't get it right.

UX design, interface design, human factors, user research, **human-computer interaction** (**HCI**), or any flavor of the art and science of interaction design is a collection of experts and expertise that can help shape this functional, engaging, usable, and fun experience. We can build successful chat-based solutions by directly applying the wealth of learnings from these disciplines to "chat" experiences or adapting what we have learned with conversational AI and graphical user experience design to fit into this new world. Using words to communicate with a computer is not new; it has just improved.

This is where you learn how to design a ChatGPT solution for customers based on company knowledge and business needs. Enterprise ChatGPT covers a wide range of experiences. One can be making a support site, a virtual assistant to help employees or job seekers, a sales engagement service that personalizes emails for sales calls, a training application, a product finder or recommender, a tool to analyze legal documents for inconsistencies, or an expert witness tool for lawyers. Code review (evaluating software written in Java or dozens of other popular languages and identifying issues or bugs) is another popular topic in tech. This book will stay away from that use case to focus on more common experiences that will impact most people, most of the time, with something important to their lives as enterprise customers. Developer productivity tools are essential, but that topic is well covered elsewhere. The learnings also apply to that space; we won't use any examples or case studies from developer productivity tools. We will start by discussing the science and art of good design in the next section.

Understanding the science and art of UX design

Every coin has two sides (okay, plus the edge!). Typically, we see two sides to UX design. Those with visual backgrounds and those who come from a science perspective. Schools are now overwhelmingly delivering visual and graphic artists to meet demand, and with conversational AI, there is some *art* to the experience but only sometimes visual elements. The introduction of generative AI impacts every facet of interaction design. The design roles will adapt or die. Adaption is the better option. As **graphical user interfaces (GUIs)** adapt to include conversational elements, the role of a visual designer will still be relevant, even if only to create the correct prompts to help them generate the look and feel that aligns with the organization's goals. The side of the equation for the science of design remains vital to requirements, understanding, and communication. Even when writing this book, ChatGPT provides some good answers related to UX. But what we cover in this book is not quickly answered by ChatGPT or any generative solutions. They help us, like all tools, move our design culture forward, but they don't know when they are wrong and still need us to decide where to apply the solutions, gather the correct data to help them form answers, and understand and improve the results.

The science of design

"Anyone can design," "Just put the button there," "I can write this copy." There is a difference between designing and making something. Anyone can make something. It might or might not work; it could work for some and not others, *"I designed this for myself, and I don't have any issues with it,"* or it could be good. We want to use the tools, expertise, wisdom, and field knowledge to ensure *design decisions* yield the highest quality product. There is a wealth of research that usually underpins quality interactions, and we want to avoid pitfalls.

When we mean research, we include controlled studies with human subjects where the team has undergone rigorous processes to return reliable, repeatable, and valid results. We then take these results and apply them to our situation. And some will say, *"That doesn't apply to this because it is different."* Well, it could be, but that is why we share these results with interaction designers to guide us to what is applicable. As ChatGPT grows and integrates with other products and features, it will become more intertwined with visual elements, forms, interactive charts and visualizations, and even typical GUIs (with buttons, tables, filters, tabs, and all the components we see in any mobile, desktop, web, or embedded experience). This will make the science and historical expertise of UX design even more critical.

Let's take an example—Hick's Law; designers know and use Hick's Law all the time. *"The time it takes to make a decision increases as the number of alternatives increases."* This is why we have menus broken up into small segments, wizards for complex processes, and debate how many buttons should appear in a dialog box. In conversational flows, we keep decisions simple to reduce the burden on the user.

Hick-Hyman Law

This law was published in 1952 in the *Quarterly Journal of Experimental Psychology*. It is an equation, $RT = a + b\,log_2(n)$, where the response time (RT) is a function of the time not included in the decision-making (a) plus a constant (about 0.155 seconds) times the *log* function of the number of alternatives to choose from (n).

```
Article: Wikipedia's explanation of Hick-Hyman Law (https://
en.wikipedia.org/wiki/Hick%27s_law)
```

We don't expect you to know or memorize this, but it is just one example of the science behind UX design decisions. Sometimes, knowing the guidelines, laws, and science helps you make better decisions and avoid mistakes by others, which you must learn to correct.

In this case, we know that a long list of choices is complex for users, and when a generative AI returns 10 to 15 choices, the effort it takes to decide goes up significantly. With this example, we can get these choices grouped into smaller logical segments and reduce them to two less complex decisions in a series. This is why we have the **File**, **Edit**, **View**, **Window**, and **Help** menus. By grouping menu items, picking the action is a less complex decision. It is also why menus fail when there are too many choices and no clear understanding of the organization of those items. Let's tell ChatGPT to return large decision trees as more logical and organized segments. We will cover this in *Chapter 7*, *Prompt Engineering*, to give ChatGPT instructions on how we want our responses framed.

How about one more classic example? Many phone calls to a business result in a voice prompt with 5-6-7-8 or even nine choices; how does a caller keep track of the right one? Do you ever have to listen to a prompt again? Maybe you got distracted and can't recall the first few choices. Do you ever use a few fingers to represent a number to remind yourself which answer might be best when multiple options are viable? This is a human working memory issue, a classic design problem.

Article: `Wikipedia explains working memory(https://en.wikipedia.org/ wiki/Working_memory)`

These human factors impact the design of many experiences—especially ones based on a lot of text. We are not going to calculate Hick's Law in this book or test on working memory. Still, one should appreciate that applying design principles will be the cornerstone of helping create a successful ChatGPT experience. Without guidance, no one should be prompted with 35 choices on the first menu. This is an unacceptable user experience. So, we could use an existing and well-organized tree and have ChatGPT (speech-to-text) determine the customer's request to skip a few levels in one step.

Design book wish list

If you are new to design and want to learn the fundamentals, there is a wealth of wonderful resources. I suggest a few non-technical first books, such as Don Norman's *The Design of Everyday Things* or Steve Krug's *Don't Make Me Think*.

Those who are familiar with these works will want more sophisticated books. I suggest Jeff Johnson's book *Designing with the Mind in Mind* to help you understand how the fundamentals of psychology are used to derive many of these guidelines and thus help you apply these principles. *Universal Principles of Design* by Lidwell, Holden, and Butler is an excellent reference book. More resources are on our book list.

Website: `Recommended Book List (https://uxdforai.com/references#C13)`

To make these experiences successful, think like a designer. Consider how users will interact, use their expectations, biases, and assumptions, and how their unique experiences will shape their future interactions. The power of the design mindset is to learn how to ensure people who use your product succeed.

To be clear, this is very different from *I will know it when I see it*. You want to apply art and style to the product and understand the scientific underpinnings of success. Some of this comes from psychology and related disciplines. Humans' cognitive and physical abilities have been unchanged over the last 50 years of the computer age. We still process information using the same senses. Our ability to react and respond is the same; only our experiences have evolved. We might now know how to type with two fingers, but there are limits to processing information, clicking on small targets, making decisions, or learning complex patterns. For example, just because today's cars have two to five times the horsepower of vehicles in the 1960s doesn't mean we can react any faster in an accident. As ChatGPT provides insights or understandings, we can overcome some cognitive or physical human limits using good UX practices.

The art of design

I bet you were waiting for us to talk about visual design here. This will apply in some places but won't be a book's cornerstone. We could have covered how to design a chat window or what visual treatment works best, but I suspect that information is in many resources independent of a book on ChatGPT. Brand identity has its subjective standards and its collection of science and expertise. Now, there will be experiences that join generative AI and traditional user interfaces, and one can apply a visual style and expertise to that experience, but that is unique to the enterprise. We want to cover the art of design that can span any enterprise here and join this with the science that supports design. And all of this is done to create something expected and comfortable for the audience.

In meetings, we hear, "*Put on your designer hat.*" This means being a customer's advocate and addressing customer's needs through how they interact with the product. This is within the basics of user-centered design. We bring customers into the process because we sometimes need to learn better. Listening to the customer means something different than what the customer says is right. It isn't, and it won't be. Many times, what the customers ask for and what they need are different. They want to work around a problem, but maybe the solution is to eliminate the problem in the first place. This is why this is also an art. You have to know where to look for problems worth solving.

Engaging users in the process will help us understand expectations and see how designs impact their behavior. And we know that generative AI will only sometimes give the expected results. Part of the designer's job is to improve this interaction, likely indirectly, through the tools we will discuss in the book, such as data cleaning, prompt engineering, fine-tuning, user research, user testing, and monitoring. There is also art we need to control and refine in generating responses with AI. As Kevin Mullet explained in a conversation for this book, "*The science describes how things ought to be done, while the generative AI describes (ideally) how they are typically done today. The designer's task (as always) is to wrangle those diverse inputs into a single coherent solution that best maps to the user's wants and needs.*" So, we need to use our depth of skills to keep our ChatGPT solutions on point.

To be fair to the process, only some decisions can be based on logic or a research method, especially regarding how you want ChatGPT to respond. Let's break down the art concept as it applies to the words used in a product's voice, style, and tone.

Let's take a large company selling surfing apparel, surfboards, and beach equipment. The website or style should be distinct from a stodgy, old-fashioned bank. Right, dude, and *dudettes*? Likewise, a bank intends to come across as reliable, safe, dependable, and trustworthy, Mrs. Customer. So, the site should speak to customers in a manner consistent with their expectations for the brand.

The overall *voice* of your conversations should reflect the brand.

The *style* of interactions should match the users' needs, wants, and expectations.

And the *tone* of the conversation can adapt or vary based on the current situation.

We will address these in detail to identify when and how to vary the tone within a style for the overall voice of a ChatGPT assistant.

There is no one correct answer to how to write. The voice, style, and tone can certainly overlap in concept. The same thing can be said in many ways and still be understood:

```
Your appointment is now scheduled for this coming Monday.
I have scheduled you to see the doctor on Monday.
Confirmed. I have a Monday appointment in the books.
```

We will discuss how to ensure that responses are clear, concise, and contain the right level of detail (maybe the customer wondered when the appointment was? Or maybe at what office?). Or perhaps they expected a calendar (`.ical` or similar) attachment to make it easier to add this appointment to a calendar:

```
Got it. Monday at 3:00 pm at the Palo Alto office. Here is the
calendar entry.
```

If a manager scheduled this appointment at the surf shop we mentioned, they would use lingo and tone that fits the customers to make the customer feel like one of them:

```
Akaw! We got you, grom. Swing by for your intro lesson on Mondo at 3
at the Santa Monica beach shop.
```

An intelligent designer will know that even if the audience doesn't know *Akaw* (awesome or cool) and *grom* (a new surfer), they can still understand the message and be a little excited to get their surf on.

Use cases for this book

This is where the art and science of design meet. And we will see this a lot in our journey. We will demand that our ChatGPT instance be clear, complete, and conversational (when needed). Using ChatGPT to create answers that present themselves in a non-conversational user interface is also perfectly reasonable. ChatGPT can process and generate data in a backend system. A UI can use alerts, buttons, or a warning dialog supplemented by ChatGPT. We call that a hybrid user experience. A combination of traditional UI elements and generative AI. Grammar or style suggestions when writing an email could be made in a suggestion window, and it doesn't have to be conversational. We will refer to these as recommenders. They provide recommendations to assist in content generation or to influence a process, such as a sales lead.

Let's illustrate these concepts with pictures. Each gives a sense of the experiences covered in the book's examples. If your use case differs but will still be a user interface, then most of what we cover will likely apply.

Messaging

Simple text messaging (SMS), voice experiences, or simple chats are two-way user experiences that can only use text or speech. These experiences have limited interactions in SMS: text, images, file uploading, and links. *Figure 1.3* shows an example of interacting with my expense assistant on the phone via SMS. It is easy to access and simple but has limits to the types of interactions it can support.

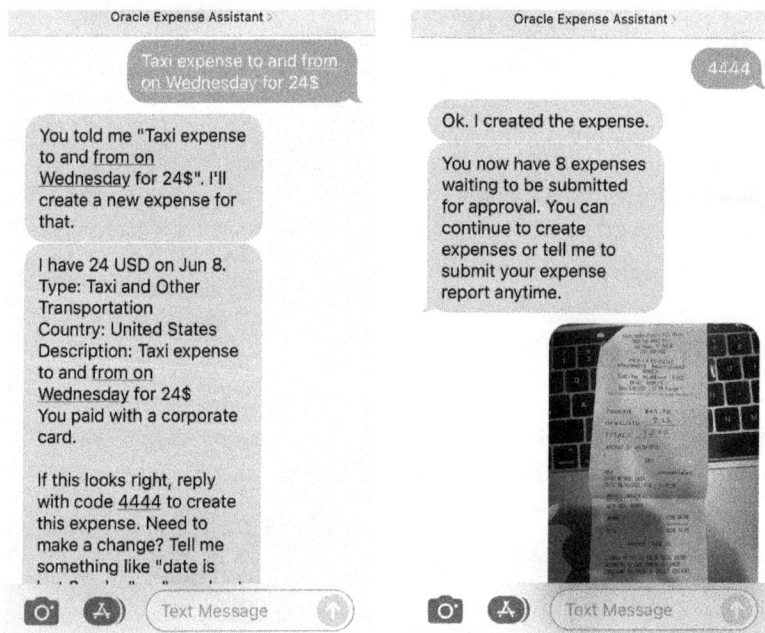

Figure 1.3 – An example of simple messaging via SMS on a mobile phone

If this was a voice experience, like in a car, or a skill in Amazon Alexa, creating well-done voice-only solutions is even more challenging than with SMS. We will discuss voice experiences in the book, but we should discuss Hybrid UIs next, as that is the future.

Hybrid UIs

Slack, Teams, and web interfaces can incorporate user interface elements with conversational text. Simple experiences can have links, buttons, charts, graphs, or forms. This hybrid experience combines LLMs and GUIs, allowing for complexities not quickly addressed by generative text or GUI components alone. In *Figure 1.4*, I show a simple example from Slack where buttons encourage the exploration of tasks.

Article: `ChatGPT AI for Slack (https://www.cityam.com/slack-to-offer-users-a-chatgpt-ai-tool-which-will-write-messages-for-them-in-seconds/)`

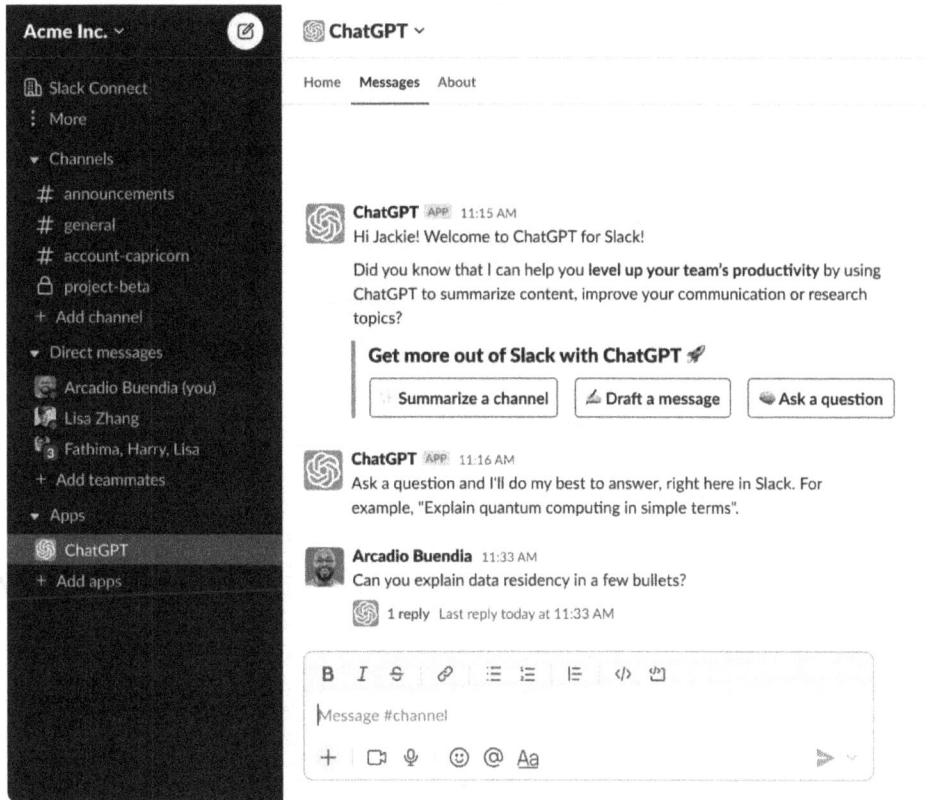

Figure 1.4 – An example of a hybrid user experience

Overall, experiences in more robust channels can take advantage of even more creative elements. For example, a web channel might support interactive charts or visualizations, shopping cart items, information, tools, and products from the enterprise. Generative conversations can refine the results or change perspective or filters, while UI components also control aspects of the view. This is by far the most robust and creative space for generative solutions.

Recommender UIs

Typically, a recommender is a recommendation as a textual prompt to encourage a specific behavior. It could be done with action buttons, but the user does not interact or converse with a system. A summary or writing suggestion tool is typically this kind of experience. Even if it has buttons to "generate email," if the user is not conversationally interacting with the generative agent, it is more like a recommender. In this Salesforce example in *Figure 1.5*, Einstein gives context to the sales process.

Article: `AI in Salesforce (https://www.salesforce.com/products/ai-for-sales/)`

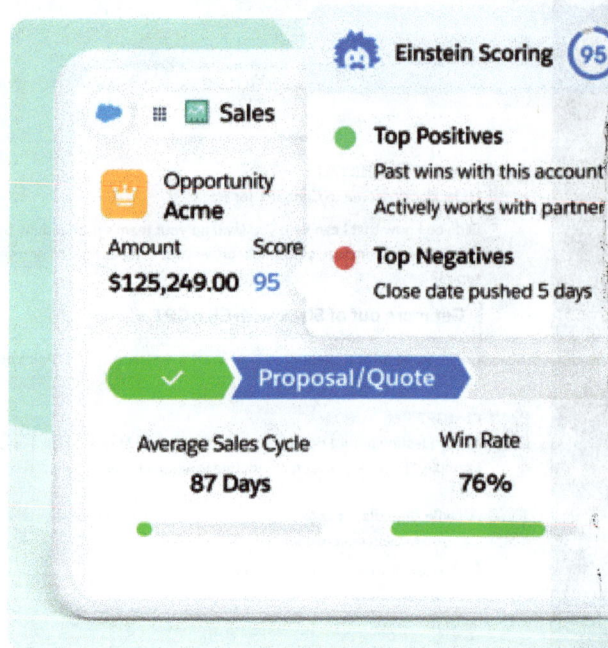

Figure 1.5 – An example of a recommendation UI

The backend uses for ChatGPT

A hidden use of a generative AI solution can be to process information and provide data, information, or wisdom to the user that is then passed to a different user experience. The user doesn't directly interact with the generative AI, so a UI doesn't exist. There could be administrative UIs or UIs that use the processed results. In *Figure 1.6*, we don't see a user interface. We see data that would be ingested, processed, and then analyzed to return clean, normalized data. We will explore this example in a case study starting in *Chapter 6, Gathering Data – Content is King*. Just realize we don't have to have a UI to get much value out of ChatGPT and generative solutions; we might use it as a tool in a much larger process.

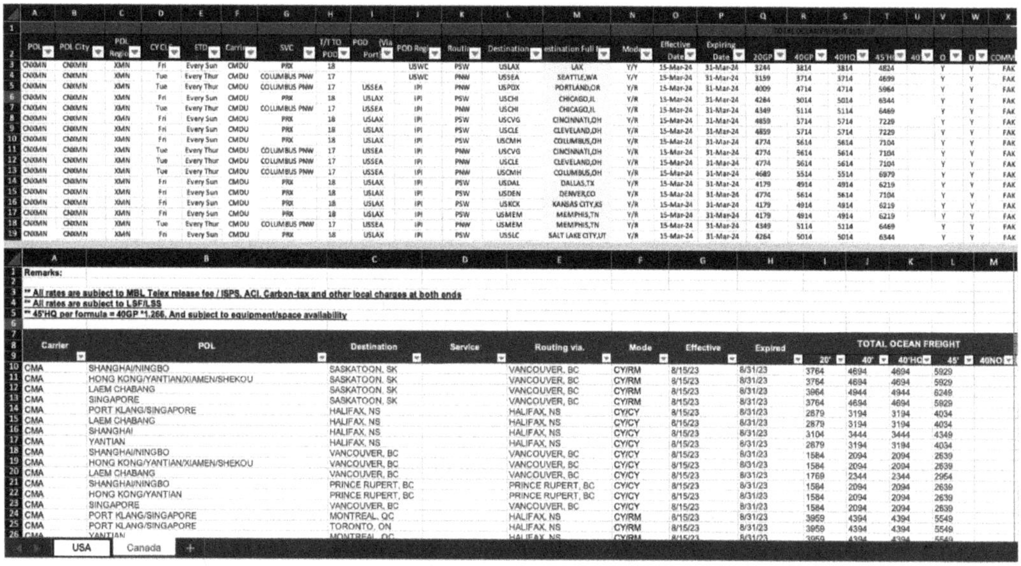

POL	POL City	POL Regio	CYCL	ETD	Carri	SVC	T/T TO POD	POD (Via Port)	POD Regi	Routi	Destination	estination Full N	Mod	Effective Date	Expiring Date	20GP	40GP	40HQ	45'H	40	O	D	COMM
CNXMN	CNXMN	XMN	Fri	Every Sun	CMDU	PRX	18		USWC	PSW	USLAX	LAX	Y/Y	15-Mar-24	31-Mar-24	3244	3814	3814	4824		Y	Y	FAK
CNXMN	CNXMN	XMN	Tue	Every Thur	CMDU	COLUMBUS PNW	17		USWC	PNW	USSEA	SEATTLE,WA	Y/Y	15-Mar-24	31-Mar-24	3159	3714	3714	4699		Y	Y	FAK
CNXMN	CNXMN	XMN	Tue	Every Thur	CMDU	COLUMBUS PNW	17	USSEA	IPI	PNW	USPDX	PORTLAND,OR	Y/R	15-Mar-24	31-Mar-24	4009	4714	4714	5964		Y	Y	FAK
CNXMN	CNXMN	XMN	Fri	Every Sun	CMDU	PRX	18	USLAX	IPI	PSW	USCHI	CHICAGO,IL	Y/R	15-Mar-24	31-Mar-24	4284	5014	5014	6344		Y	Y	FAK
CNXMN	CNXMN	XMN	Tue	Every Thur	CMDU	COLUMBUS PNW	17	USSEA	IPI	PNW	USCHI	CHICAGO,IL	Y/R	15-Mar-24	31-Mar-24	4349	5114	5114	6469		Y	Y	FAK
CNXMN	CNXMN	XMN	Fri	Every Sun	CMDU	PRX	18	USLAX	IPI	PSW	USCVG	CINCINNATI,OH	Y/R	15-Mar-24	31-Mar-24	4859	5714	5714	7229		Y	Y	FAK
CNXMN	CNXMN	XMN	Fri	Every Sun	CMDU	PRX	18	USLAX	IPI	PSW	USCLE	CLEVELAND,OH	Y/R	15-Mar-24	31-Mar-24	4859	5714	5714	7229		Y	Y	FAK
CNXMN	CNXMN	XMN	Fri	Every Sun	CMDU	PRX	18	USLAX	IPI	PSW	USCMH	COLUMBUS,OH	Y/R	15-Mar-24	31-Mar-24	4774	5614	5614	7104		Y	Y	FAK
CNXMN	CNXMN	XMN	Tue	Every Thur	CMDU	COLUMBUS PNW	17	USSEA	IPI	PNW	USCVG	CINCINNATI,OH	Y/R	15-Mar-24	31-Mar-24	4774	5614	5614	7104		Y	Y	FAK
CNXMN	CNXMN	XMN	Tue	Every Thur	CMDU	COLUMBUS PNW	17	USSEA	IPI	PNW	USCLE	CLEVELAND,OH	Y/R	15-Mar-24	31-Mar-24	4774	5614	5614	7104		Y	Y	FAK
CNXMN	CNXMN	XMN	Tue	Every Thur	CMDU	COLUMBUS PNW	17	USSEA	IPI	PNW	USCMH	COLUMBUS,OH	Y/R	15-Mar-24	31-Mar-24	4689	5514	5514	6979		Y	Y	FAK
CNXMN	CNXMN	XMN	Fri	Every Sun	CMDU	PRX	18	USLAX	IPI	PSW	USDAL	DALLAS,TX	Y/R	15-Mar-24	31-Mar-24	4179	4914	4914	6219		Y	Y	FAK
CNXMN	CNXMN	XMN	Fri	Every Sun	CMDU	PRX	18	USLAX	IPI	PSW	USDEN	DENVER,CO	Y/R	15-Mar-24	31-Mar-24	4774	5614	5614	7104		Y	Y	FAK
CNXMN	CNXMN	XMN	Fri	Every Sun	CMDU	PRX	18	USLAX	IPI	PSW	USKCK	KANSAS CITY,KS	Y/R	15-Mar-24	31-Mar-24	4179	4914	4914	6219		Y	Y	FAK
CNXMN	CNXMN	XMN	Fri	Every Sun	CMDU	PRX	18	USLAX	IPI	PSW	USMEM	MEMPHIS,TN	Y/R	15-Mar-24	31-Mar-24	4179	4914	4914	6219		Y	Y	FAK
CNXMN	CNXMN	XMN	Tue	Every Thur	CMDU	COLUMBUS PNW	17	USSEA	IPI	PNW	USMEM	MEMPHIS,TN	Y/R	15-Mar-24	31-Mar-24	4349	5114	5114	6469		Y	Y	FAK
CNXMN	CNXMN	XMN	Fri	Every Sun	CMDU	PRX	18	USLAX	IPI	PSW	USSLC	SALT LAKE CITY,UT	Y/R	15-Mar-24	31-Mar-24	4264	5014	5014	6344		Y	Y	FAK

Remarks:

** All rates are subject to MBL Telex release fee / ISPS, ACI, Carbon-tax and other local charges at both ends
** All rates are subject to LSF/LSS
** 45'HQ per formula = 40GP *1.266, And subject to equipment/space availability

Carrier	POL	Destination	Service	Routing via.	Mode	Effective	Expired	TOTAL OCEAN FREIGHT 20'	40'	40'HQ	45'	40NO
CMA	SHANGHAI/NINGBO	SASKATOON, SK		VANCOUVER, BC	CY/RM	8/15/23	8/31/23	3764	4694	4694	5929	
CMA	HONG KONG/YANTIAN/XIAMEN/SHEKOU	SASKATOON, SK		VANCOUVER, BC	CY/RM	8/15/23	8/31/23	3764	4694	4694	5929	
CMA	LAEM CHABANG	SASKATOON, SK		VANCOUVER, BC	CY/RM	8/15/23	8/31/23	3964	4944	4944	6249	
CMA	SINGAPORE	SASKATOON, SK		VANCOUVER, BC	CY/RM	8/15/23	8/31/23	3764	4694	4694	5929	
CMA	PORT KLANG/SINGAPORE	HALIFAX, NS		HALIFAX, NS	CY/CY	8/15/23	8/31/23	2879	3194	3194	4034	
CMA	LAEM CHABANG	HALIFAX, NS		HALIFAX, NS	CY/CY	8/15/23	8/31/23	2879	3194	3194	4034	
CMA	SHANGHAI	HALIFAX, NS		HALIFAX, NS	CY/CY	8/15/23	8/31/23	3104	3444	3444	4349	
CMA	YANTIAN	HALIFAX, NS		HALIFAX, NS	CY/CY	8/15/23	8/31/23	2879	3194	3194	4034	
CMA	SHANGHAI/NINGBO	VANCOUVER, BC		VANCOUVER, BC	CY/CY	8/15/23	8/31/23	1584	2094	2094	2639	
CMA	HONG KONG/YANTIAN/XIAMEN/SHEKOU	VANCOUVER, BC		VANCOUVER, BC	CY/CY	8/15/23	8/31/23	1584	2094	2094	2639	
CMA	LAEM CHABANG	VANCOUVER, BC		VANCOUVER, BC	CY/CY	8/15/23	8/31/23	1769	2344	2344	2954	
CMA	SHANGHAI/NINGBO	PRINCE RUPERT, BC		PRINCE RUPERT, BC	CY/CY	8/15/23	8/31/23	1584	2094	2094	2639	
CMA	HONG KONG/YANTIAN	PRINCE RUPERT, BC		PRINCE RUPERT, BC	CY/CY	8/15/23	8/31/23	1584	2094	2094	2639	
CMA	SINGAPORE	VANCOUVER, BC		VANCOUVER, BC	CY/CY	8/15/23	8/31/23	1584	2094	2094	2639	
CMA	PORT KLANG/SINGAPORE	MONTREAL, QC		HALIFAX, NS	CY/RM	8/15/23	8/31/23	3959	4394	4394	5549	
CMA	PORT KLANG/SINGAPORE	TORONTO, ON		HALIFAX, NS	CY/RM	8/15/23	8/31/23	3959	4394	4394	5549	
CMA	YANTIAN	MONTREAL, QC		HALIFAX, NS	CY/RM	8/15/23	8/31/23	3959	4394	4394	5549	

Figure 1.6 – Spreadsheets processed in a backend ChatGPT solution

I suspect these four concepts cover almost all uses of textual generative AI. The most interesting for us are ones with richer or robust UX, while something like a pure backend solution leaves little discussion to improve the user experience. But we do have examples! Sometimes, work starts as a backend solution, and then, with the need for controls and feedback mechanisms, these experiences come to the forefront. None of this work can be done in isolation. It takes a team.

It takes a village to create superb UX

If you are serious about creating world-class experiences (and I hope every reader is!), consider the resources that will help get the most value on the design side. As shown in *Figure 1.7*, the collection of specialists you need goes beyond the typical software team. Let's explore their contributions briefly, as this is not the usual team organization for traditional software development.

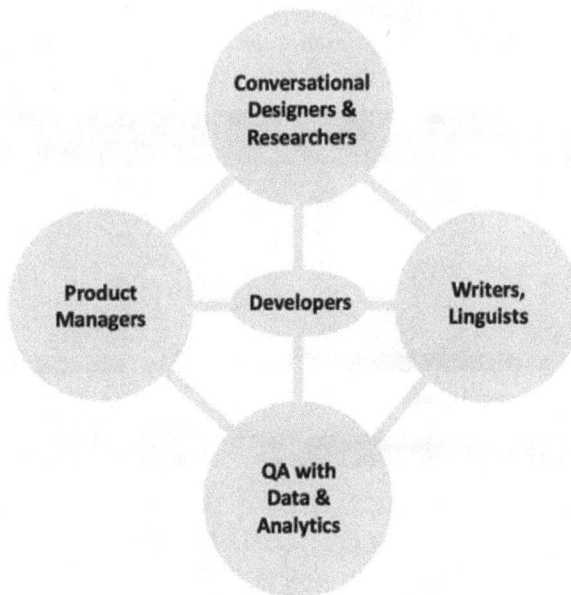

Figure 1.7 – It takes a village to build a solution

Writing comes into play fundamentally because we have an assumption in the enterprise design space that there are materials, FAQs, articles, manuals, help, installation processes, and bug fixes, which might not be easily accessible to the Internet. This material will be added to a private ChatGPT instance so that a paying customer will get value. We hope the business employs writers and editors to create high-quality content. How this is written, ingested, and represented in ChatGPT will reflect strongly on how well it was written originally (quality in will help with quality out). Designers must help understand the goals, design tools, and even integrations that might optimize the user's experience. Should a customer answer 15 questions individually or use a form to see the chunks of questions and answers contextually? We must design the experiences to match the problem we need to solve (the use case). In complex environments, where technical language or even multiple written or verbal languages are expected, linguists will be part of the team to help train and improve communication. Writers and linguists are also crucial in setting the style and tone of the conversation. We will dive into this in more detail later.

Finally, let's not leave off *user research*. Besides reviewing logs and analyzing the product's use, researchers can work with customers to learn more about how they want to use the assistant, where and when they will likely call on it, and how they will interact with it. These learnings are valuable for plotting a strategic direction and can help fix tactical issues. We will provide a set of heuristics that can be used to put on your research hat to understand where problems will arise and how to classify them.

This is just on the design side. We know that a team of engineers, product managers, quality engineers, and others will be involved in the journey. They can also benefit from this book. Do share it. Great ideas can come from anywhere, so always keep learning. The best way to keep learning is not to relearn something we already know from the history of conversational AI.

So, we now know who should be involved in this process, and we understand that OpenAI's ChatGPT is foundational, but there is still one missing piece. Why is there a need for a custom version of ChatGPT? Because proprietary data and custom answers are unavailable in a worldly-trained ChatGPT instance. Let's explore how to add unique content by setting up a personalized ChatGPT instance.

Setting up a customized model

So now we have a sense of the people and tools we need. One more piece of the puzzle is the content that will make a ChatGPT solution valuable. There are many ways to include content in an LLM, such as ChatGPT. However, we assume a focus on unique, proprietary content hidden behind a secure paywall or available only to authenticated users. If company answers are already out in the world and answered by the basic ChatGPT, ask yourself, *How would my custom version of ChatGPT provide value?* We can help get that answer by focusing our content discussions on building a private model and including company data without sharing it with the world.

The idea of *enterprise* assistants is precisely that. Ensure company data is only exposed to customers and that adding it to an instance of ChatGPT does not expose the data to the world. Only some people were this careful when ChatGPT was first introduced. Don't add sensitive data to a model used by competitors or to be a doctor and accidentally break patient confidentiality. The enterprise market demands security. Look at this article to go a little deeper:

Article: `Feeding Sensitive Data to ChatGPT (https://web.archive.org/web/20240119052608/https:/www.darkreading.com/cyber-risk/employees-feeding-sensitive-business-data-chatgpt-raising-security-fears)`

Let's make a simple request and see how adding data dramatically changes the landscape for answers. I will set up a ChatGPT assistant in the Playground for this example. I am selecting the gpt-3.5-turbo-1106 model because it supports "Retrieval" and allows me to upload my files. Retrieval means I can share content with the AI so it can answer based on the information I provide. This sets enterprise solutions apart from our day-to-day use of public conversational assistants. An enterprise solution knows things that are specific to the enterprise. For this example, everything can be done from the Playground user interface. We don't need to set up or run any code. You will need a free account to run this. Don't upload content you are not comfortable sharing.

Newer models, like Chat GPT 40-mini, are always becoming available; use the latest models in your playground. In my example, I included a 200-page document (a draft of my Master's thesis on interface design). This material is not easily accessed online, so we can see how results change based on its inclusion.

Uploading this document took only a few seconds, and the model was ready to answer. Of course, adding thousands or a million files is very different. Still, I want to share a sense of this power and ChatGPT's innate ability to prioritize uploaded content over the base model. Follow along using my example content, as shown in *Figure 1.8*.

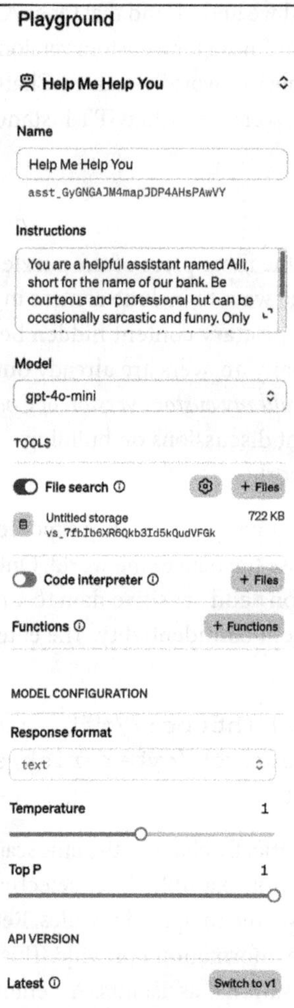

1. Sign up for an account.
2. Website: OpenAI `https://chat.openai.com/`
3. Head over to the Playground.
4. Website: OpenAI Playground `https://platform.openai.com/playground`
5. Choose **Assistants** from the top menu.
6. Select a model that supports Retrieval.
7. Test the model by asking question before uploading documents.
8. Create **Instruction** (this is prompt engineering) to ask it to "focus on answers only gathered from the documents".
9. Upload documents and retest – asking questions that only are known within the document.

Figure 1.8 – Setting up the Playground to do a simple test

We can compare the answer provided by the base model to the same question answered after adding the `Full Thesis.pdf` file, a draft of almost 200 pages.

GitHub: `Chapter1-Full_Thesis.pdf (https://github.com/PacktPublishing/UX-for-Enterprise-ChatGPT-Solutions/blob/main/Chapter1-Full_Thesis.pdf)`

As shown in *Table 1.1*, we compared the results from a generic model to a model with the document attached. Very different results. We can drill down into those differences.

What type of inconsistencies are there in UI designs?	
Generic Model	**Model with Retrieval and one file**
Inconsistencies in UI (user interface) designs can occur in various ways. Some common types of inconsistencies are: 1. Visual inconsistencies: These involve inconsistencies in the visual elements of the UI, such as font styles, colors, layouts, and spacing. For example, buttons or icons that are different sizes or have inconsistent spacing could create visual inconsistencies. 2. Interaction inconsistencies: These occur when the behavior of UI elements, such as buttons, links, or menus, is inconsistent across different parts of the application. For instance, inconsistent hover effects or different ways of navigating between screens can create confusion for users. 3. Naming and labeling inconsistencies: (removed for brevity) … 4. Functional inconsistencies: (removed for brevity) 5. Platform-specific inconsistencies: (removed for brevity) 6. It is essential to strive for consistency in UI designs to provide users with a familiar and intuitive experience. Consistent UI designs improve usability, reduce cognitive load, and enhance the overall user experience.	In a study regarding inconsistencies in user interface (UI) designs, various types of inconsistencies were identified, including physical location inconsistencies, syntactic inconsistencies, semantic inconsistencies, and feedback inconsistencies. The study found that subjects had an easier time identifying physical location inconsistencies in the interface, and that the number of physical location inconsistencies seeded in the different versions of the application was higher than other types of inconsistencies. Additionally, it was observed that different groups of subjects found varying numbers of inconsistencies in their versions of the interface, even when the seeded number of inconsistencies was almost identical across groups[1]. This suggests that inconsistencies in UI designs can manifest in different ways, including visual, interaction, naming and labeling, functional, and platform-specific inconsistencies as mentioned earlier, but the study further highlights specific types of inconsistencies identified in a research context. If you need more detailed information or other specific types of inconsistencies, feel free to ask!

Table 1.1 – Comparison of answers once custom data is added

Now, this is pretty good, even without reading the 200 pages of the document. Here are some highlights at this early stage of our discussion:

- It prioritized the answer from the file over the generic *Internet* answer.

- It summarized it somewhat, including referencing it as a *study*.

- Although not shown, it had difficulty showing tables and detailed results.

- It was easy to extend ChatGPT with novel content.

- It was available *instantly*. This should indicate that one can leverage these models to make them custom and unique for customers.

- Notice the results included a reference citation [1] but didn't provide a reference. This is a little bit of a hallucination. We will cover this in the book.

- Notice that the first answer is a numbered list, while the second is a paragraph. We can only get consistent results if we specify how we want the results (see *Chapter 7, Prompt Engineering*).

We did this example, and I encourage you to try this yourself. This shows the basics of asking ChatGPT to prioritize enterprise data over what it knows in the base-trained model. We will go much further with this in the coming chapters.

> **OpenAI supports basic retrieval**
>
> On their knowledge page (December 2023), OpenAI said, "*Retrieval of content in this example is optimized for quality by adding all the content to the model calls. In the future, ChatGPT will introduce other strategies to help optimize the trade-off between quality and cost.*"
>
> Here is one of those strategies:
>
> Article: `File Search from OpenAI (https://platform.openai.com/docs/assistants/tools/file-search)`

If you read about OpenAI retrieval, it has significant limitations and costs that might hinder an enterprise solution. Other parts of generative AI solutions can address scale. This is covered in *Chapter 6, Gathering Data – Content is King*.

Summary

At this point, this only scratches the surface. We will go into how to develop a user research plan and apply standard research methods to the conversational world, then dive deep into user interface patterns, guidelines, and heuristics to help guide and improve the quality of an enterprise ChatGPT solution. Once we have something, it is time to apply our care and feeding approaches to monitor and improve the quality of customer experiences, extending it further into more customer use cases with the quality they expect. We will wrap it all up in a nice bow so this newfound expertise can be shared with your friends in development and product management. No person is an island. It does take a team.

Next, let's cover basic research and tools we can use to understand our users and how we can learn what our customers need. A few suitable research methods can go a long way to bridging the gap for unknown unknowns…

> *"Reports that say that something hasn't happened are always interesting to me, because as we know, there are known knowns; there are things we know we know. We also know there are known unknowns; that is to say we know there are some things we do not know. But there are also unknown unknowns—the ones we don't know we don't know. And if one looks throughout the history of our country and other free countries, the latter category that tends to be the difficult ones."*
>
> *– Donald Rumsfeld (Defense.gov news transcript: DoD news briefing – Secretary Rumsfeld and Gen. Myers. United States Department of Defense. February 12, 2002)*

Let's go!

References

The links, book recommendations, and GitHub files in this chapter are posted on the reference page:

Web page: `Chapter 1 References (https://uxdforai.com/references#C1)`

2
Conducting Effective User Research

We all want to dive in and start working on things immediately—that's human nature. There is more value in our efforts with a plan. And if that plan includes a solid understanding of users, any effort will be more than returned in value. This is where user research comes in. Let's use expertise in this field to understand the user's needs, listen to the problems, and then drive solutions. Explore a few essential methods to understand where to go with ChatGPT projects. There is a well-known adage in development circles that change gets exponentially more costly at every step of the development process. So, why not take the proper first steps and reduce expensive change later? This doesn't mean you can't adapt and improve.

The world of LLM is all about refinement and incremental improvement. Let's start on the right foot to reduce complexity and cost. We can do this by knowing more about how customers will use the product, what needs will be the most critical, and how deep the solution needs to go.

We will not give a survey of all the possible user research methods out in the wild. But we always encourage deeper exploration. Let's cover a few main topics that give the best value for an investment in research:

- Surveying UX research methods
- Understanding user needs analysis
- Creating effective surveys
- Designing insightful interviews
- Trying conversational analysis

Surveying UX research methods

This section will focus on some popular UX research methods.

I want to respect all the possible methods out there. We tend to pick techniques to help a specific problem, and one can't apply all methods to all situations. But reminding yourself of options is good. There is more out there than covered in this chapter.

Article: `Overview of research methods (https://www.nngroup.com/articles/which-ux-research-methods/)`

The takeaway from that article is to notice where our review methods appear on Christian's landscape map from the Nielsen Norman Group article, as in *Figure 2.1*.

I circled three methods we will cover and highlighted **usability testing**. The ones we will cover in this chapter provide the most value for helping decide what to build and start building a collection of customers to engage with over time. Usability testing is also essential; it will come into play when we have something to test. Once we have a product, we will discuss methods and tools to help evaluate the quality of LLM solutions.

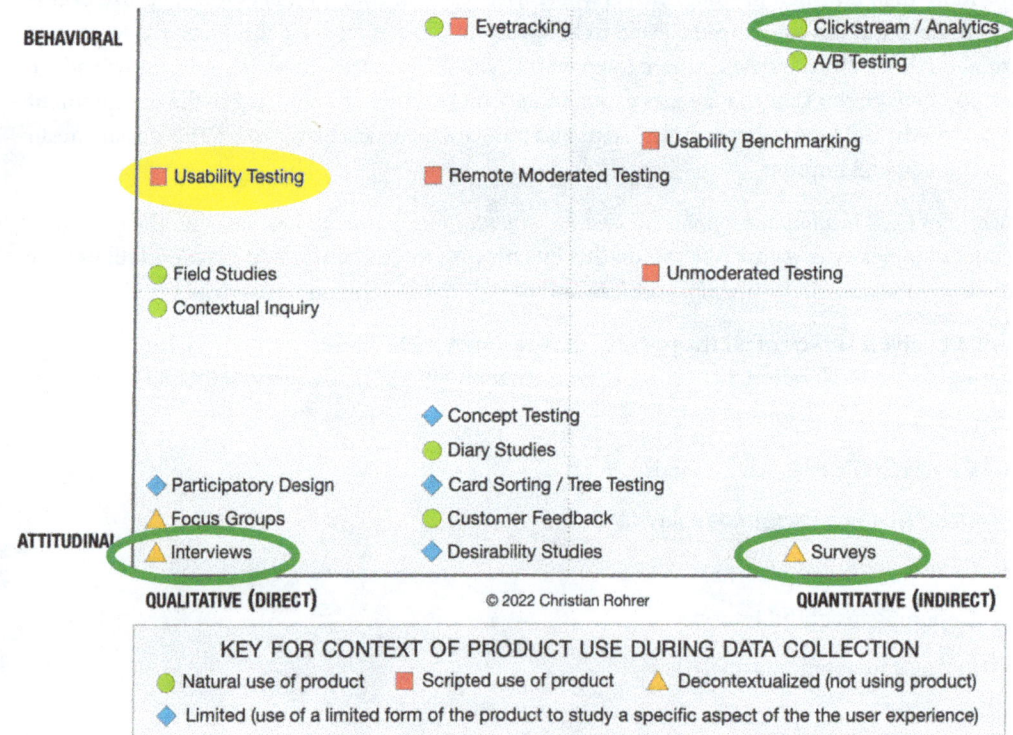

Figure 2.1 – The broad landscape of user research methods

> **More on UX research methods**
>
> Try some research methods books for a deeper exploration that can help design generative AI solutions.
>
> Book: A deeper dive into user research methods (https://amzn.to/3zYtzN1)

We spend the effort researching to understand the user's needs. We don't ask them what they want; we learn what they need and use that to drive design and feature decisions. This is a needs analysis, which we will discuss next.

Understanding user needs analysis

Let's review the basics of **needs analysis**. The needs analysis helps us identify, evaluate, and prioritize the user's requirements and needs. We want to understand the users' outcomes and ensure we meet the challenges they will see. The product manager typically leads this product definition, but an interdisciplinary team with a designer and researcher helps improve the outcome.

We want to accomplish something significant by introducing ChatGPT into our company's resource bag. Let's ensure we know how this will solve *real* problems for the right customers and at the right time during their user journey. Later in the book, we will go deep into ensuring that we service the right customers with a service they use with some frequency and solve significant problems. For now, let's get the context for what is needed and gather some knowledge on the gaps in our product that a ChatGPT solution might solve:

1. **Define the purpose**: What goals do we want to achieve by introducing an LLM solution? A service example could be to reduce costs by introducing a chat assistant to solve simple problems or gather information before handing it off to human agents. But this is not a goal for the user. The user aims to solve a problem as quickly and accurately as possible. And not to make things worse by solving the wrong problem or making matters worse. That is their take from their boss's perspective. A needs analysis for Sales might show that the time to close a deal is 30% longer than the industry average. A salesperson whose commissions are based on closing deals should want to help move their process along, so the goal of reducing close time helps them and the company.

2. **Identify stakeholders**: Identify and involve stakeholders, including end-users, clients, decision-makers, and others interested in the product or service. This is important because the methods we discuss can be applied to customers and internal decision-makers.

3. **User profiling**: Create user profiles or personas representing different target audience segments. Be careful not to create a caricature of a customer. Avoid definitions that distract from connecting the user and the system. For example, if the persona or profile includes that they get a coffee at a local shop on the way to work, let's ensure this is relevant to the use case. Later, when we want to tune the messaging to the user's individual needs based on experience level, previous use of the tools, or technical understanding, we should be able to use this knowledge in the profile to tune the dynamics of the conversational style and tone. If getting coffee is only there because you think it is cool, drop it.

4. **Conduct surveys and interviews**: Gather information directly from users through surveys, interviews, or focus group discussions. Ask open-ended questions to understand their goals, challenges, preferences, and expectations. Ask specific questions to segment and quantify results easily. This is useful for customers and internal stakeholders.

5. **Observe user behavior**: Conduct usability testing or observe users interacting with *existing* systems or prototypes. Analyze how they perform tasks, identify pain points, and follow any patterns in their behavior. We will see some of this in our log analysis discussion. A generative AI solution isn't necessarily up and running, so use other channels to analyze user behavior and gather that data.

6. **Analyze feedback and support data**: Review customer support channels, user reviews, blog posts, and any other sources of information that can provide insights into user issues, concerns, or suggestions. This is a time-consuming and hard-to-quantify task. Analyzing existing log files is also time-consuming but easier to quantify. We will drill down into data sources like a chat log file.

7. **Define user stories**: **User stories** are narratives that describe a user's interaction with the system, focusing on the desired outcome and the context in which the user's goals occur. Let's briefly explore this and save the bulk of it for the next chapter. Using cases or case-driven design and development is prevalent in customer-centric companies. This is encouraged as a more systematic and tactical than the profiling approach discussed above in the *3rd item*. Some do cartoons or visually write up a story like a comic book. These are very popular and allow a broad audience to grasp the goals quickly. This example from Chris Spalton reminds us that we don't always have to start with specifications. We can also use our creative and visual side to explain our story. *Figure 2.2* shows Chris's example of a UX concept expressed as a storyboard.

Article: `Using comic strips to show UX concepts` by Chris Spalton (`https://uxdesign.cc/using-comic-strips-and-storyboards-to-test-your-ux-concepts-cccad7ac7f71`):

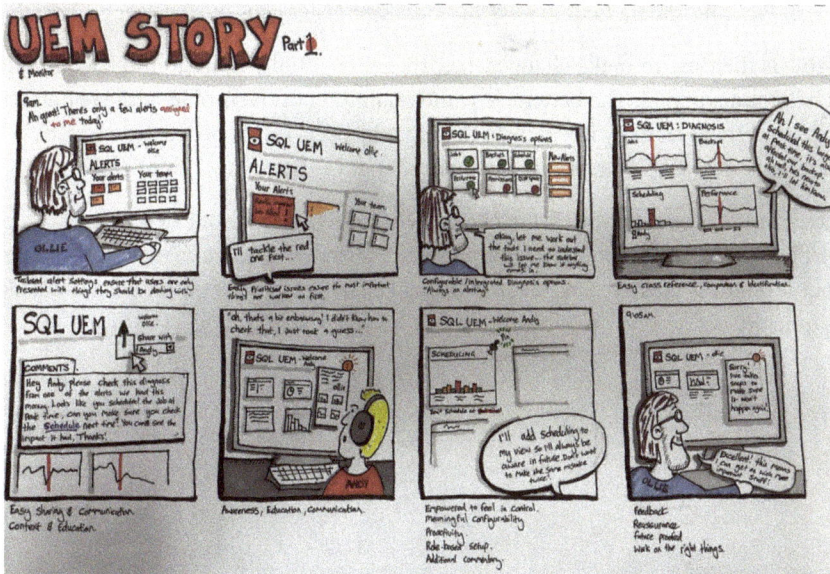

Figure 2.2 – Example of a UX story as a comic strip

8. **Prioritize user needs**: Prioritize the identified needs based on their importance and impact on the user experience. This can "drive focus" during the design and development process. We will drill down into this heavily by providing a repeatable tool to address this significant issue.

9. **Create use cases**: Develop use cases that describe the step-by-step interactions between users and the system. This helps visualize how users will accomplish specific tasks. Alistair Cockburn's *Writing Effective Use Cases* is the definitive guide when teaching use case design (*Chapter 3, Identifying Optimal Use Cases for ChatGPT*).

    Book: `Writing Effective Use Cases (https://amzn.to/3YnbGSp)`

10. **Document research**: Compile and document results from the user needs analysis. This documentation serves as a reference for the design and development teams.

11. **Iterative process**: User needs may evolve, so analysis should be considered an iterative process. Regularly revisit and update user needs to align with changing requirements and user expectations.

> **Note**
>
> No matter how it is implemented, a ChatGPT solution will still require care and feeding throughout the foreseeable future, so plan on interactions at a regular cadence when in production. It could be weekly or faster. This is reviewed in *Chapter 11, Process*.

By thoroughly understanding user needs, you can tailor products or services to meet the target audience's expectations, improving user satisfaction and overall success. It is unlikely anyone will use or need all the steps listed. *This is why we can apply design thinking to our problems. Don't reinvent the wheel for every project. Pick and choose tools to solve the most significant gaps in your team's understanding.*

Remember, this is the time to make changes. It costs exponentially more to improve quality once a product is built and shipped. The better early understanding of the problem, the better decision-making will be, and the less costly it will be to the organization. This is not to say we won't continue improving solutions after shipping, but incremental changes at later stages cost more.

Figure 2.3 shows how to consider the cost of change when moving from one development step to the next. We don't get it all right in requirements, and with a ChatGPT solution, we expect to iterate to get it right, but it reminds us there is value in learning at every step because the cost of learning and repair grows quickly.

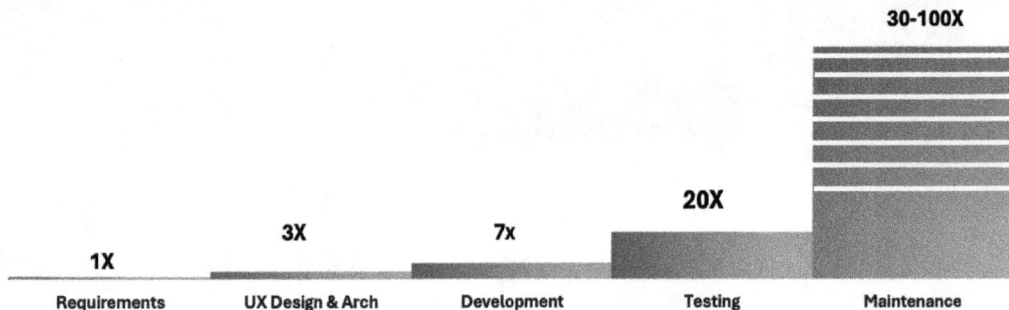

Figure 2.3 – The cost of change rises at every stage

I prefer to focus on a few essential methods and expose some additional tips and tricks to make them effective when researching a generative AI solution. Let's dig into the first tool in our bag that sometimes helps us refine our needs analysis: a survey.

Creating effective surveys

Of all the methods available, why choose a **survey**? There are challenges with building a survey, but when a new project starts, there is sometimes little data to help form a design *space* around users' needs. Instead of relying solely on customer feedback, visionary leaders and design innovators anticipate needs and create solutions that customers might not have imagined. Steve Jobs, co-founder of Apple Inc., was attributed by saying, "People don't know what they want until you show it to them." But for some of us, when we start, we need a clue where to go, what problems to solve, or help deciding which problems to solve have the most value. We do not suggest asking the customer what they want; we want to ask questions to help us understand what they need and where current products and services don't meet their needs. We don't ask them how to solve the problem. Let me give an example of how to ask the wrong question.

Jim is cooking in the kitchen when Kelly comes in from a run. They both reach for the orange on the counter. They both want the orange. Jim suggests cutting it in half. Kelly says that won't work. Kelly asks, "What do you want to do with the orange?" Jim replies, I am cooking and need the rind. Kelly says, "Great, you can have the entire rind, and I will take the insides for eating since I just got back from the gym."

This is more than just about negotiation and conflict resolution; it is essential to understand the use case. If the design team knows how someone wants to use a product, you can help design a better product. We don't ask "what do you need" in a survey. We focus on understanding intent. Once we have some generative output to share, we might adjust the styles and tone to one that resonates better. It is common to run a survey after releasing software (or by embedding a survey within the software or service) and ask for feedback on their experience. In any case, ask quality, unbiased questions that you know apriori (ahead of time) how the results should be analyzed.

Writing surveys is a science and expertise all on its own. Even experts can benefit from the science in the field. We have resources for that.

Book: `Questionnaire Design` by Bolton and Bolton (5th Edition, 2022) (`https://amzn.to/3AemWWY`).

Surveys for conversational AI

Surveys are well suited for broadening an understanding of customers in a new or novel space. We can only offer generic questions and sage advice. Questions specific to your space will have to come from you. But there are questions we should not ask. We cover some of this in the checklist. For example, don't ask, "Do you want to use a conversational AI to answer questions?" or "Can I offer you recommendations based on either an algorithm or a generative AI analysis?". These questions return us to a time when we asked our customers if they wanted a faster horse. The horse wasn't the answer; a car, something they could not imagine, was the solution.

Focus questions on customer problems, their frustration with solving them, or the tools and techniques they use to interact with the company. This is a method for gaining insights into how they experience the brand and go deeper into why and how. With recommendation UIs, you can explore style, tone, and content in a survey. Probe users concerning different recommendations and how they might react to them, which ones would be more valuable (by giving multiple examples and asking to pick the one that is the most useful), when the recommendation would be most helpful, or even how to share it (should we email the users when we know it, or only show it in the UI when they come to the deal page?). And this is an excellent opportunity to ask why. Have them fill in details on why they made their selections and take the time to organize and assess this feedback. Do this research before making recommendations because users might only accept future improvements if we hit the mark. Start on a sound footing. This research will shape how to form prompts, gather the data for analysis, and generate results. We can dig into issues to watch out for when developing a survey.

Survey checklist

We dive right into tricks of the trade, assuming some survey expertise; grab our resources if needed. If you have the basics, proceed, as this is a collection of lessons we sometimes need to remember when building an effective survey. A poorly designed survey can be worse than no survey at all. It can give a sense of false hope and guide the team into wrong assumptions. The village of people it takes to build a solution will all be in an excellent position to contribute to the survey. Conversational writers, designers, researchers, product owners, and managers will all find value in the results of a well-done study. So, be sure to engage them to help create the questions. Here are some checklist items to ensure an effective survey.

- **Only ask questions knowing how they will be analyzed**: Plan to analyze every question. This forces a purpose behind every question, and when it comes time to trim, it is easier to judge its value.

- **Don't ask leading questions**: Let the user form their opinion. Ask, "How satisfied are you with your support experience earlier today?" Don't ask, "Considering how improved our excellent support process and staff is with handling concerns, how satisfied were you with your recent interaction?"

- **The simpler the questions, the easier the user can answer:** Multiple-choice (a, b, c, d) and simple-choice (a or b) questions are easier and faster to analyze. Use them and rework open-ended questions when possible.

Simple choice questions should be direct. Try to avoid yes/no questions.

We avoid yes/no questions in UIs because they require more cognitive effort. Here's an example from banking:

```
Do you wish to cancel your transfer? [ Yes ]   [ No ]
```

Instead, we would use this:

```
Do you wish to cancel your transfer?
[ Cancel Transfer ]    [Continue Transfer ]
```

Survey questions are similar. For simple choices, try to use descriptive words so the selection is straightforward and can stand alone without reading or re-reading the prompt. Don't use this:

```
Were you happy with your support call today?   [ Yes ]   [ No ]
```

Instead, use this:

```
Were you satisfied with your support call today?
[ Satisfied ]   [ Not Satisfied ]
```

This is less leading and more direct. Decide what kind of results are needed from the survey.

- **Choice questions should be Venn diagram complete**: This is my funny way of saying make sure question choices cover all the possible outcomes. This avoids someone attempting to answer a question and not seeing a response that matches their use case. Take a look at this:

```
What is your current age?
    Younger than 20
    20 to 50
    Older than 70
```

Oops! The question missed a group of users. This is an obvious example, but don't be surprised how often this happens. With a long list (say, 5 to 10 choices) covering 90+ percent of the answers, including an "Other" option, they can optionally fill in a text answer for the details of the other answer. Manual entries do require more effort to analyze.

Results sometimes need normalization. If you asked a question about the customer's favorite streaming service and didn't include Disney, for example, they might choose "Other" and then type in answers such as "Disney," "Disney Plus," "Disney+," The Walt Disney channel." Results have to be tabulated into one answer. Try to be complete in the answer options to avoid this problem.

- **Include a Not Applicable choice when needed to complete the possible use cases**: This will help ensure better results by not forcing users to select the wrong answer to a required question.

- **Consider what questions are required**: Most questions are typically needed so users can't accidentally skip entire sections.

- **Create balance in the list of choice questions (single select)**:

Let's use an example to ask for an age range:

```
What is your current age?
    < 20
    20 to 30
    30 to 50
    50 to 60
    60 to 70
    > 70
```

A few issues are of interest. Firstly, we might be trying to exclude a particular group (maybe only adults older than 20 are needed), so that group has 20 years in it, and that might be fine. Because we were interested in the older groups, we broke those segments down into 10-year segments but used 20 years as a group for the younger group. Be mindful of these decisions. Use consistent groups, especially in the middle of a range, and go beyond the range with those consistent groups. Let's use a 70-80 group and then an 80+ group. So, when the data is graphed, it is segmented better. Then, the graph won't show a significant spike in the 30-50 group, which might only happen because the group is twice the size of the other groups. Finally, it is expected to do a cross-tab on data like this (for example, comparing age range with income to verify the right audience segment is being targeted and then chart that). So, consistent group sizes will make that chart more appealing. But if the customer is 30, which choice do they select? We don't want to have overlapping values. Use these larger groups on the edges of the data, assuming they are small, insignificant groups (a few percentages or less).

I would revise this as follows:

```
What is your age?
      Younger than 20
      20 to 29
      30 to 39
      40 to 49
      50 to 59
      60 to 69
      70 to 79
      80 or older
```

This is good as a drop-down menu. Everyone knows their age. They don't need to compare answers on the screen and can quickly pick from the list. And we are not asking their age, so I changed the prompt. It takes up less space and is a cleaner user experience. I removed those greater than and less than arrows and used English instead. There is no reason to make them work hard to understand symbols like < or > to pick a simple answer. I selected ranges such as 20 to 29, not 21 to 30. This is consistent with how we talk (someone in their 20s or 30s).

- **If a drop-down menu defaults to one answer, then the user is not required to select an option, which can bias results**: So, in these cases, ensure the software you use defaults to an unselected state, as shown above.

- **Data can always be merged later when asking a specific question (such as age). If the question uses groupings, you can't change the groupings later**: Grouping is nice when you want to correlate data. For example, with technical support customers, there is interest in knowing how many products they own. Don't ask this if this is known from their account information. It can be correlated with the number of times they have used the site or the number of posts on each product forum. This information might help explain their expertise level. Use this information to adjust the level of detail, style, or tone in generative answers. In a GUI, the system can offer expert advice instead of a novice UI. It can also adapt further to give novice generative advice only for the user's new products and more advanced advice for products they are known to use regularly. And then adjust based on their use of that advice.

- **Display all the choices for multiple selections and keep the entire list visible on the screen**: With numerous options, especially if there is a limit to the number of selections, they should all be visible. If not, the user might hit their max and then see other answers, which causes them to need to rework their selection, and they might not bother. So, by ensuring choices are all visible, they can scan and decide which ones to pick.

- **Keep the survey short**: When fully designed, try to then reduce it by 30%. Look for places where asking the same question again is unnecessary or asking questions you don't need. There are cases for asking the same question again, in a slightly different way, to validate the first time it was answered.

- **Use branches in a survey to focus specific users on specific questions**: If user answers go in one direction, we don't need to ask them additional questions in a different direction. That is unnecessary. For example, ask questions about the iPhone only to iPhone users. Please don't ask them Android questions that are not applicable.

- **Resist the urge to collect unnecessary demographics (age, location, maybe gender)**: Don't ask if it won't add value to the analysis.

- **See if a survey mechanism exists, such as a quarterly customer survey**: In that case, add questions to the existing process and get results without the overhead of running a new survey. Of course, this will be on their schedule and be limited in some ways, but it could be an option.

- **Use an anchored Likert scale with a middle value for ratings**: Opinions, perceptions, and behaviors can be asked using a multi-point scale, typically with labels (anchors) for each or some values. I suggest a 1 to 5 range or 1 to 7 at the most (not more, except for Net Promoter Score). Anchors are the labels at each end of the scale (Poor to Great, and labels for each step if they are reasonable to define, such as for the common UX questions around ease of use: Very Difficult, Difficult, Neutral, Easy, Very Easy). We see these in post-use one-question surveys, such as "Please rate your experience today with our chat assistant."

Survey Monkey's primer is straightforward to digest for some basics.

Article: SurveyMonkey's primer to Likert Scales (https://www.surveymonkey.com/mp/likert-scale/)

- **Power tip for Likert Scales**: Vary what is 1 with a collection of scales. Some 1s can be for the bad or poor concepts (Very Difficult, Never, Not Recommended), and some can mean positive or good options (Very Easy, Always, Strongly Agree). This forces the user to read and review each scale. If they don't notice, the data is invalid, but likewise, if they read and mark everything with the same choice, there is no value either.

- **Test the survey with pilot users (friendlies)**: Verify the wording, grammar, and flow. Check all branches and review the results to confirm they are valid. Get them to be frank. Encourage feedback.

- **Consider distinct surveys for customers and internal stakeholders**. Stakeholders might deeply understand customers' needs, gather stakeholder feedback, and keep this data separate from customer results.

- **Consider adding a final question about participating in future user research activities**: This can generate an additional source of customers for interviews, user testing, or participatory design sessions.

- **Don't bug customers**: Companies have rules. Sending one email and a follow-up is likely okay. But we don't suggest sending more than that. Drop it if a customer doesn't have time to reply to the second request. Also, track user responses (many tools allow this), only sending the follow-up to those who have not yet participated.

- **Don't expect more than 10-20% return rates for a survey**: It is possible to incent customers if the company allows it. It is expensive and time-consuming to send every participant a gift card. It can cost less time and expense to enter customers in a raffle and distribute only a few significant incentives (I gave away iPods back in the day; a $100 gift card or more, depending on the complexity of the survey, is reasonable). This can increase the return rate somewhat (but also requires the survey to include name and contact information). If 5% return surveys by emailing 100 participants, there won't be enough subjects for valuable results. We will cover the number of participants shortly.

- **There are no good rules for how many respondents are needed for reliable data**. Since some survey questions are subjective and qualitative, 30 to 200 answers could be required for a specific item. However, a more extensive survey with branching and multiple segments of an audience might need 1,000 or more respondents to get enough results on each branch. For statistically significant data for quantitative results, math found in statistics books is required to calculate the necessary sample size of respondents.

- **Some survey examples:** *Figure 2.4* shows how questions and the UI interact. Learn how to use a survey tool's capabilities to get the best results. I strongly encourage using one of the robust third-party tools that have been around for a while. Internal survey tools at large companies can lack the robustness and flexibility needed for deploying a quality survey.

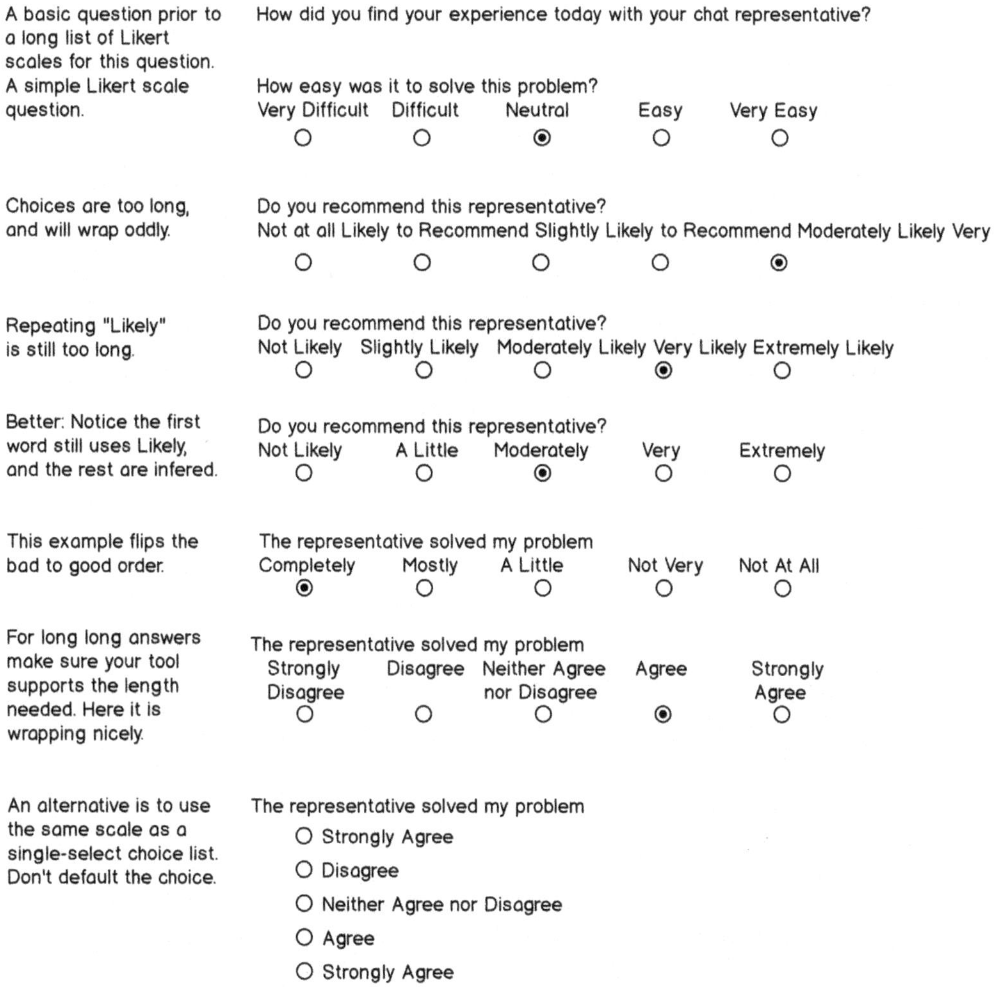

A basic question prior to a long list of Likert scales for this question. A simple Likert scale question.	How did you find your experience today with your chat representative? How easy was it to solve this problem? Very Difficult · Difficult · Neutral (selected) · Easy · Very Easy
Choices are too long, and will wrap oddly.	Do you recommend this representative? Not at all Likely to Recommend · Slightly Likely to Recommend · Moderately Likely · Very (selected)
Repeating "Likely" is still too long.	Do you recommend this representative? Not Likely · Slightly Likely · Moderately Likely · Very Likely (selected) · Extremely Likely
Better: Notice the first word still uses Likely, and the rest are infered.	Do you recommend this representative? Not Likely · A Little · Moderately (selected) · Very · Extremely
This example flips the bad to good order.	The representative solved my problem Completely (selected) · Mostly · A Little · Not Very · Not At All
For long long answers make sure your tool supports the length needed. Here it is wrapping nicely.	The representative solved my problem Strongly Disagree · Disagree · Neither Agree nor Disagree · Agree (selected) · Strongly Agree
An alternative is to use the same scale as a single-select choice list. Don't default the choice.	The representative solved my problem ○ Strongly Agree ○ Disagree ○ Neither Agree nor Disagree ○ Agree ○ Strongly Agree

Figure 2.4 – Examples of Likert survey questions

- **If clarity is needed, ask ChatGPT**: It has a wealth of knowledge about creating surveys. Use it to understand how to set up a survey process and how to edit questions to be clear and concise. However, as we will see in this book, the results from generative AI can also miss the mark, so make sure you understand its results. Here is an example to try.

```
I want to ask in a survey how the agent performed in a support
session. How would I ask this question and use an anchored
Likert scale to collect responses?
```

Now, let's apply this knowledge to learn how to craft an effective survey.

Case study on an effective survey

I took a survey on my phone based on an email for a product I use. They gave me about 20 Likert scales. I don't want to screenshot the vendor because there is no reason to embarrass them. I can set the stage. It was a one-page survey focused on my use of their online service. There were no questions on the page, only labels such as Internationalization, Reporting and Auditing, Performance, System of Record Integration, and Web Interface on a scale from 1 (Less) to 7 (Excellent). We won't mention the company, but many use their products. We will call them CircleRize.

Let's discuss the positive and negative things learned from this case study.

The positive things include the following:

- Before this page, they asked a Net Promoter Score question (1 to 10 scale). We will cover Net Promoter Score in more detail in *Chapter 10, Monitoring and Evaluation*. This single market research question determines whether a customer would recommend a company or product.

- They use a good range of 1 to 7 for questions.

- They labeled the 1 and the 7 anchors. It is good to know their meaning.

- They have titles for the items.

The negative things are as follows:

- Those anchors? They used *Less* and *Excellent* labels. These are not opposites. They need to be balanced and clear. I don't know what Less means as an anchor with the word Excellent. Do they mean Poor?

- The questions don't have a *Not Applicable* option. I am forced to answer questions I don't understand, such as the one on System of Record Integration. They can do better by segmenting the 20 questions into groups and branching based on product use questions to eliminate confusion. They should also use language I understand.

- There were no questions on the actual survey page. It was a page labeled Feature Ranking. I needed clarification by looking at the page label. I first thought they wanted me to rank each item in importance (Web UI is more critical than Internationalization). Then I realized they were asking for individual ratings. It needs better wording to be clear.

- I don't know how to answer statements such as Internationalization without a question to give it context. Do they want to see if I use it, need it, or even notice it is supported? No actual question was asked, just generic labels such as Internationalization. I have no idea what they wanted.

- What they expect to learn from this is unclear unless they plan on asking the questions regularly. For example, a survey can ask about performance to measure improvements or degradation. These questions must be repeated in subsequent surveys to see differences over time.

- The email subject for this survey was "Review your CircleRize plan, get a $25 gift card!". I first read this as "check to see if the details of your CircleRize subscription with us are correct" instead of their intended "Provide feedback on your experience using the CircleRize plan." I would suggest words matter. They should label the email clearly to reduce confusion and increase participation.

There is more than meets the eye when throwing out some simple questions. The results from this survey, likely sent to all customers like me who were looking at various plans, are worthless. What is worse than the time and energy spent putting out this survey and wasting customers' time is that they didn't know better and will likely use this data to make decisions. User testing or someone with more experience could have caught some of these issues. It's hard to tell why these issues appeared. Use your resources to create effective surveys. It is a lot of effort and won't give a lot of return if everyone clicks on a score of four (the middle) because they don't understand the question.

We have discussed what to ask and what not to ask, but specific examples are more helpful. We don't know your use case, but we can offer suggestions on how to use ChatGPT to solve customer issues.

Here are some questions typically asked in a survey and how they might be analyzed.

- Demographic:

  - **Examples**: Age, gender, education level, years of product use, expertise

  - **Analysis**: Style or tone can be associated with expectations for support, level of expertise, age, and other demographics

- Multiple choice:

  - **Examples**: What three attributes of a service are most important to you?

    Which products give you the most headaches?

    A list of competitors' products that they already use.

  - **Analysis**: Judge user goals; accuracy is likely one, but availability in the context of their use is more important than, say, integration with another service. The proper selections are needed so a lot of data doesn't go into the overflow *Other* category. The survey can explore other products that customers use and like to learn about possible advantages for competitors.

- Likert Scales (1 to 5, 1 to 7 scales with labeled anchors):

    - **Examples**: How satisfied were you with the support call today?

      What is your level of experience with product x?

      Comfort level with giving confidential information to an agent or bot.

      Satisfaction with the level of service, responsiveness, availability, accuracy, and cost.

    - **Analysis**: Find opportunities for improvement for support and gauge overall quality over time as improvements or changes occur. Determine tradeoffs between 24/7 support and level of service.

- Net Promoter Score:

    - **Example**: On a scale of 0 to 10, how likely are you to recommend [company, product, or service] to a friend or colleague?

    - **Analysis**: This is a quick snapshot of customer sentiment that can easily be compared to other companies or services or used as a benchmark for improvements. *Chapter 10, Monitoring and Evaluation* will cover this in more detail.

- Open-Ended questions:

    - **Example**: What do you like about your favorite streaming service?

      Tell us about your best experience using product Y.

      Tell us any horror stories from using product X.

    - **Analysis**: This is an opportunity to benchmark against the competition; open-ended questions help go deeper into one area of interest. It could be a follow-up to the multiple-choice questions exploring what competitive products they use. Ask them specific questions. This can be correlated with the demographic question above to identify where particular demographics are likelier to enjoy different services. These require more effort to analyze, classify, and correlate.

- Quantitative data:

    - **Example**: How many hours a day do you spend on our site?

      How much did you pay last month for streaming services?

      What is your monthly car payment?

      When was the last time you emailed us for support?

      How long is too long when on a tech-support phone call?

- **Analysis**: The question about hours a day differs from what it seems. There can be analytics on their time spent on your site; this is more about perception. It can be helpful to compare actual usage data if they report less than they pay or if they report more. It can also be cross-tabbed with information on the types of activities they do on the site. This might reveal if some activities are more time-consuming than others and provide insights into how to invest resources to reduce activities that take a long time. With customer data, you can correlate perceived time, cost, and use with actual time, price, and use. This can explain if customers' perceptions match expectations.

- Trends:

  - **Example**: Which of the following online payment services have you used?

  - **Analysis**: A survey can capture competitive information. This data can be used to research partner opportunities or see how certain trends are viewed within specific demographic groups.

- Behavioral:

  - **Example**: What was the primary reason for visiting our site?

    If you were to tell a friend about your experience with support today, what would you say?

    How likely are you to post something about product Y on social media?

  - **Analysis**: Hopefully, this is a simple single-choice list, but it can branch into more details (e.g., whether they were having product issues, looking for a new product, downloading something, or providing a review).

- Comparative:

  - **Example**: Place these five services in order of value to your needs.

    Which of these pieces of sales information is the most valuable?

    Rank these attributes from most to least desirable.

    Which phrasing of these statements feels most natural to you?

  - **Analysis**: Asking customers to rank features is better than asking them to rate them. Ranking only places features in order of importance. It doesn't help you know how important they are. Ranking questions are hard for a user to answer. Limiting ranking to five items, but if some items are considered equal, forcing a ranking can cause frustration. This type of question can be used to try different styles and tones. See if there are correlations with demographics or other information to create customized personas for the conversational style. There is a lot of psychology in writing effectively.

Wow! I hope that wasn't too much to process without giving all the training necessary to be a survey expert. Grow expertise, use resources, partner with other experts, and do what it takes! Please return to this when it is time to survey and check off the items to ensure a reliable study with results the team can trust.

As mentioned, surveys are a great place to meet people worthy of an interview. What a nice segue into discussing our next user research method: interviews.

Designing insightful interviews

Interviews might not have been your first choice, but we encourage exploring interviews as a valuable tool for understanding use cases. Use a survey to gather general information, and do a click analysis (or log analysis) to get actual behavior. An interview is vital to get a cohesive feel for the customer. It has its downsides: it is hard to scale, results are more open to interpretation, and it is sometimes challenging to get the right mix of customers. The upside is the feeling you get for the customer and the empathy one can embody as the customer advocate during the design process. This will enrich the project. In surveys, consider finishing with questions about participating in future user research. Although this is self-selecting, it is fruitful almost all of the time.

There might be time for five to ten interviews, but most teams will unlikely invest in more than 20. Targeting the right audience has value. The cost of setting up the interviews is likely more significant than the time required to conduct the interviews (typical sessions run for 30 to 60 minutes). Customers are busy, and incentives might not be an option when working with existing customers. They tend to need to reschedule because this is a low priority for them. Even with the downsides, I still recommend the interview process.

Don't expect demos or visuals to be shared to get feedback when dealing with conversational solutions. Ask them to have the existing products or web support site handy during an interview. Walking through some use cases creates a discussion around enhancement opportunities (focus on their needs, not their solutions). Don't constrain them to what are perceived as AI issues; if they go somewhere else, take their feedback, pass it on to a team (or file a bug or enhancement request on their behalf), and then bring their discussions back into focus. Create a relationship where they trust that you are trying to help the company help them to do their job or task.

Put together a framework for interviews that supports gathering the needed information. If you are trying to do a conversational assistant around support, focus on understanding the complexities they experience when trying to get answers. If you are building a recommendation engine, look for places to discuss how they currently get recommendations, who gives them, and how much they trust the information. Maybe the idea is to replace or improve existing algorithms that offer sales information (likelihood to close, timeline for shipping custom solutions, promotional tie-in offers, product recommendations). Then, focus on understanding when they need this information, what channels help or annoy them (SMS, email, automated phone calls, or on the website), and how the placement of the information in the process or frequency of that presentation might influence their opinion.

There are plenty of examples of scripts for interviews. Adapt these templates to inject the appropriate content into an interview script.

Article: `Example user Interview Script` (https://guides.18f.gov/ux-guide/
interview-script/)

Let me walk through some tips and suggestions for interviews.

Defining research objectives

Clearly outline the goals and objectives of the interview. For example, with a survey, we want to know why we ask these questions beforehand. Keep the interview focused on the results suited for the interview process. Please don't turn it into a live survey. If you need demographic information, get it out of the way before. This pre-interview allows tailoring the interview based on the results.

The interview can use samples, demos, or visuals. But the goal is to get them talking and giving feedback, insight, knowledge, and wisdom. Use the opportunity to go deeper and adapt to what they say and how they say it. There can be nuance in how they answer that might reveal hidden truths. Explore existing or previous customer interactions, competitive product use and feedback, style and tone discussions, and a wealth of use case examples, like in surveys. However, unlike a survey, go deeper, easily adjust or drill down the line of questioning, impromptu sign into a website, show a different example, and try to get more understanding of and empathy for the customer.

Selecting participants

Define the target participant group based on research objectives. Ensure the selected participants have the relevant knowledge or experience related to the research topic. If they are existing customers, consider the customer profile (long-term customers, new customers, those with lots of products, those only with a specific product, those who file a lot of trouble tickets, technical, non-technical, those who self-serve through a community website, etc.). Set up a process to find and interview the right audience. Sometimes, it is the survey that will resolve this issue. Here are some tips to help with this part of the process:

- Schedule 20-30% more than needed for interviews of 10 to 20 participants. If someone cancels, you don't have to start the recruiting process again, and results will not be delayed. Don't cancel if you get lucky and everyone shows up; it is not respectful. Either follow the process or look for opportunities to extend understanding with different lines of questioning. Maybe drill down further into areas exposed by subjects where there wasn't time with previous subjects.

- It would be great if your company had someone to recruit and schedule. Expect recruiting to take longer than the actual interviews. So, account for this in the schedule. There are also outside firms that will assist in recruiting.

- Don't book interviews back-to-back. Allow an hour in between. This allows overtime if the customer is okay with it and gives time to write up additional notes while they are fresh.

- The most interviews I have done for one project is 20. There is no rule for figuring out the correct number; with vast subject areas or many different customer profiles, more interviewees are warranted. However, as with discount usability methods, overlapping feedback will start to dominate, and you don't want to be only 10% of the way through interviews and notice the same results over and over. If this was a user test, we have data to suggest a small number, such as five, is sufficient for some tests. However, an interview is less structured than using a piece of software, so consider the cost-benefit tradeoff. Read this article to see how to apply discount usability methods, a practice for iterative small studies useful for context even with the interview process.

Article: `Discount usability methods` (`https://www.nngroup.com/articles/discount-usability-20-years/`)

> **Pro tip**
>
> To improve the rate of people showing up, call (an actual phone call) the day before, reminding them of the scheduled time. They may appreciate the personal touch to confirm the meeting, and if they can't show up, they might tell you right then so you are not sitting and waiting. When it is time, join the web conference five minutes early (I assume most sessions are via a web conference; if they are in person, good for you!), and send them a message with the link one more time. "I joined a few minutes early to make sure I am ready for you to join. I look forward to our session in a few minutes."

Develop a structured interview program

Create a **detailed interview program** that includes standard questions and optional follow-ups. Questions should be clear, unbiased, and designed to elicit the necessary information. Please encourage them to think aloud and to expand on their reasoning. Recognize that some customers don't have anything to say on a subject. Be prepared to skip ahead or branch to the topics of interest. No rule says everyone has to answer every question or line of discussion. However, when doing the analysis, it is nice to say, "6 out of 8 customers expressed the same lack of support for XYZ without being prompted." That is a powerful statement.

In the script, include tips to be reminded to prompt customers to think aloud or expand their thinking. If they are already doing it, great, but when in the zone, you might forget; this is a good reminder.

Keep biases to yourself. We can easily express where we want the customer to go with their feedback. Sometimes, there are cultural issues with providing honest feedback. That is, some people will only give positive suggestions. Feel out the audience. Give them prompts that probe positive and negative discussions. This will help adjust the approach to tease out the most valuable feedback from that participant.

Pilot the interview process and program

Conduct a pilot with a small group of participants to identify any issues with the interview process. This should include technology, sound, and quality checks, as well as checking the entire process to test timing and scheduling issues. Ask the questions as expected to be asked. This will verify if the scheduled time is sufficient and if the questions sound natural when spoken out loud. Record the session. Take note of where something should be changed or review the recording. It is a pilot. Make and correct mistakes now. Use these learning to update and edit the program and try again:

- It is okay to pivot the program even after some interviews are conducted. You won't get as much data for something the interview shifts away from, but if you realize that a different tactic is necessary, make the change. Focus on more concerning areas.

- Just because one user gives excellent feedback doesn't mean it will be heard from more than one user. Proceed with caution if you pivot too soon.

- ChatGPT can edit the program to give it a more natural and conversational feel.

Conduct the structured interviews

Follow the interview program, asking questions in the predefined order. Be neutral in tone and avoid leading questions that might bias responses. Record responses accurately through note-taking, recording, a note-taker, or another method:

- Before starting the interview, ensure participants understand the purpose of the research and agree to participate voluntarily. Generally, we read a short statement that addresses the purpose, how the information will be used (video or audio recordings, quotes, and profile information), and how participant confidentiality will be respected. We also informed them that they could stop and withdraw from the study. A user researcher probably already has this boilerplate information available to them. If you are a designer or product person, use a researcher to run the study.

- If you're screen recording, help the user hide or remove confidential or unrelated materials before starting the recording. This will help build trust and allow the recording to focus more on the project.

- If you are the interviewer, have a confederate take notes. It allows more focus while in the interview. There are a variety of user experience tools that will enable synchronized note-taking. Interviewer verbal cues such as "That is a very interesting point," "We should make a note of that," and "Thanks for your observation" can help the note-taker know what is essential to you and engage the participant.

- Allow other insiders, owners, or advocates to watch or join the call. Be clear that they are silent observers during the interview.

- Those silent observers can chat on a private channel. They might have insights into an answer you don't know or can provide prompts with helpful follow-up questions.

- Probe for clarification and encourage participants to elaborate on their responses. We want their insight, which is more valuable than simple answers readily available from a survey. The interview is a time to *fill in the blanks* and to understand their needs, why, when, with what, how, and who.

- Ensure consistency in how questions are asked and how responses are recorded. Audio or screen-sharing recordings are great to go back and review. Account for at least 2x the time of the recording for review.

- Review the recording to ensure the quote is correct and in context. It is wrong to quote a customer when they say, "This is the best!" but it was entirely out of context (or said sarcastically).

- This is worth repeating a few times: avoid introducing personal bias into the interview process. It is ok to emphasize a specific feature or task, but let the customers speak. Don't push them one way or another. Instead, put them at ease to address complex problems and give honest feedback. If the interviewer feels too invested in the solution and wants to defend it against scrutiny, your interviewee will sense this and not share fully.

- One way to avoid bias is to conduct the interview with a researcher who is not invested in the results. If you are the designer or product manager, let them run the interviews, take notes, and help behind the scenes. This is one of the most valuable tips we can share.

- Be an active listener. Use verbal neutral interjections such as "I understand" and "got it," and reiterate what they said. Use nonverbal cues (nodding and eye contact) to show you are listening.

Record and document findings

Record detailed notes or transcribe interviews for analysis. Generative AI tools are an excellent resource for transcription. Some web conferencing tools have this built-in; turn it on or do it after the fact. This can make it easy to mark up critical elements of the conversation. Document any patterns, trends, or notable insights during the interviews. Once you have this annotated data, turn to the analysis. Some tools allow note-taking with timestamps. This is a great way to log events to return to later. Even if you are alone and don't have time to take notes, type a placeholder (like ***) and return to the recording to capture that insight. The more notes, the better. Ignore off-topic stuff later. Work with an interdisciplinary team to evaluate items that might be off-topic for you but are in scope for other parts of the organization. It is better to catch and discard than not capture and then later realize there might be something to that information.

Data analysis

Analyze the collected quantitative (for closed or demographic questions) and qualitative (for open-ended questions) results. Combine them to form more robust results. For example, what if three out of four novice users didn't think to ask for shipping information, while all four expert users explicitly wanted shipping information? Because this line of discussion was explored, it was revealed that the experts know that the company doesn't ship quickly, and that has been an issue in the past, while the novice users are expecting Amazon-like shipping and didn't think and didn't have enough experience to know that wasn't the case. Only by segmenting this data by the novice/expert role would we recognize how to solve this problem:

- Expect to take at least three times the clock time of the interviews to document the findings (2x) and then do the analysis (1x). If it took a week to gather 10 hours of interviews, it will take the next week to complete this analysis.

- Even small results (found by one or two customers) might be valuable, so be sure to capture all data. In *Chapter 4, Scoring Stories*, we explain how to prioritize based on the number of customers who can be impacted. Even though only one or two customers reported something, it doesn't imply that only a few would benefit. Profound insight can come from one person; you have to recognize it.

Report findings

Present research findings in a clear and organized manner. Use the User Needs Scoring method (*Chapter 4, Scoring Stories*) to prioritize the results. Review the most to least valuable results in order. This way, when an item is discussed, the attendee's attention is focused on the most valuable pieces of feedback:

- Include participant quotes and critical insights to support conclusions. Be sure to quote correctly and in the context, the participant intended.

- Share out-of-scope findings with the appropriate teams.

- File bugs or enhancements into the company tracking tool. Assign them to the appropriate person for resolution.

- Include feedback from colleagues participating in the process.

- Include issues or hiccups from the study to follow up for yourself and the team.

- Have a retrospective with the team to learn and document what to do differently and better next time.

Summary of the interview process

Conducting good interviews is a skill. It takes practice, expertise, and good people skills. Especially for novice interviewers, do a few extra practice sessions before the first official interview. Ask a design researcher to participate in the pilot interviews and give feedback. There is an art and science to doing this well. The Nielsen Norman Group article has some helpful content. They even offer classes on how to perform successful user interviews.

Article: `User interviews (https://www.nngroup.com/articles/user-interviews/)`

Also, once we get a ChatGPT solution running, revisit this process to interview those who have used the solution and those who have not or intentionally not used the ChatGPT-based solution. This applies to any implementation, not just conversational solutions. Once a solution is available, there will be different topics to explore.

Surveys and interviews are only two methods. We encourage using the correct methods to uncover user attitudes and expectations. Later, you might do user testing or a heuristic evaluation. More on those methods in *Chapter 10, Monitoring and Evaluation*. However, we have some context from which to work during this research phase. So, we covered two methods that directly engage customers, focused on their attitudes, with primarily qualitative results. Now, we pivot to behavioral and attitude research with clickstream analysis.

Getting started with conversational analysis

The final type of research we want to cover is **conversational analysis**. We don't have a ChatGPT solution yet, so how can I have anything to analyze? Maybe not, but let's consider other sources. Are there existing trouble tickets or service requests? Maybe there are live-person chats or phone channels. Even reviewing community posts in forums is a good source. Perhaps an existing conversational AI doesn't include ChatGPT integration. This is an excellent source of existing data. Conversational analysis isn't valuable for backend solutions or even recommendation systems. Use other research and needs analysis tools there.

At one company, we kept metrics on the quantity and type of support calls. Unsurprisingly, password reset requests and account activations were broken out. We precisely knew the percentage of calls for these services and could then calculate savings to a change in a process that improved that task. For example, everyone has done a password reset via a web form. After providing some additional private information, and these days, a phone number to send a 2-factor authentication code to, an email is sent with a link to reset the password. It wasn't always this way. The cost of a phone call to do this versus a web form makes it easy to calculate cost savings. A list of service types broken down by percentage is a valuable tool for use case analysis to determine possible opportunities for a ChatGPT-driven solution. As we will see, some use cases are better suited for a generative AI solution.

Consider all sources of customer interaction; there can be data worth investigating. And yes, once a conversational AI with ChatGPT is running, that becomes the primary source of learning to continue the care and feeding process.

Reviewing logs for clues on gaps in support, sales, service, or other features has advantages and disadvantages.

Some upsides include the following:

- It is accurate information about the current state of affairs
- It is likely already grouped and segmented, as existing people and teams support these interactions
- Existing AI tools can help triage and organize this data (we will share one example from HumanFirst in a few minutes)

- It is helpful for people new to the space to understand the support pain points

- It could be a great source of knowledge to feed the solution

There are a few downsides as well:

- It doesn't represent where the solution should go

- It is a lot of data to process

- It is time-consuming (some pieces can be automated)

- It might be self-selecting and focusing the user on only what it can do, not what the user came to do

Conversational analysis is a valuable tool. We will work to address some of the downsides, but I also want to focus on what to do with the data and how to use this research tool to establish the organization's needs. Let's use these methods to drive the requirements, priorities, and a roadmap for improvements.

The best logs to analyze would be those closest to the new ChatGPT solution. For example, existing Chatbot logs are a perfect place to look for existing needs and understand the failures of the current solution. The second best would be chat logs with human agents. These have much more conversational noise, and the interactions are not directed at solving a problem (we call this chit-chat). The analysis of these logs takes more time to filter. Other locations would be community social interactions or public support sites. They are even more challenging to parse. Sometimes, support people monitor those sites and offload challenging interactions into private channels while some community members answer common questions.

In these cases, the correct answer is needed to analyze the logs. We call this **ground truth**: information we know is factually accurate. Sometimes, this is easy. The user gets the answer, tries it out, and reports back. Sometimes, you don't have that feedback. Knowing the space or working with experts can help determine how well the current solution works. A benchmark is needed to understand how well the improved solution performs.

There are automation tools to help organize and parse logs and usage information. Companies such as humanfirst.ai (`https://www.humanfirst.ai/conversational-ai`) provide tools to manage data. This is faster than manually tagging and labeling. It takes examples and then offers lists of similar phrases, as shown in *Figure 2.5*. This works for many use cases. We will review the manual version of these processes so you know how to tag and organize data and will understand if automation is performing well.

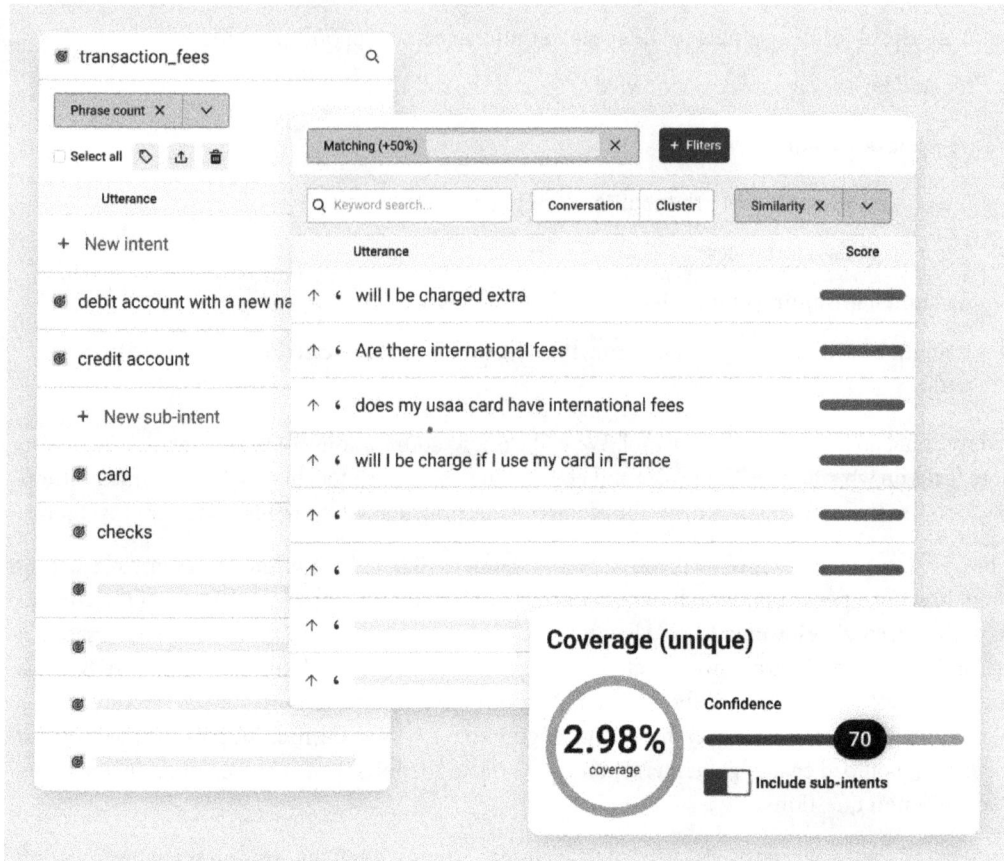

Figure 2.5 – An example of the HumanFirst tool that classifies input into collections

Seek the necessary approvals to access and analyze the material. Sometimes, chat logs can contain **Personally Identifiable Information** (**PII**), which makes security teams worried about distribution. We agree. So, work with them to mask this information, as we don't need access to this PII. When appropriate, work with them to prevent this information from being stored in the first place. We have seen some reasonable solutions that use automated systems to make a payment, authenticate, or mask sensitive data, so there is never a human in the loop with access to something they should not.

Here is my proposal for measuring and creating a benchmark to judge future improvements and a tool to identify gaps in current processes that become candidates for the ChatGPT solution.

1. Collect a recent log file, hopefully a chat log file, as discussed. Gather at least 1,000 interactions (a question and a response are one interaction).

2. Mask any PII as needed.

3. Tag each interaction as a success or failure. Subclassify the interactions to understand the issues that need to be addressed.

4. Analyze the results for improvement opportunities (or places where the ChatGPT solution might solve the existing problem).

5. Score the 1,000 interactions and use this score (% success rate) as an improvement benchmark.

To do all this, we want to teach skills to tag interactions and classify the opportunities for improvement. Once this is done, the score should be easy to measure. An expert in the space can, even with very technical material, judge the success or failure of each turn with about 99% accuracy. Even novices can correctly judge 90% of interactions. So, with limited expert availability, ask for their help on that missing 10%. Over time, novices will reduce the 10% gap. Let's get started with how to tag a log file.

Tagging a log file should focus on each interaction

Each back-and-forth is a single interaction. If a question is asked, it should be answered. We can analyze each question if multiple questions were asked before getting an answer (like from a community forum). The answer might cover none, some, or all of the questions. Each answer might be a success or failure, each counted individually.

Figure 2.6 is a sneak peek at a conversational log we will use in our drill-downs. We will provide a spreadsheet template to use with your data.

	Text	Choices	Source	Classification	What	When	Where	Utterance Issues	Notes	Usage Intent	Date	Time	User Id
104	manager keying maternity		USER	Failure > Utterance	Failure	Current	Utteran Mismatch			Manager Action	8/1/23	9:58:16	3001750310360
105	For detailed information, please review our Maternity Leave Policy. To estimate your leave and pay, use our Maternity Pay & Leave Questionnaire. Maternity Leave Policy: https://www.gov.uk/maternity-pay-leave Maternity Pay & Leave Questionnaire: https://www.gov.uk/maternity-pay-leave		BOT								8/1/23	9:58:16	3001750310360
106	eye test		USER	Failure > Utterance	Failure	Future	Utteran Unrecogni			Benefits	8/1/23	10:01:29	4440003502698
107	Hey there! You can find all the <a href="https://xxxx.hcmcloud.com/fscmUI/objType=CSO_ARTICLE_CONTENT_KM&objKey=docId%123456789&action=EDIT_IN_TAB">HR Forms</a> in one spot on our Knowledge Base. You can bookmark it by clicking the star icon, making it easily accessible under 'Favorites'.		BOT								8/1/23	10:01:29	4440003502698
108	£100.00 uplift payment		USER	Failure > Utterance	Failure	Current	Utteran Mismatch			Pay Questions	8/1/23	10:02:11	4440003502698
109	Deputy employees are eligible for an in-charge payment when: - They oversee an outlet (Monday to Sunday) for at least 4 consecutive hours, covering either the opening or closing time. - They do not claim an additional responsibility payment for the same period. - They are not officially in managerial roles. For more details, please refer to the Acting Up Arrangements Policy https://www.elft.nhs.uk/sites/default/files/2022-01/acting_up_policy.pdf		BOT								8/1/23	10:02:11	4440003502698

Figure 2.6 – A snapshot from a conversational log being analyzed

Masking tip

Sometimes, teams are uncomfortable with pure log files because of the PII. If this data still exists in the file, one suggestion is to mask it. Typically, the data to be masked is *numerical*, such as social security or credit card numbers or people's names and emails. Some countries might have specific rules around PII, so understand your region.

Hopefully, you have automated scripts for masking user data. Tools from vendors like IBM, Oracle, ServiceNow, and Salesforce have data masking features. There are third-party tools readily available. If not, we will help with a simple and practical approach. Do it manually and then turn it into a script:

1. Numerical data can be substituted or masked by simple Find and Replace features in any spreadsheet. Look for patterns for numbers (Social Security, Phone, Amounts) and replace them with 1s.

2. The @ sign can be found and used as a marker to replace email addresses with something generic in email patterns. I suggest keeping the same patterns in the file and not deleting the now-masked data.

3. You might have data around a specific type of masked data that could be useful in automation or process improvements. For example, how often do customers provide a PIN or security code to a human agent? This can give a sense of the scale of the problem, how much effort is involved, and if there is an opportunity to resolve this with automation or a generative AI solution.

4. Dates can be manipulated. Consider whether they must maintain a sequence for the data to be valid (such as when charting progress).

5. Consider whether masking should use random values so someone can't quickly reverse engineer the original data.

6. I have rarely found that masked data impacts understanding.

Table 2.1 depicts some examples of masking.

Type	Original	Masked
Social Security # (SSN)	123-45-6789	*-*-6789
Credit Card Numbers	1234-5678-9012-3456	--****-3456
Names	John Doe	J*** D**
Email Addresses	john.doe@company.ai	j***.d**@company.ai
Phone Numbers	(555) 123-4567	(***) ***-4567
Addresses	5622 Main St, City, Country	5622 Main St, C***
IP Addresses	192.168.1.1	192.***.*.*
Usernames	Johndoe123	J*******
Employee IDs	EMP-00123	EMP-*****
Date of Birth	07/15/1975	01/01/1980

Table 2.1 – Typical masking examples for PII

Once the data is masked, consider outsourcing or using cloud analysis tools. By reviewing and analyzing the logs, we can define success and failure.

Define success and failure categories

Each row in a spreadsheet that represents a response (from the human agent, a bot, or someone on a community site) is either helpful or not. We classify interactions into success, failure, unclassifiable, or ignore. However, there is value in a more robust breakdown depending on the log analysis and the kinds of fixes that issues might necessitate. Problems can be aggregated and allocated to different teams for resolution. *Consider collapsing categories if the same team is going to resolve similar items or ignore some of these if the type of data collected makes them unnecessary.*

Think about the user's intent, which should match to an answer. The content presented is the response to the user, and it is organized and presented in a structure, such as a bulleted list or just text. We can determine if we gave a good answer, an answer that needs some improvements, or one that missed the mark for various reasons. This forms the system of classification.

The goal of categorization is not just to see where problems exist but because this classification can be used to evaluate the effort needed by different teams to improve the overall quality. Picking a category is intended to be done in a manageable amount of time. The time in the log file is well spent to get an overall sense of quality and then have the specific issues to address, and they also can create new test cases for future iterations.

Do not take these categories as perfect; add or remove some and annotate others based on the value of classification and what the team will fix. Some may be done by the prompt engineers, others by the fine-tuning squad, and with a voice interface, the test-to-speech team will work on improvements. Moreover, identify when the knowledge base articles need improvement. Here are the detailed categories.

Category – Success

The question was answered. We are happy with it being complete, correct, and up to date. It is much less costly to classify successes than any issues. Classifying the first 1,000 responses with a 60% success rate will take longer than later iterations with a 95% success rate. Just keep at it and judge each interaction on its merit.

Category – Qualified Success > Response > Content

Qualified Success, but the response has a *content* issue.

Even though the answer was correct, it contained additional content, which might cause some issues. The content could include unrelated or immaterial information or too much or too little content, which might relate to a knowledge base problem. For example, including the phone number when asked for an address is reasonable, but if the answer also includes the company's mission statement, then it would be immaterial. A recommender including references to something unrelated would be a content issue. This needs to be addressed in the data and training.

Category – Qualified Success > Response > Structure

Qualified Success, but the response from the system or agent has a *structure* issue.

The structure includes examples where the order of results matters. The answer could repeat steps multiple times, put something in the wrong order, or contain results that should be excluded. The structure category contains formatting issues where a table or a numbered list should have been used. This could be a content team problem and be addressed with prompt engineering or more complex use of multiple models.

Category – Qualified Success > Response > Grammar

Qualified Success, but the response has a *grammar* issue.

Any spelling, grammar, or editorial issues would qualify. Typos might come from source documents, while formatting, such as correctly using no space before a colon, might be challenging because this might come from the basic model training. If they can be reproduced, point them out in a training session with the model and tell the model these are wrong and how they should be fixed. The editorial team will likely solve this.

Category – Qualified Success > Response > Voice and Tone

Qualified Success, but the response has a *voice and tone* issue.

This includes the correct answers, but communication should be in a more appropriate tone. Long-winded answers can be classified here. If the voice is inconsistent with the context (for example, happy when talking about a sad subject or too serious when handling something light-hearted), include it here. Linguistics and design should address this issue by prompt engineering (*Chapter 7, Prompt Engineering*). Review articles to see how they might have been written. Documentation is critical and covered in *Chapter 6, Gathering Data – Content is King*.

Category – Qualified Success > Response > Provisionally Handled

Qualified Success, the response was *provisionally handled.*

This is a unique section for handled answers, where a better answer is planned for the future. It allows you to track things that are coming. For example, the assistant will handle real-time weather reports in the Q3 release. In that case, temporarily linking to a weather service until the weather API is available is a way of provisionally handling the use case. This is for the Product Manager to track and prioritize these issues.

Category – Failure > Utterance > Mismatched

Failure: the input was *mismatched* to the wrong answer.

There is a mismatch between the question and the answer. This is where we asked one question (How fast is a jaguar?), and the AI provided an answer to the *wrong* question. Did the user mean the animal or the car brand? Or it didn't have a clear answer, giving the proper instructions but for an incomplete question. For example, it answers the question of how to reboot a phone, but without confirming the model of the phone, it gives the wrong instructions. These are critically important issues that the AI team needs to address.

Category – Failure > Utterance > Unrecognized > Current

Failure: the input was unrecognized, but we have a *current* response.

It means we (humans or AIs) didn't understand the question, but we could have if we had asked another way. Sometimes, this is due to custom language or company-specific initialisms when the model isn't fine-tuned (explained in *Chapter 8, Fine-Tuning*) with these extra details. If this is generally unrecognized, log it here; if it is specific to a custom word or phrase, use the custom category. With generative AI on the front end, this will be rare. The AI is a people pleaser, so it will try to provide an answer. But if ChatGPT is inside another tool or process, the output might cause a downstream system, response, or recommender to fail. It could be a content or AI issue.

Category – Failure > Utterance > Unrecognized > Future Development

Failure: the input was unrecognized; we should support this in the *future.*

These are the basics of enhancements or new features. As mentioned, the ChatGPT instance will want to answer no matter what, so your system will likely return a mismatched answer. Depending on the data, it is ok if all of these get tagged as a mismatch. If you are running a traditional chat solution or logging against a human agent answering, use this for times when the system or human can't answer because it isn't supported. Typically, these items are for the backlog and are addressed by the Product Manager. This backlog can be reviewed, scored, and prioritized. This is usually done on a regular cadence as part of sprint planning. We will cover a process for these items in *Chapter 4, Scoring Stories*.

Category – Failure > Utterance > Unrecognized > Customer Unique – Future

Failure: the input was unrecognized but should be handled by *another team* in the future.

Similar to the future intent, if this is something that another team or sister organization (or where multiple ChatGPT instances are joined together, and this instance doesn't support this intent, but another one should), log the problem here. Another team's Product Manager should handle these new requirements.

Category – Failure > Utterance > Unrecognized (Missed) Entity

Failure: the input needed to be recognized. A *custom* or company-specific word needed to be understood.

We segment out this type of issue only because this is typically a more direct issue to fix. Fine-tuning or using **Retrieval Augmented Generation** (**RAG**), which we will explain in *Chapter 6, Gathering Data – Content is King* can solve this problem, but first, try prompt engineering. If there is a company thesaurus, put it to good use. At Oracle, there are many initialisms and shortcuts for named products (RDBMS, Oracle DB, Oracle Server Enterprise Edition, DB12, Oracle Server, Oracle SQL Database, OraDB, and other names all mean the Oracle Database). The system should understand terms that are unique to the company and culture. This is also a good reminder in general. Human support agents will likely know much of this. An AI must also trained to learn this material. The AI team should address this training issue; the writers on the team should have the data.

Category – Failure > Utterance > False Positive

Failure: *misunderstood* (a false positive).

The answer is wrong, but someone or something thinks it is correct. This is important for human answers as well. Humans tend to be overconfident, so it is essential to recognize if their truth is the ground truth. This can lead to a better understanding of the quality of the solution. There is a fine line between this and *mismatched*. It might be pedantic, so merging the two classifications is okay. The AI team typically addresses these. We include this for completeness.

Category – Failure > Utterance > Irrelevant

Failure: the input was unrecognized and *irrelevant*.

If we have customer responses during live logs, such as "Are you there?" or "Hello????", they should be cataloged, but they are primarily irrelevant from a model or use case perspective. We see them in live agent chat logs. The agent might be handling four or five chats simultaneously; thus, delays are inevitable. A conversational ChatGPT agent won't be distracted by the thousands of other simultaneous interactions. Garbled or garbage messages will also go here.

Category – Failure > Interaction

Failure: the failure was due to the type of user experience provided or something structural in the response.

The answer might be the correct content, but it could require a UI component to filter the choices based on the product's version. In this case, the channel doesn't support a drop-down menu to make the selection. Or a button is rendered in the answer to start a process (such as uploading a file), and nothing happens when the user clicks the button. I saw this recently, and the problem was browser-specific. Solutions need to understand the context of use. You need to know that it works on a robust channel with interactive elements. Conversational AI must adapt to the channel's limitations if the same generative solution is used on a less robust channel. The engineers prompt engineering and development work to do. Or there can be unique instructions for each channel. Instructions, a form of prompt engineering, are discussed in *Chapter 7, Prompt Engineering*.

Category – Failure > Response > Content

Failure: The response content is wrong or is the right content but still has inaccuracies or incorrect details.

Sometimes, these take time to determine. Did we return the correct document or agent response but lack some additional context to give the proper details? Or it could be that the document is accurate, but the facts are wrong. This differs from *mismatched*, where the document might be valid but matched to the wrong question. Here, something we wrote needs correction. The content team should update the document.

Category – Failure, No Engagement

Failure: there was no response.

This is more common with social and community posts and less common with conversational AI, which always tries to provide an answer. For some systems, no response could have been classified as an error, but we include it for completeness. Product Managers can decide if these should be considered.

Category– Failure > System Error > (further classifications, if needed)

Failure: *system error* due to timeout, error message, infinite loop, or no response.

Typically, this is found when working with a chat solution or integrations that might be down. Of course, we don't ever want our customers to see these. They are not common, but they are worth capturing, logging, and getting fixed when they occur. A well-designed user interface might suppress some of these errors, so scan system logs for errors. For example, if a knowledge base connection is down, and a ChatGPT solution ignores that it is down, then the answers to questions might be wrong, and it might not be evident in the log of this problem due to a system error. Monitor resources or set up alerts to notify the right administrators when a downed service is detected. Development or operations should handle these failures.

Category – Unclassifiable

Sometimes, we don't know what the customer was trying to do, and likely, they came and went in the conversation with so few interactions we can't piece together enough to infer their goals. Unclassifiable utterances will probably not get addressed. This will only be frequent with a new analyst. Provide support from experts who have more experience in classifying this data.

Category – Ignore

This could be chit-chat, test data, or garbage interactions. If no other category fits, and you don't want to deal with this, put it in this dumping ground to ignore. I found that in enterprise logs, this is very rare (less than a fraction of 1% of interaction).

We are looking for groups of classifications so that we can continue to improve the models. Some of this might go back directly into existing solutions as ERs or enhancements, but it also forms a basis for what is wanted for the new solution. If the new ChatGPT solution replaces or supplements these tools, use these benchmarks to know if future iterations are better. Aggregating these into three buckets (Success, Qualified Success, and Failures) makes an excellent chart for tracking progress.

Developing categories for analysis

In this phase, we are focused on learning what is working and what is not. We use this to understand existing content sources and input them into our conversational AI process. You might also be fixing an existing system while building the generative solution. Categories might be of value for that effort. When we have a generative AI log, we can use the same process, maybe with different categories, to analyze how we are doing. By collapsing interactions into groups around a specific issue, we can determine how many customers are impacted by issues. We can see priorities for our problems (and solutions) based on our understanding of the severity. We will teach how to score these findings in *Chapter 4, Scoring Stories* to create and maintain a backlog of issues.

Once we deploy our ChatGPT solution, we can re-run this analysis on its use and use the results to drive further improvements.

> **Pro tip**
> Use categories that make sense for the data. This is a tradeoff. The more categories, the more time each classification takes. With fewer categories, more explanations and details are needed to understand the problems.

"What gets measured gets managed"—Peter Drucker (quality maven and world-famous management consultant). By creating measurements, we can communicate the quality and improve it. Improvements become increasingly complex over time. It is one thing to go from 67% to 97% success; it is much harder to go from 97% to 99%.

Let's see how we can do this in practice.

Trying conversational analysis

I have placed two files into GitHub to explore. The first is a case study conversational analysis for a human resource tool. One can imagine the kinds of questions that come to HR within a company. When is my check coming? How much vacation do I have left? I need to change my last name. Can I get a discount at the company store? Within the file, explore our reasoning for the classifications for hundreds of examples and see how we summarize and aggregate the data. Use this as a basis for data analysis.

GitHub: `Analysis Log File (https://github.com/PacktPublishing/ UX-for-Enterprise-ChatGPT-Solutions/blob/main/Chapter2- ConversationalLogAnalysis-2024.xlsb)`

That brings us to the second file:

GitHub: `Blank Log Analysis File (https://github.com/PacktPublishing/ UX-for-Enterprise-ChatGPT-Solutions/blob/main/Chapter2- ConversationalLogAnalysis-2024-empty.xlsx)`

This is the empty case study file for your data (from one of the sources we discussed earlier), so run a data analysis. *Using a tool that automates this process won't give you a good feel for the user's experience.* Doing some of this manual analysis forces thinking from the customer's perspective with their experiences and an eye on improving. The file contains various spreadsheet tricks, such as pivot tables, which can be hard to maintain, analyze, and generate summaries. There are videos and documentation on these subjects online that can be used to start using pivot tables. I provide the files to show what is possible with standard tools. Please use them to get started or find similar support from other tools.

Exploring the examples from the case study

The examples from the GitHub file help us form a common understanding of how to classify interactions. Recognize that expertise around the material being classified will go a long way to doing a good job. Someone entirely new to the classification material might only be able to annotate 80% of the responses correctly. This number can rise to 99% if you are an expert.

I selected examples from the GitHub files in *Table 2.2* as success examples and *Table 2.3* as failure examples because they should be familiar to anyone who has worked at a large enterprise. This file is what would be expected from a conversational AI log. Each prompt will have a response. Multiple questions posed at once might not get all the answers, system errors can occur, and because of the model, some entities might not be trained, thus missing. This type of log file will expose us to most of the classifications explained. Other files, like social media posts, won't have this variety.

Row #*	User Prompt	Failure	Reason
514	I can't find the Bullying policy which explains the steps I need to take to raise a bullying grievance (sic)	Qualified Success > Response > Content	Correct response, but additional content and direct action would have been better for this instead of just a link to the policy. This is a critical but infrequent interaction. It requires immediate attention.
1113	2024 holiday	Qualified Success > Response > Grammar	The answer was not well written and asked them to say 'Book absence' so it knew what to do and didn't do it.
1546	how do I change a job title on HR Cloud	Qualified Success > Response > Structure	The user can, when allowed, change their discretionary title. There were a lot of links and options, and the structure of the response could be improved.
1588	Raise a Service Request	Success	Started the SR process.
* In the spreadsheet sort by Row # to see the conversations in the correct order.			

Table 2.2 – Examples and explanations of successful interactions

Order the spreadsheet by user ID, date, and time to see a user's complete thread. This allows an entire conversation in order without other interactions being in between. To sort the file in the original order, sort by row number.

Log files follow different formats. There is no standard out there. The critical information is what the user said, and the system responds with the date, time, and user ID. Extend this with the labeling information as we did. Notice how I build a menu of classifications in the **Menu** and **Tips** tabs of the spreadsheet. Then, when we include this column in the log sheet, we can quickly and repeatably select from the list of classifications. We also further classified the data by the intent of the user. Since those intents roughly group to different teams, we can share items related to absences, HR details, benefits, and so on with the corresponding teams.

Table 2.3 is just a sample of the types of prompts we see. Explore the examples in the GitHub spreadsheet for many more. This is challenging science. Even when reviewing this data, I saw one place where it was better to reclassify the utterance. The more you become an expert, the easier classification becomes. The more straightforward the classification, the better one can judge issues.

Row #	User Prompt	Failure	Reason
141	Balance (sic)	Failure > Interaction	They were trying to find their vacation balance. A typo in their entry prevented success. The wrong response was returned.
219	what is the service level for actioning promotion requests	Failure > System Error > No Response	The system did not respond.
604	Payrise (sic) not showing on HR Cloud	Failure > Utterance > Mismatched	Should have gone to the current pay stub details for a pay raise. Could also consider if it should compare the old and new salaries to reveal if it sees a change.
750	car allowance mileage claim	Failure > Interaction	Should have started an expense, or at least reviewed if existing claims were in progress.
915	what do I do if I need a sick day	Failure > Utterance > Mismatched	The Absence answer probably isn't correct. They needed help to actually take sick time.
1341	where do I send my P45	Failure > Utterance > Mismatched	A P45 is a UK form showing taxes paid this year. We returned a final salary result, which is not only unlikely but disconcerting if they are still working at the company.
2578	I want to promote	Failure > Response > Content	It responded the user didn't have permission. This is interesting because it is possible that the user is not a manager, or we need more context.
3851	date of birth	Failure > Utterance > Unrecognized > Current	Only by seeing the thread do we know they actually want to view a team member's date of birth, not edit or confirm their own. We didn't recognize this.

Table 2.3 – Examples and explanations of failed interactions

The spreadsheet shows more than the **Log Data**, **Menu**, and **Tips** tabs. Explore the other tabs that summarize results. *Table 2.4* explains the other tabs:

Tab	Purpose
Success Summary	Charting and summarizing the overall quality of the interactions being analyzed. Useful to compare results over time to chart improvements.
Log Data	This is the raw data tab. Some additional summary and analysis columns are generated. We suggest at least 1,000 interactions.
Utterance Count vs Success	This measures the quality of the interaction based on prompt length. LLMs benefit from longer prompts. With traditional conversational AI shorter prompts are more accurate. Three to four words returned the highest success rate in this example.
Daily Users	This is useful to trend in longer periods of time. Stickiness is valuable. People returning and using the skills more and more. Use the daily chart to mark where events occur that trigger increased usage, such as pay day, a merger, or a product announcement.
Repeat Users	This is of particular interest if the same users return. This means they are getting value and continue to see value. It is dependent on logs supporting unique identifiers over time. Systems with authenticated users should always have this identifier. This can track trends in repeated usage. This metric should be going up.
Unique Users	This gets to marketing and growth. Is the word getting out? Do new people engage? Flag dates for announcements or email campaigns here to find correlations.
Intent Classification	This is very specific to the data. In the case of the example, these are the baskets of important tasks and areas the log covered. This is a valuable chart to help us decide how to invest in improvements. Use the failure rate on the intents with the most usage as important. A high failure ratio for a well-used intent is worth investigating.
System Errors	This is also specific to the service. Hopefully, there are no errors. If they occur, catalog them, and get them addressed by the right teams.
Conversational Buttons	This is specific to the data. The conversational AI had sets of buttons that might appear depending on the flow. When they were presented, did users click on them? What buttons were not used? This is helpful in this case for removing buttons that didn't have value and for creating good labels for buttons that were of value.
Menu and Tips	The drop-down menu that is used in log data. Some analysis suggestions and notes are also included. The drop-down menu can be customized and reduced to make sense for the type of data being analyzed.

Table 2.4 – The tabs in the conversational log analysis spreadsheet

We will now focus on grouping issues to manage this enormous influx of issues. In this case, we explain the process assuming a conversational log. However, you might need more time or energy to generate bugs if the logs are from message boards, human agent logs, or service request threads.

Generate enhancements and bugs from groups of issues

So, grouping by specific issue is typically straightforward but takes some time. For example, in the case study, we tagged the user interactions by the feature and content we expect to resolve that line of questioning. Utterances such as "career break" and "sabbatical" should be grouped to form one issue to resolve. If five entries around the style and tone are too negative for a line of questioning around bank balance, then it would be better to have one issue logged with five supporting examples. We group to get a sense of volume around specific topics. This also helps because no one wants to see bugs on 1,000 interactions; maybe there are 80 issues worth reporting, and each issue is a group of 2 to 50 similar interactions.

For example, in one log analysis, we had 19 interactions around how the system handled direct deposit (automation for paychecks to go directly into their bank account). The system supported the answer, but the model didn't recognize the indirect words (paper check, savings, new account, bank deposit) for the direct deposit interaction. So, one tag for direct deposit issues allowed us to collect all the related examples and form a corpus to train the AI.

Score results

A benchmark to compare against when a new solution is deployed is good. Scoring is pretty straightforward. Take the total number of interactions measured, maybe a few that can't be classified, and sum up the successes and failures. Create a percentage from the data. Use this as a benchmark and run the same analysis on the ChatGPT solution for each significant iteration. We will explore other tools for charting your success as well. But in all cases, always measure performance. And now there is a benchmark to exceed as the ChatGPT solution matures. Score results, and comparing at every milestone.

The intent is to share progress within the organization, showing how each assistant improves. It could compare different assistants or the same assistant on different channels. Use this information to help meet quality goals. *Figure 2.7* shows actual data from three assistants. The **Minimum Viable Product** (**MVP**) line is where to enter live customer testing, and the **High-Quality** bar, at 97%, is where we would be happy. Product C version 3 was in development then, hence the dotted line. Product B came late to the game, so it shows the first and second iterations starting at release four for Product A. And iterations do not have to mean releases. If quality decreases, like in Product A, don't release that version.

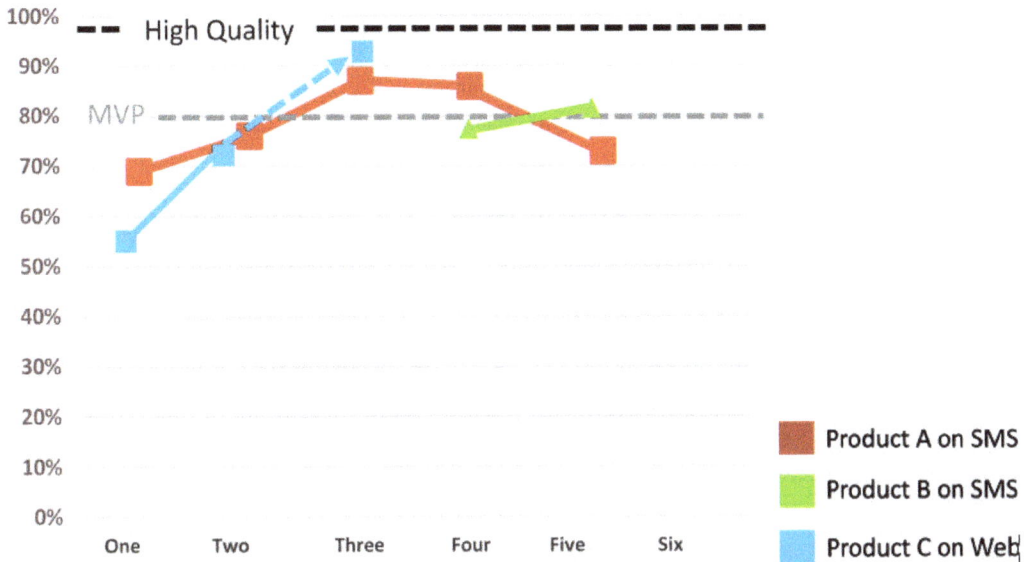

Figure 2.7 – Comparison of assistant scores across iterations

Results

With a simple spreadsheet summarizing the results, a range of quality-related questions can be answered:

- What is the success and failure rate of the current solution?
- Identify and rank the issues using the highest to lowest user needs score
- Document the number of interactions for each issue
- Chart growth in usage from external events or triggers
- Recognize and calculate stickiness – the likelihood of repeat users

Identify and segment issues to identify the work to be done:

- Within the scope of the current solution (it should work)
- Missing from the current solution (it should be added)
- This is unrelated to the current solution (some other solution should solve this)
- Not relevant (gibberish, out of scope)

This work will form a design vision for how the ChatGPT solution should work, goals for the kinds of answers it will need to answer, and even how to establish the right style and tone for responses. This kind of analysis offers a wealth of learning beyond the numbers and data.

Summary

So, through any method, we can start to understand user needs. The survey and interview processes are related but very different. Along with log analysis, this is a good start. You might realize, as we all have to at some point, that there is more work than time. So, how does a team decide what to do first? It seems apparent to do the most important things first, but that is only sometimes the case. And there is some math behind that to help you believe it. In *Chapter 4, Scoring Stories* we will explore how to prioritize research and customer needs (user needs). This method works on any list of activities, projects, tasks, upgrades, patches, or new features. I suspect that to build an effective ChatGPT solution, you will likely focus on creating or curating content to include in the model. What should you do first? This is where **Users Need Scoring** and the concept of **Weighted Shortest Job First (WSJF)**, an Agile method, comes to good use. We need to take all this great information collected and put it to good use. We are focused on ChatGPT, but it can't always provide a good solution. When we use it, and for what purposes will it come into play? We have one more stop before exploring scoring. We need to identify the optimal use cases for enterprise ChatGPT solution.

In this chapter, we tried to set up learning about what was done and what we can learn from customers about what to do. In the following chapters, we will focus on getting it done. To put your role as a UX or product leader in perspective, I would quote Peter Drucker again: *"Management is doing things right; leadership is doing the right things."* So, now it is time to use this research and feedback from internal and external sources to determine what use cases can be identified for our solution.

References

The links, book recommendations, and GitHub files in this chapter are posted on the reference page.

Web Page: `Chapter 2 References` (`https://uxdforai.com/references#C2`).

3
Identifying Optimal Use Cases for ChatGPT

A use case is a description of the possible sequences of interactions between the system under discussion and its external actors, related to a particular goal.

- Alistair Cockburn

There is a wealth of explanations around writing use cases. I suspect we have all created something we call **use cases**. We will first clarify a robust definition for a use case to work from the same understanding. Then, we can go into examples within use cases to look for opportunities to insert generative AI into the mix, while considering where a ChatGPT integration might be challenging or where issues might arise. By the end of this chapter, you will be armed with the ability to take the research tools from *Chapter 2, Conducting Effective User Research* document use cases in this chapter, and be ready to prioritize use cases for a backlog in *Chapter 4, Scoring Stories*. Experts in use case development will benefit from learning how to apply ChatGPT to the use case process. I encourage all readers to be knowledgeable about use case development, as it is not just for designers!

In this chapter, we will cover the following topics:

- Understanding use case basics
- Aligning **large language models (LLMs)** with user goals
- Avoiding ChatGPT limitations, biases, and inappropriate responses

Understanding use case basics

These five essential use case components have been consistent over the years:

- **Primary actor**: Who is the person or type of person who will use this case?

- **Preconditions**: What has to have happened for this use case to be used?

- **Triggers**: Why does this use case happen now?

- **The path**: The steps in the process and the system response to those steps.

- **Extensions**: Exceptions and variations to the primary path should be considered and accounted for in the process.

> **How to pick the proper use cases**
>
> I suggested Alistair Cockburn's book in *Chapter 2, Conducting Effective User Research* as the best primer on use cases because he explains how to write them effectively. We should also be sensitive when picking the proper use cases. It's possible to write a well-formed use case that does not represent significant business value. Cockburn's book covers the proper construction of use cases and provides examples. Consider printing out the one-page summary from *Chapter 1, Recognizing the Power of Design in ChatGPT*. It has plenty of great suggestions to help write accurate use cases.
>
> Book: `Writing Effective Use Cases (https://amzn.to/3YnbGSp)`
>
> Book (PDF): `Writing Effective Use Cases (https://www-public.imtbs-tsp.eu/~gibson/Teaching/Teaching-ReadingMaterial/Cockburn00.pdf)`

Use case or user stories

Some people have heard of **user stories**, and the overlap with use cases is significant. User stories are typically more general and treated as a narrative. As we explained in *Chapter 2, Conducting Effective User Research* we tell an actual story with the storyboard cartoon. At the same time, a use case is more descriptive of the interactions. Storytelling can set the mood and integrate the context and actor, while the use case defines this upfront. Writing a story from a use case is easier because what should happen and the order of events is known. Going from a user story to a use case is more complex because of the need to address the details of each step. The focus here is on the use case approach to provide details to estimate and understand the value of the interactions and the engineering effort needed to address the problem. It would be easier to ask ChatGPT to create a story from a use case than to make the use case from the story.

We can create use cases and use the process to identify steps suitable for ChatGPT. We can also learn enough about when not to apply ChatGPT.

Here's a step-by-step guide:

1. Understand how to write compelling use cases.

2. Identify the use cases and any needed user stories.

3. Break down the use cases into steps.

4. Identify places in those steps where ChatGPT can provide support (and where it would not be suited).

5. Work with the development team to explore the value of ChatGPT to solve specific steps in the problem. Prioritize the ones with the most value and least cost (more on this in the next chapter).

6. Ignore problems that are not well suited to ChatGPT.

We will use a sample use case to explore how to include ChatGPT in the process. First, we can review how ChatGPT might be engaged within the use case.

Establishing a baseline with ChatGPT

Considering all the activities customers do and even within the company for employees, this can add up to many existing use cases. Some will be ripe for improvement by integrating a generative solution such as ChatGPT. Conversely, avoid applying ChatGPT when it is of limited or no value.

Throughout the book, we cover different ways of understanding what ChatGPT will be good at and what it won't. Because we expect solutions to merge enterprise data with one or more generative models (and some specific models, as discussed in *Chapter 6, Gathering Data – Content is King*, are better at particular tasks), it is good to understand what it can and can't do.

In this context, there might be a chat interaction, recommendation UI, enterprise workflow, or even a backend solution, but ChatGPT might only apply part of the problem. ChatGPT was asked what it was good at and provided the results in *Table 3.1*.

Good At	Not Good At
• **Guidance through logical processes**	• Real-time interaction
• **Breaking down complex procedures into steps**	• Controlling physical devices
• **Providing explanations and examples**	• Highly specialized knowledge
• **Adapting responses to user queries**	• Rapidly changing information
• **Tailoring guidance to specific needs**	• Sensitive or controversial topics
• **Structuring and explaining workflows**	• Nuanced emotional or interpersonal dynamics
• **Assisting with troubleshooting and tasks**	• Providing professional advice or judgment
• **Offering personalized assistance**	

Table 3.1 – ChatGPT's opinion of its capabilities

This book covers some of these items from *Table 3.1*. Challenging areas like sensitive, emotional, or relationship advice are only briefly addressed. Most enterprise solutions will need to handle the problem of specialized knowledge by using a model that can be adapted to support enterprise knowledge. We will cover this in subsequent chapters. Even ChatGPT fails to explain that more value might be gained from applying ChatGPT at a process step, optimized and tuned to do one specific task in a sequence. ChatGPT doesn't have to do everything (as we say, don't *boil the ocean*). That is the beauty of creating a use case to break down a process into a series of steps. It is easier to compare step details to the capabilities of ChatGPT.

Let's take an example. Customers commonly deploy patches or firmware updates for a variety of products. A mobile phone gets regular updates, and enterprise software gets patched, even in the cloud. In the cloud, consumers rarely see this as patching is done behind the scenes, typically without an outage for critical systems. Someone is working on this problem on our behalf. With personal devices such as phones, TVs, and even a toaster, the user manages the process: download software updates, schedule it for later (maybe while sleeping), or pick and choose when and what update to take. For the enterprise, there might be dozens of patches and updates to take and test before deploying them. This is the big brother version of updating a phone.

Here is an example use case of a simplified enterprise system consisting of a step, intent, and system response (processes for patching can be far more complex because of customization, integrations, and legal or regional requirements):

- **Primary actor**:
  - System administrator, patch manager, or by operations

- **Preconditions**:
  - The user is signed into the system
  - The software patches are available

- **Triggers**:
  - Recommended available patch details are emailed to the actor

The following section will expand on this basic patching use case by identifying where ChatGPT can be applied. This is brainstorming: find opportunities in the use case where gaps exist and identify what ChatGPT can do for customers at each step.

Example use case for a ChatGPT instance – patching software

This use case is for a traditional graphical user interface. Even in a single use case, there might be multiple applications of a generative enhancement to adapt the **user experience** (**UX**). We consider the mantra that *the best user interface is no user interface*. Though an overarching goal might be to automate everything, for this example, we presume a complex scenario where human interaction is necessary. But automation of steps along the journey is possible.

We can expand the details of an enterprise patching example. Large enterprises frequently manage their software deployments. Even offloading this work to the cloud just transitions the problem to the cloud service provider. The problem is still the same for those vendors. This software is 100 times more complex than our personal updates, containing 1,000 to 10,000 possible patches, with hundreds of instances of that software running in various departments and on multiple development, test, and staging servers. Knowing precisely what is patched and how that might impact the computers becomes essential. It is a complex and time-consuming process. This is partially why we see so many security breaches in the news. It is not because the software is insecure but because the processes are complex and fraught with possible error. In most cases, vendors have already provided patches to fix security issues, sometimes years earlier. Companies are reluctant to patch because of the complexity and sometimes the risk involved in bringing down a piece of software, which means their systems are vulnerable.

Here is a simplified version of this complex process to help us work with this use case. Notice how each step in *Table 3.2* resolves to actions or experiences that the system needs to address. UI will be designed from this use case, but *we are not defining the UI in the use case*. The high-level needs of each step are documented. There are better times to determine the exact information, layout, and details for patching with the UI design. There is an entire flow and details just for the patch detail's view; the primary use case does not cover this. The same would be said for the plan concept introduced in *Step 3*. A plan is a collection of patches we can test and deploy simultaneously. A plan would have more than just the name. A separate use case can help define the metadata for the plan (who has worked on it, dates, contents, state, and deployment details). Review the steps in the use case to understand the types of interactions we might need to handle at each phase. This use case is about viewing patch recommendations, deciding which patches to include in a plan, and then scheduling and deploying those patches. Based on our understanding of what ChatGPT is good at, we can think about how it might help us in these steps.

The last column of the table identifies opportunities for ChatGPT. For those more experienced with use cases, there are extensions to this use case that address conditions that are not typical but expected at each step. I have removed and placed extensions to this use case in *Table 3.3* to make it easier to read. Know that there are many ways of presenting a use case. This is just the one we used for convenience.

Step	Description	System responsibility	ChatGPT opportunities
1	Recommended patches are shown.	Displays the patch recommended for the system.	Make better recommendations from existing patches and known bugs.
2	The user selects one or many of the recommendations.	Displays details about the patch and options available.	Recommend collections of patches that work well together.
3	The user reviews and selects recommendations to add to the plan.	Patches are added. Completion of the task is communicated. The plan updates.	
4	The user repeats the review process until all patch decisions are completed.	Display details, updates, and the number of patches in the plan.	
5	The user starts the plan.	Take the user to the plan.	
6	The user reviews the plan, checking for conflicts and issues.	Displays the plan contents and status. A conflict check is run.	Predict the results of the patch plan, and the implications for missing or conflicting patches.
7	The user deploys the plan on a test instance.	The system allows the selection of a test instance.	
8	Run evaluation of test instance.	*(Results and feedback are not in scope for this example.)*	Generate tests from existing cases. Identify gaps in testing.
9	Adjust and evaluate the plan based on test results.	*(Cycle through previous steps with a new plan and track the versioning of the plan.)*	
10	Schedule to deploy on the production instance.	Schedule and notify the user of success or failure.	Make deployment recommendations.
11		*Patches deployed; system available at new patch state.*	Generate test case. Identify possible issues.
12	Schedule and deploy to other instances.	*Add more instances to a plan, such as production instances.*	Monitor for abnormalities.

Table 3.2 – An example use case for a patching system

The primary use cases have multiple places with opportunities to insert ChatGPT. If there is a wealth of data on what patches have been installed and the resulting tickets from those versions and releases, a model can be trained to suggest a collection of patches that work better together. The LLM can propose a patch plan or at least find issues that could result in a conflict or alternative collections of patches that resolve a conflict. This model could be used before, during, and after the process. This is by no means trivial. It is just an opportunity suited to ChatGPT; it is good at providing examples and test cases. After the fact, different data from monitoring software performance could be used to train ChatGPT to understand or even predict abnormal conditions. Patching, testing, and monitoring can have different training models with varying prompts, APIs, and results. In this case, uses of the LLM are identified in the table. We break down the steps to a level of detail so that it is easier to imagine a LLM solution. It still might be significant work to make it happen, but the problem is constrained. It can be evaluated to see if there is valuable data that will allow the model to perform successfully.

A case study in *Chapter 6, Gathering Data – Content is King* presents an example of deploying a dozen tuned models to perform specific tasks in a life cycle. Each step of the process uses a different model. In this case, some patch recommendation steps might not be exposed directly to the customer but be used by other models to refine or support suggestions to the customer.

Additional sequences that supplement the process or are not the likely path are called **extensions** to the use case; some extensions are included in *Table 3.3*. Usually, the primary flow is the **happy path**, which most people are likely to follow to achieve success. Extensions supplement the happy path with other necessary functionality. *The happy path is not the only path*. A robust enterprise design must consider edge cases, error conditions, slow connectivity and performance bottlenecks, accessibility, internationalization, and other issues.

In this case, there are a few extensions to point out. Adding to a plan is standard and typical, but without any plans, the first thing that happens (even if automated) is a new plan, which would be created like a new folder on a desktop. On a computer, the folder is created as `untitled`. A well designed system automatically selects that text so the user can immediately type and name it. The use case does not describe the behavior at this level. This is good, a product manager can deliver the use case, and the designer can pick this up and develop the UX design. Based on the main flow and general understanding of ChatGPT capabilities, there are places where ChatGPT might be included to offer entire plans or make suggestions to resolve conflicts.

Step	Description	System responsibility	ChatGPT opportunities
3a	Create and name a new plan.	Create a new plan. Selected patches are added to the new plan. Message response for the completion of the task and a link to the plan.	
3b	The user selects an existing plan and adds recommendations.	Plan is updated and displayed with new details.	Suggest plans that might overlap.
6a	The user is notified of a conflict.	Take the user to a conflict resolution step.	Offer suggestions to resolve conflict.
6a1	The user saves as a named plan (useful to apply the same plan to other instances).	Save a unique named plan.	
10a		Alert the user to deprecated patches; the patch cannot be deployed.	Offers suggestions.

Table 3.3 – Extensions to the patching use case

Suppose data about what patches have been installed in the field and resulting trouble tickets from those versions and releases are available. In that case, a model can be trained to suggest a collection of patches that will work together. The model can recommend a patch plan or suggest an alternative patch collection that resolves a conflict. Conflicts are patches that don't play well with other patches. This is by no means trivial. It is just an opportunity that is suited to ChatGPT.

Prioritizing use cases based on usage and value – human resource example

It is critical to understand the importance of every flow. If 10 or 20 flows each have five or ten percent of usage, these flows represent almost all usage. However, if primary flows represent only a tiny portion of the interactions, this is called the **long-tail effect**. If some flows and paths serve few people and are infrequently used, there will be a long-tail distribution for these flows. This means that any single flow will only have a small impact on the user community. When there are a few cases representing most interactions, improvements to these flows or tasks have more value than those with little usage. Long-tailed use cases will not improve the customer experience as dramatically as the ones with the most use. The difference in value between these can be quantified. And if the cost of developing one with the most use is the same as the one with the least use, the choice is obvious. This will be explained in *Chapter 4, Scoring Stories*.

A common distribution (for the statistic geeks, this looks similar to a one-tail normal distribution; real-world data doesn't always look exactly like a normal curve) looks something like *Figure 3.1* by sorting use cases most frequently used to least frequent. This dataset is from a human resource (HR) system, so it is not a perfectly smooth distribution. In this example, the few top cases represent the most usage (like the 80/20 rule). In this case, 80% of the usage comes from 20% of the features or capabilities.

A One-Tailed Distribution Example

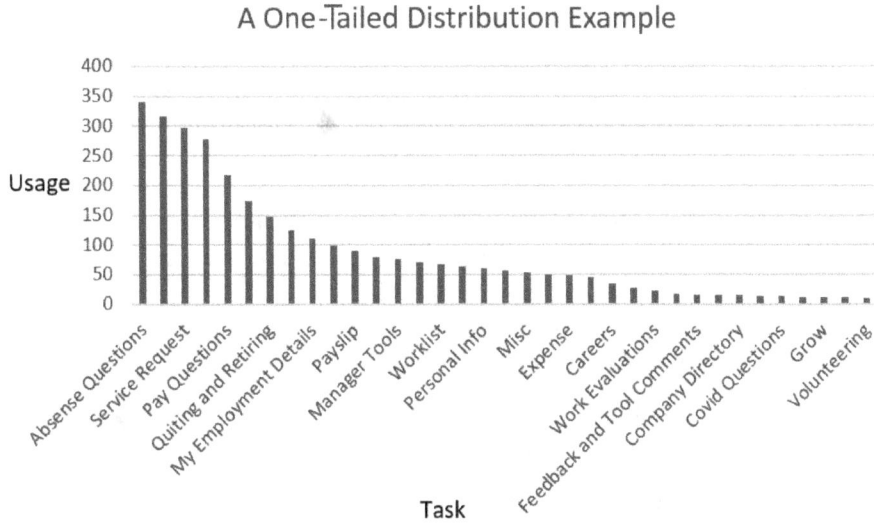

Figure 3.1 – A typical one-sided distribution, a few cases have the most impact

Fewer big-ticket use cases appear in a long-tailed distribution, which extends further into the less frequently used use cases, as demonstrated in *Figure 3.2*. It takes more than 40% of the use cases to get to 80% of the usage.

Long Tail

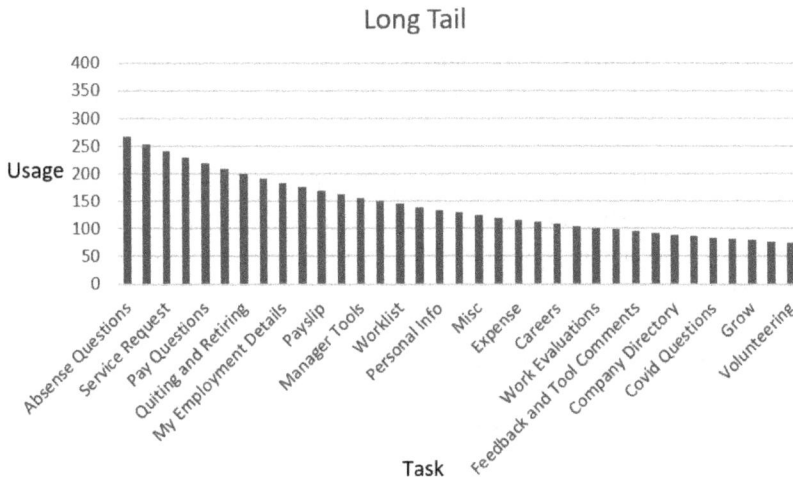

Figure 3.2 – An example of a distribution with a long tail – many cases have similar usage

Prioritizing based on the 80/20 rule – Amazon shopping

Features on Amazon.com are something we can explain with no introduction. This shopping website has features that represent an 80/20 distribution. Features such as searching for a product, adding it to a cart, and checking out represent the majority of use. Small features make up the rest, such as creating a wish list, registering for an account, or adding a new address. However, if you switch perspectives to looking at products, the most popular product would be less than one percent of sales. Even popular products based on search traffic, such as the Apple Watch, Crocs, AirPods, air fryers, and iPads, might total six million searches, but they represent the top of a long tail of product searches. Compared to the volume of searches of over 200 million, even the top seller is not even one percent of sales. This is a long tail. This article discusses these five search terms out of over 274,000 keyword searches to make the long-tail problem appear even more impressive.

Article Top 100 most searched items on Amazon (https://www.semrush.com/blog/most-searched-items-amazon/)

The Amazon example demonstrates that the most value will likely be derived from a few features (like the add-to-cart function). Likewise, focusing on content (like items to purchase at Amazon) might create a long-tail problem. In either case, if AI is added for something on the *right* of the curve, it will provide less value than something on the *left*.

Returning to *Figures 3.1 and 3.2,* we can apply the Amazon example to the HR data. Notice that the *Absence Balance* item (about vacation time remaining for the employee) has almost 350 interactions in *Figure 3.1*. It would only be over 250 in a long tail (*Figure 3.2*). No one use case stands out in a long tail situation. Spending the same effort on this use case in a long-tail environment would return a different reward.

It is helpful to know the distribution of use. A more normal distribution means value will likely come from improving a few critical use cases, while the long tail means the impact will be more challenging without covering more use cases.

But even small cases can be relevant. Signing up customers is a small use case by frequency compared to the usage after registration. Still, no one would argue that getting signups correct and reducing friction in the process isn't essential. Now that we have covered use cases, we can still consider the impact of a user story.

Creating a user story from a use case

Recall that the concept of use cases and user stories overlap. The form is different. Let's switch back to the patching example. It provides a robust set of steps that afford various opportunities for applying ChatGPT. Running ChatGPT on the user story from *Table 3.1* produced the story accompanying the use case:

```
I will give you a use case that includes the user's steps and the
resulting system response. Please summarize this into a user story
```

```
using storytelling techniques. Then, summarize the story into a book
synopsis.
```

The ChatGPT response to this use case was as follows:

```
In the magical kingdom of Systemia, the wise and diligent IT wizard,
Merlin, sets out on a quest to secure the kingdom's digital fortress.
Armed with the magical tool "PatchMaster," Merlin navigates through a
series of enchanting chapters to review, select, and deploy patches
across the kingdom's instances.

The story unfolds as Merlin discovers a captivating array of patch
recommendations, meticulously builds a Patch Plan, and navigates
through the patching odyssey, ensuring every decision is made with
precision. The wizard crafts a masterpiece - the Plan - checking
for conflicts and issues to guarantee a flawless defense against
vulnerabilities.

Testing the waters, Merlin bravely deploys the Plan on a test
instance, adjusting and evaluating based on results, showcasing the
wisdom of iteration. A grand deployment ensues, scheduled and executed
with magical finesse, fortifying the kingdom's defenses.

In a grand finale, Merlin expands the kingdom's shield by deploying
the Plan on multiple production instances, declaring victory as the
patches are deployed and the kingdom's instances stand secure at their
new patch state.

The epilogue reflects on the enduring legend of PatchMaster, as
Merlin's wisdom ensures the kingdom's digital realm remains resilient
against the forces of vulnerabilities. "Patch Perfect" stands as a
testament to the enduring power of thoughtful patch management in the
ever-evolving landscape of technology and security.
```

This generated story from ChatGPT was not edited in any way (grammar errors included). I want to use this to make a few points. First, ChatGPT has value in our design process. This story is creative and enchanting and could be customized further, giving more directions in the prompt to match a specific style and tone. Second, it is easy to go from a use case to a user story, even though editing the output may be necessary. This could be provided to a cartoon or sketch artist to develop the storyboards for an engaging marketing spin. Additional ChatGPT refinements could be made to flush out the story beyond the happy path. It could be asked to break down the story into eight storyboard cells that would be more readily visualized as a storyboard.

It would be best to recognize that it is easier to identify opportunities for ChatGPT using the step-by-step approach in the use cases than in this storytelling format. The audience for a user story is different. The use case has details that the development team needs, but the story is for upper management, other teams, and marketing.

Prioritizing ChatGPT opportunities from the use case

This patching example is useful because there is an actual cost to apply a ChatGPT solution at every step, including design and development, and the actual cost to run production, stage, and test LLMs.

Don't expect to deploy ChatGPT in all these opportunities, so in *Table 3.4*, let's re-list the steps identified as places for ChatGPT support and score each to see which can unlock the most value for customers. The team can then spend the time and money on the most valuable steps.

Each of these possible stories is scored on three metrics. *Chapter 4, Scoring Stories* teaches **User Need Scoring**, the product of the three scores. The higher the score, the more value to the customer. In the next chapter, we will also factor in the development cost. From this analysis, we have three stories with the most value.

Steps Identified for ChatGPT Support	A Customers Impacted (1 to 3)	B Frequency of Need (1 to 3)	C Severity of Issue (1 to 4)	User Need Score (A*B*C)
Step 1: Make better recommendations from existing patches and customer bugs.	2	1	3	6
Step 2: Recommend collections of patches that work well together.	2	1	3	6
Step 6: Predict the results of the patch plan, and the implications for missing or conflicting patches.	2	1	4	8
Step 8: Generate tests from existing cases. Identify gaps in testing.	2	1	2	4
Step 10: Make deployment recommendations.	2	1	1	2
Step 11: Generate tests from existing cases. Identify gaps in testing.	2	3	2	12
Step 12: Monitor for abnormalities.	2	1	4	8

Table 3.4 – Opportunities for ChatGPT within the UX (top two items are bold)

Let me break down the scores. All of these stories impact the same number of customers. Only some of our customers do patching, so we ranked this a two on a three-point scale. Three is the highest and is reserved for tasks that most customers do. It is not surprising that patching is done infrequently. Individuals update their phones a few times a year; enterprise customers are more hesitant to update. Each of these steps is done about the same number of times and typically done quarterly, so the frequency of this need is a 1 (again, on a 3-point scale where a two would be most of the time, while a three would be all the time). But test cases are different. It is an ongoing process, so it scores a 3 on the Frequency scale. It is also the one place to catch potential problems before a customer encounters them.

The significant differences in the user score come from the severity. Think of this as the inverse of bug severity. A severity 1 bug is a big deal in most companies. Everyone should agree that this needs to be fixed now, and it should be scored the highest, so it is 4 points. A 3 would be a significant problem, while a 2 is not so much, and a 1 is the least severe issue (typically a severity four bug).

We document the explanations of the scores so multiple people will learn the process and be able to score similar problems. Since scores range from 1 to 4 points, it keeps the explanations and scoring simple. We want people to be able to repeat scores, share, and learn a common method. Teams develop more complex processes, which can also get in the way. This is discussed in the next chapter.

The list can be prioritized based on these scores and values. We will explain in depth how to score stories and, importantly, how to incorporate engineering's cost into developing these solutions. We can continue our journey knowing we have at least three stories (within *Steps 6, 11*, and *12*) with significant user value. We will use these same stories in *Chapter 4, Scoring Stories* to explain scoring in more detail.

So, we started with an easy example. I am an expert in this space, making this example easy to explain. Business expertise is critical to solve problems with LLMs. If you are not the expert, create a team that balances customer empathy with business expertise.

This patching case is a known set of tasks, and we could quickly identify where ChatGPT solutions might fit. But what about when it takes more work? You may have to explore or do research to determine the best fit. Where do the use cases come from? Whose goals are we trying to achieve? Do you know enough to recognize where ChatGPT should be considered? Challenging questions. I recommend becoming an expert in your product area. The more known about the product or service, the more the user's goals are understood, and the better ChatGPT capabilities are understood, the easier it will be to recognize opportunities and solutions. Each new version of ChatGPT is different and can cause changes in the plans. As we will explore in *Chapter 6, Gathering Data – Content is King*, different models can be used for other tasks. We can't say which model is better; this changes every few months.

Let's examine how to define those goals. Remember that these are the user's goals, not the organization's needs.

Aligning LLMs with user goals

Let's say we have a customer or role in mind and want to define their goal. This book concerns the user's goals, not marketing, engineering, or sales. It is up to product leaders to figure out valid user goals and how valuable the goal is to the overall product. Viability is also essential and typically provided by engineering. Compelling use cases can come from anywhere; the head of the company might have a vision, a customer could complain, the product manager gets an idea from a customer, the QA engineer recognizes a shortcut, user researchers uncover a critical workaround, or designers imagine a solution to a pain point for customers. Be less concerned with who comes up with the solution and more focused on how much value it provides. Don't get hung up on which vendors LLM is needed. As OpenAI and the industry evolve, a broad range of models with specific capabilities and costs will be available. This fit-to-purpose allows building solutions with a suite of LLMs in one product.

Applications of ChatGPT

There is a world of solutions waiting to be solved. In the space of enterprise solutions, there are solutions that ChatGPT can help with. Let's throw out a collection of use cases that can be improved with ChatGPT in the mix. Don't be shy; ask ChatGPT for suggestions specific to the business. Drill down into detail with it. The process might not highlight what is most important for the company, but generating ideas starts the process. Having an extensive collection of ideas is okay, but use the skills provided to prioritize based on value.

If you draw on what others have done with ChatGPT, it is essential to understand if they have done it *well*. Here are some areas that ChatGPT can handle. There are thousands of ideas out there worth exploring. Use this to spur some thinking. Rethink these examples into your company's business and practice. Consider the following ideas to help with a brainstorming session:

- Finding critical information in an extensive knowledge base for specific solutions for specific versions of products.

- Offering suggestions to sales reps on what factors help close customer deals.

- Diagnosing service issues, walking customers through the steps to resolve, sending repair agents into the field when all else has failed, following up with changes to the service appointment, and closing the request when completed.

- Consider if other modes of interaction, vision-based analysis, monitoring for sounds, or sensor data (temperature, pressure, air quality, proximity, light, vibration, optical, level, motion, or speed) support a process, a recommendation, or a coaching opportunity.

- Recommending products and solutions based on previous purchase cycles (in the loop directly with the customer or as a support angle for a sales rep).

- Translating for world languages to broaden support channels.

- Generating code, autocomplete for tasks, and regression test suite creation.

- Helping to file human resources (HR), finance, accounting, service, sales, and other requests or forms. Expense reports, end-of-month balance sheets, service tickets, sales orders, vacation requests, or helping to match candidates to jobs (it can't be worse than the tools out there today).

- Validating and ensuring contract conditions are met before completing a transaction using blockchain.

- Brainstorming with a user to explore and diagnose alternatives to problems.

- Summarizing, drafting, writing, or editing material.

- HR has many use cases around scheduling interviews, coaching new employees, helping to write job descriptions, analyzing exit interviews for reasons for leaving, determining the best employee to coach new employees (watch out for bias), and detecting bias in those processes. HR is always doing surveys that can be analyzed with LLMs.

- Financials is a hot area – forecasting, market trends, and advice based on spending patterns, goals, and tolerance. Some company financial reports, daily stock prices, and volatility commentary are already autogenerated.

Many of these could be handled by a 24/7 virtual assistant who supports the business and supplements or replaces trivial and simple human-agent interactions.

Let's pull a random company out of a hat and see the wealth of AI possibilities. I picked the 2,000th largest company in the world (so we are inclusive of large enterprise companies of many sizes). This search landed on The Gap, an iconic retail organization that sells clothes via many brands, such as Old Navy, The Gap, Athleta, Banana Republic, and Intermix. It is at the end of the Forbes 2000 list of global companies. Let me give an example of how to brainstorm with ChatGPT by generating a prompt to help brainstorm

```
Our company is called the Gap; it includes many brands that sell
apparel in retail settings in stores worldwide. We deal with a very
competitive and dynamic space, handling customers in stores, handling
returns, promoting and advertising, designing clothes, manufacturing
products worldwide, and dealing with logistics and over 70,000
employees while managing a significant web presence. Provide me with a
numbered list of detailed important use cases that ChatGPT can inject
into to give my customers a more effective customer experience.
```

I have truncated the descriptions, except for the last one, to show how it framed the user benefit. Run this exercise for your business and see the results.

- In-Store Customer Assistance: Implement ChatGPT-powered kiosks or mobile apps….
- Virtual Fitting Rooms: Develop a virtual fitting room(to)receive personalized recommendations….
- Returns and Exchanges Assistance: …responses to common return and exchange inquiries….
- Personalized Product Recommendations: …analyze customer preferences based on past purchases, browsing history, and interactions to generate personalized product recommendations….
- Live Chat Support: …provide instant responses to customer inquiries….
- Fashion Advice and Styling Tips: …customers can input their preferences, body type, and occasion, receiving personalized styling tips….
- Supply Chain Transparency: …provide real-time updates to customers about (order) status….
- Customer Feedback and Surveys: (collecting) customer feedback and conducting surveys….
- Multilingual Support: …provide support and assistance to customers in various languages….
- AI-Powered Visual Search: search for products using images, enabling them to quickly find similar items or complete looks….
- Customized Loyalty Programs: Utilize ChatGPT to analyze customer data and behavior patterns to create customized loyalty programs and

```
        incentives tailored to individual customers, rewarding them for their
        engagement and purchases, and fostering brand loyalty.
    •   Integrating ChatGPT into these use cases will not only streamline various
        aspects of your business operations and significantly enhance the
        overall customer experience, driving customer satisfaction, retention,
        and, ultimately, revenue growth.
```

We could continue. It did a great job exploring opportunities for ChatGPT. These examples benefit from what ChatGPT is good at. When good enterprise data is passed to an LLM with content and intelligent control via prompting and security, it can deliver value at many customer engagement steps. *Chapter 6, Gathering Data – Content is King* explores bringing enterprise data into the fold.

When doing competitive research, it might be clear that other companies are building similar LLM solutions, and it could be a game of catch-up. For example, clothing recommendations will be the norm. A company must have this in the future. In any of these deployments for cloth recommenders, there will be failures. Most failures will come from not caring and feeding the LLM; some will fail because they picked the wrong use case, and a few will fail, likely because LLMs are imperfect. Active monitoring, iterative design, and fine-tuning are staples of success and will help avoid failure. We will take all of *Chapter 9, Guidelines and Heuristics* to discuss monitoring to ensure an understanding of success and future improvements. We will draw on these earlier chapters to refine and improve. Stay tuned. I hope we instill the idea that not *every* problem is solved with ChatGPT. Focus on the most valuable pieces of a problem that can be delivered sooner.

Examples of generative AI outside of chat

Try to think outside of the box. Great ideas can always come from very different fields. We have Velcro® because an engineer noticed the burrs that clung to his clothes after a walk with his dog (in 1941), Post-it® notes because of a failed attempt by Spencer Silver to create a super-strong adhesive (in 1968), and Super Glue® by Dr. Harry Coover, at Eastman Kodak (in 1942), from a failed attempt to create a clear plastic. So, it is always good to think outside of the box. I always suggest looking at problems in the context of the whole business. Sometimes, we get caught solving issues in the weeds when we need to look up and see the entire forest. As I mentioned with the patching example, sometimes the solution is to remove the UX and automate the whole practice. That is the ultimate simplification. This list of innovations that ChatGPT can address grows every day. However, I put these out here because they might connect experiences and the company's needs. Innovation can come from anywhere. Could any of these widely varying spaces help in a quest for solutions? Let's see:

- **Vision analysis of objects for detection, sorting, counting, or analyzing**: Are there inventory, shipping, handling, or retail issues? Amazon has already implemented stores where customers put stuff in shopping bags in their cart (or take it out when they change their mind), the store calculates the cost, and they walk out the door. There is no checking out.

- **Image creation for the visual arts, marketing, or social media**: Although this space is much bigger than the example, an artist, with prompting, can generate images without licensing issues for media. There might be issues with LLMS training on unlicensed material. Something we will discuss later.

- **Musical composition**: I am impressed by how ChatGPT calls Mathematica to create a note-for-note melody. I can't speak to how catchy the jingles might be…

- **System simulation and testing**: It is very trendy to create virtual simulations of processes or building of buildings (we hear the term digital twin), so look to see places to use ChatGPT to simulate or test conditions.

- **Prototyping and user interface design**: This was the book we should write, but I didn't think the world was ready for it. Use the models as tools. Someone out there will write this book for us.

- **Drug discovery**: Google it. If drug companies can search for new compounds based on research and results, I bet a goldmine is in enterprise data. Mine it.

- **Finding new solutions to age-old problems**: With the right tools and framework, a generative solution can solve problems with unknown answers.

We have plenty of uses for ChatGPT, but it will not always be the right solution. Here are some places where ChatGPT might not be the best fit.

Avoiding ChatGPT limitations, biases, and inappropriate responses

A puzzle piece might not fit a puzzle because it is from a different puzzle. We love to take a new tool and apply it to every solution. A hammer has a lot more uses than just pounding nails, but that doesn't make it the right tool for the job. ChatGPT's unique ability to predict what to say next will not always fit a specific use case perfectly. There are plenty of spaces where there might not be a good match. Or, as shown in the enterprise patching example, it's not necessary to apply ChatGPT to every step to get value. Not only should the use cases be the most important, but they should also be ones where ChatGPT integration makes sense. Here are some areas where it might be challenging to apply ChatGPT to a solution and some creative thinking to help if these situations apply.

Lack of real-time information

Foundational models have cut-off dates for the ingestion of new material. A new model might have data only from a year prior or earlier. This problem will improve over time; we have seen the ingested material dates as recent as less than a year. However, if there is a lag and a customer needs that time-critical information, it must provide it. If the company releases a 2.0 widget, new documentation can update a custom model with that information at launch. However, if the LLM solution depends on keeping third-party details up to date, ingesting up-to-date third-party data is also needed. It won't matter if the third party posts the information on their website since the foundational model won't be that current. Since the model updates are limited, Retrieval Augmented Generation (RAG), explained in *Chapter 6, Gathering Data – Content is King* can ingest this data and be up to date. Recognize that new testing and validation are required with changes to the model.

Complex or specialized topics

Legal and medical advice are the two areas that could cause issues. However, many opportunities exist when building enterprise versions of these services. They don't have to be the answer to the legal or medical questions; they can help refine the space or look for connections that might be too difficult for a human to process (such as possible side effects or interactions for a patient with 15 different medications). A support tool in the legal space could be to review mountains of legal files to highlight information related to specific expertise. Expert witnesses have to review depositions to form their conclusions. Still, with a ChatGPT tool that can be trained, it can identify important information in hundreds of pages of depositions and speed up the process by a factor of 10. It also comes back to the lack of real-time data if it is expected to know everything. Create high-quality solutions by carefully monitoring the quality of the results by experts, incrementally improving, documenting, and refining success.

Long-form content generation

It is common for enterprise use cases to need long-form content, such as writing a book. We have knowledge and technical documentation needs, but this is not an effective use of ChatGPT. ChatGPT can help edit material, shorten tests, tighten language, or even do essential translations, but there are limits. Don't expect to put out long material like this without technical editors. ChatGPT can drift away from the topic, introduce unrelated content, and repeat phrasing. It might even repeat topics. I have seen good examples where a ten-step guide had two steps labeled slightly differently but identical to earlier steps – creating and maintaining the organization and hierarchy found in complex and lengthy documents taxes the long-term memory limitations of generative AI. Also, facts are not what they used to be. Especially with technical documents, we always have additional testers run through the processes and tasks in the document to verify that the solutions work. ChatGPT might first identify and point the user to the correct pieces of well-authored documents. Then, the user can trust the results to be accurate if they recognize that the document applies to them. For example, we don't want to show a valid step-by-step solution to the problem in Microsoft Windows when we know they are using an Apple Macintosh.

Long-term memory

A ChatGPT virtual assistant might be good from one comment to the next, but not for a long duration. A conversation that spans a week will not likely maintain the context or information to continue that conversation. However, some workarounds could be developed. Consider if long-term attributes or the conversation could be saved and processed to remind the LLM of previous discussions. A collection of contextual details could be built up over time. Even a historical list of earlier elements of a conversation could be stored. It is exciting to give an LLM the appearance of long-term memory.

Vendors are working on providing foundational models better long-term memory. We have seen this in ChatGPT 4o, for example. Imagine a directory service within a company:

- The directory app provides people's names, phone numbers, email addresses, and titles.

- Many people have asked the app many times. It can look at search history to understand popular people and shortcuts for people's names (Ben for Benjamin).

- If a user returns a month later, they don't want to scroll through the history (or maybe the UI doesn't have this data).

- So, the user asks, "What was *his* phone number again?"

- Once the AI responds with a name, the user responds, "No, it was someone else," and the app returns a short list of other possible names.

- By storing people's history, it looked up and provided that context in the prompt (or via other solutions). This recreates the history and presents the appearance of long-term memory.

- Context about this user and their organization might allow the LLM to provide names that were not previously used but are likely answers. "I don't see that we talked about Ben before, but maybe you mean Benjamin Buttons. Here are his details…"

Thus, long-term memory could be constructed to provide context to new conversations. Each time it is constructed, we can confirm that the user has permission to access this information. This is one method to protect sensitive information. Sensitive information takes many forms in an enterprise.

Sensitive information

Understanding what is private and keeping that information confidential is a real challenge. Recall that an LLM doesn't understand what it is saying; it is good at predicting. We have seen time and time again stories about chat models being forced to spill information that should be sensitive. A workaround is not to use that information directly in the model. For example, if the app needs to collect credit card information, develop a secure service to pass this information between the user and the credit card backend. It doesn't have to be done within the conversational model. Doing so with a GUI element might support building more trust with the user. Another place we can build trust and have issues with LLMs is with biased thinking.

Biased thinking

The model is only as good as its training data. If the model has biases toward a way of thinking, is harmful to a user group, or contains offensive material, it can have unintended consequences on results. This presumably comes from the base model. Be cautious of this with proprietary data. It might not be a personal or cultural bias, but how a product or a competitor's products are discussed can impact results. If the foundational model has biases, it might take some engineering to protect users from these using instructions to the LLM. But no matter how good the instructions are, some things are hard to teach, like emotion and empathy.

Emotion and empathy

We discussed the history of **ELIZA**, the original chatbot, in *Chapter 1*, *Recognizing the Power of Design in ChatGPT*; thus, using AI for psychological support is at the root of our AI journey. But be careful here. It lacks empathy and emotional understanding. There may be ways ChatGPT can be used to support the mental health community, but AI is not ready to answer 911 (or 988 mental health) calls (these are the emergency phone numbers in the USA). Use guardrails, test extensively, and monitor results where humans might be in a sensitive state. Even if a conversation with an LLM might be perceived as empathetic, this can go wrong quickly. It is only doing what is likely and not based on the nuances of empathy or emotion. In some sense, it is unethical for the LLM to sound empathetic.

Ethical and moral guidance

Most enterprise solutions usually do not delve into ethical or moral dilemmas. Much of this comes back to the Trolley problem when making a decision that can adversely affect two parties. The problem is based on a situation where a runaway trolley is about to kill people near the track, or it can be diverted and kill other people on a second track. How do you make such a decision?

> The Trolly problem
>
> This article describes the moral and ethical dilemmas described by the Trolley problem.
>
> Wikipedia: `The Trolly Problem (https://www.britannica.com/topic/trolley-problem)`

Autonomous cars must consider these decisions, so commercially deployed systems already have to handle these hard decisions. ChatGPT is not built to make ethical decisions; it will use its predictive model to form its answer. Traditional machine learning already has this problem, and LLMs are not immune. Most popular models avoid answering ethical questions. Enterprise solutions that need to delve into these areas can be impacted by safeguards built into the models. Carefully apply best practices to prevent these conditions. Related to ethical and moral issues is how to handle critical decisions with significant consequences.

Critical decision making

We have talked about ethics, morality, security, and empathy. These might come into play when making time-critical decisions with profound consequences. Depending on business decisions based on the model, and the model has been trained on the diverse internet data, will it give the same conclusions that a trusted expert would provide? It is essential to monitor this interaction closely, train it on company-specific data to avoid biases, and provide guidance and guardrails to the recommendation engine. Remember, it is basing its results on patterns it has seen. *Chapter 6, Gathering Data – Content is King* covers how ChatGPT works with enterprise data to make the LLM respond with business-savvy knowledge. It isn't going to answer with a human's moral compass and is not in a position to decide on the value of its judgment based on the expectations of its human user. Maybe less challenging would be something more geeky, like programming.

Programming and debugging

This is one area that is getting a tremendous amount of visibility. I think it's mostly because geeks love geeky things. Also, getting ChatGPT to do work is fantastic. However, if you don't know what you don't know, be very careful with its responses. Production-level code requires more robustness than what ChatGPT can provide. LLMs can write test cases against the code the model wrote. It can easily knock out simple scripts and templates in many common languages. And it is not just programming. Different models can be used to generate test cases and give the human in the loop some support. I suspect this area will mature quickly in the next few years. The more it can learn about good patterns and quality code examples, the more critical this area will become. Just know and expect it isn't perfect. But these programming languages are sometimes better than world languages.

Translation accuracy

Most models handle translation but are only sometimes successful in technical or scientific areas. The training models have traditionally been very English-centric. Since they were trained on available web resources and 55% to 60% of the web is in English, the sophistication of translations varies considerably based on the training available.

There are significant hurdles to overcome for technical and company-specific vocabulary. Test and verify how these terms might be handled in other languages or used without translation. Test and verify cases of multiple languages being used in one interaction. Additionally, brand names are translated into only some cultures.

Speaking the user's language has so much value. We have a UX saying to *bring the tool to the user, don't make the user go to the tool*. Translating languages is the same. Customers would appreciate working in their native tongue rather than forgoing support or communicating in another language. The quality of the language mirrors the training available. Expect better support for Spanish, French, and German, while it can be challenged in Russian, Arabic, and Hindi, especially with idiomatic dialects, formalities, and technical language. Again, monitor results and invest in training technical jargon and unique company vocabulary. Some world languages have smaller training sets, grammar, syntax, and phonetics complexities, overlaps or confusion with another language, and a limited human feedback loop. There is no substitute for a great teacher.

Educational substitution

Human trainers and training or educational material have a long history of success (and failure) in various areas. The best class I have ever taken was a three-day hands-on class in the 100-degree heat of the desert. We rarely sat down, and even with ten years of college, I learned more per hour than in any classroom. Don't expect ChatGPT to have the expertise to train or educate students effectively. It doesn't know how the student reacts and will be challenged to adapt to their needs. Hands-on skill training is difficult and even more challenging (but not impossible) to analyze and share with an LLM. For example, a training application might teach and coach how to change out a faulty sensor on an airplane engine. The LLM can coach the user, while cameras can detect the situation and guide the user to perform the steps in the correct order, point to the right tool (using a laser pointer), and identify the location to apply the tool. Assisted maintenance is a vast area, and given the cost of mistakes, investments in this space will be significant. So, it is coming, but it will be hard, and accuracy will be critical.

Don't force-fit a solution

Recognize that for every bottleneck to which ChatGPT is applied, ChatGPT might offer suggestions on how to avoid or mitigate issues. However, don't try to force-fit ChatGPT into a solution. There seems to be a rich set of opportunities to apply ChatGPT in almost any vertical market. Focus on the use cases that provide the most value and match the unique capabilities of a generative AI solution.

ChatGPT might be used in a space not covered in this book. Try asking it, reviewing it, and diving deeper into its response:

```
What problems would ChatGPT have with identifying legal challenges if
I gave you a collection of depositions to review?
What issues would ChatGPT have with providing a summary of knowledge
articles on resolving complex technical problems with airplane engine
repair?
```

Summary

Developing user stories can sometimes be easy. If requirements are apparent, straightforward, and ideally suited to ChatGPT, write those up and get to work. We also realize that some things need to be clarified. Use research, discussions with internal stakeholders, and your keen mind for problem-solving to find LLM opportunities.

Start thinking about the opportunities best suited for a generative solution in your business and explain why some problems are unsuitable for ChatGPT.

Hopefully, this chapter will create excitement around introducing our next topic for scoring and prioritizing stories. We will jump into that next since we only briefly explained User Needs Scoring. This will allow us to prioritize the backlog features and use cases we have explored up to this point.

It will also be helpful to know this when we get a ChatGPT solution up and running and want to test or monitor it.

Once we have a solution, we can reuse some of these skills to verify what we did and continue to care for and feed the solution for continuous improvement. We will then have data and some confidence that we are doing the right things.

References

The links, book recommendations, and GitHub files in this chapter are posted on the reference page.

Web Page: `Chapter 3 References (https://uxdforai.com/references#C3)`

4
Scoring Stories

One of the secrets of successful development efforts is to work on the proper work. The agile method helps break down effort into manageable pieces, but some concepts, like Estimating Poker (we will explain), focus on the *cost* of doing the work, not the *value*. We want to share a method to systematically break down stories so they can be prioritized based on the value to the customer. ChatGPT solutions require a lot of decision-making. We need to prioritize those decisions. Adding or editing knowledge, doing more testing, adding a new integration, changing the models, improving fine-tuning, or refining prompts must all be prioritized. It might seem obvious, but work on the work that provides the most value as soon as possible. Items that hold little value or benefit a small subset of customers could wait. We have a way of putting this backlog of work into an order the team can understand.

We will discuss the concept of *User Needs Scoring* in-depth and tie it to the development cost. That will set up *Chapter 5*, Define the Desired (User) Experience.

This chapter will cover the following main topics:

- Prioritizing the backlog
- Creating more complex scoring methods
- Real-world hiccups with scoring

Prioritizing the backlog

First, some context and history. A **backlog** means the collection of known work to be evaluated by the team for development. The closer an item on the backlog is to being worked, the more detail is needed to understand and scope it. This is why we have different approaches to managing items on this list of work. Sometimes, a large team will have many sprint teams, each managing its backlog of work. However, since all teams are marching toward the same goals, it is essential to understand that everyone is working on the backlog with the most value. We will be better off having a repeatable and consistent understanding of what work to do first. **Weighted shortest job first** (**WSJF**) is one of these ways.

WSJF

The agile concept of WSJF helps one decide which job/task/project/feature/content should be given resources to complete first.

We learn that the most critical tasks don't always come first because their value depends on the development cost. Sometimes, doing something different on the backlog can provide better value for the development investment (in the USA, we call this getting a *good bang for your buck*—a buck is another term for a US dollar).

I had a great Agile teacher who explained this the best and will try to do it justice. WSJF prioritizes work based on the **cost of delay (CoD)** *to the customer*. Customers don't get any value from features that don't ship. Shipping the most valuable features sooner minimizes the CoD. However, for two features of equal value, deliver the one with the least cost first. Why? Because the customer gets the value from the product sooner rather than waiting for the more costly feature.

Let's examine *Figure 4.1*, from the **Scaled Agile Framework (SAFe)**.

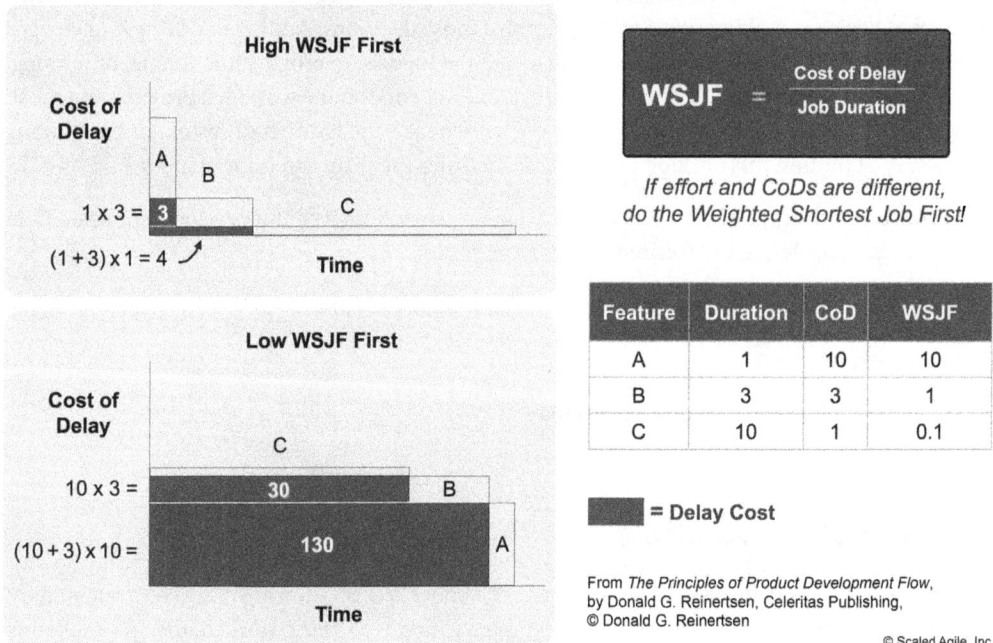

High WSJF First

$$WSJF = \frac{Cost\ of\ Delay}{Job\ Duration}$$

If effort and CoDs are different, do the Weighted Shortest Job First!

Feature	Duration	CoD	WSJF
A	1	10	10
B	3	3	1
C	10	1	0.1

■ = Delay Cost

Cost of Delay

$1 \times 3 = 3$

$(1 + 3) \times 1 = 4$

Time

Low WSJF First

Cost of Delay

$10 \times 3 = 30$

$(10 + 3) \times 10 = 130$

Time

From *The Principles of Product Development Flow*, by Donald G. Reinertsen, Celeritas Publishing, © Donald G. Reinertsen

© Scaled Agile, Inc.

Figure 4.1 – Explaining the CoD

Article: WSJF (https://scaledagileframework.com/wsjf/)

In the SAFe, this example is based on item C having a small CoD. However, because of its long time to complete it, its development blocks getting value from delivering B and A, represented in the *Low*

WSJF First graphic with their `1.0` and `0.1` WSJF values. Consider that CoD is an unrealized value that can't be recovered. Instead, take the *High WSJF First* example, where A is done, B, and then C. Now, because A contains a high CoD and is first, there isn't as much remaining CoD. Customers get value sooner. Less value is wasted, like in the *Low WSJF* example, where we delivered C first.

This allows for comparing features in the backlog needed by the generative AI product. Let's try an example with two features: password reset support and ordering a replacement part. Suppose they are of equal value to the customer (and we will go deep into how to calculate their value later). Which one should be done first?

Password reset takes 45 days to develop, while the replacement part feature takes 90 days. The team can only work on one feature due to staffing. If the password reset feature ships, customers get 45 days of value before the replacement part feature would have shipped. They get value from this feature starting on day 46.After 135 days and the completion of both features, the customer enjoys 90 days of value from the reset feature. Had they completed the replacement part feature first, they would have only received 45 days of value from both features combined. By changing the order of development, the customer effectively doubled the value.

I find it challenging to wrap my head around negative words in the CoD. The CoD happens because the work is holding off delivery to the customer and can't provide value to the customers until it ships. Doing this effectively assumes a cost of development estimate. We can all agree that estimating is hard. All development teams know this. The relative cost *is reasonable* to compare with consistent costing across features. We will make estimation errors. For Agile Scrum teams, use the retrospective at the end of each sprint to continually improve estimating costs.

> **Note**
>
> The CoD is focused on the cost to a consumer. The size of the CoD is a function of the delay in the customer getting value. Delaying a more valuable job to deliver a less valuable job incurs a more significant CoD.
>
> *CoD is the money lost by delaying or not doing a job for a specific time. It's a measure of the economic value of a job over time.* – SAFe website

Figure 4.2 from the Scaled Agile website shows that the CoD model has some user-centric elements.

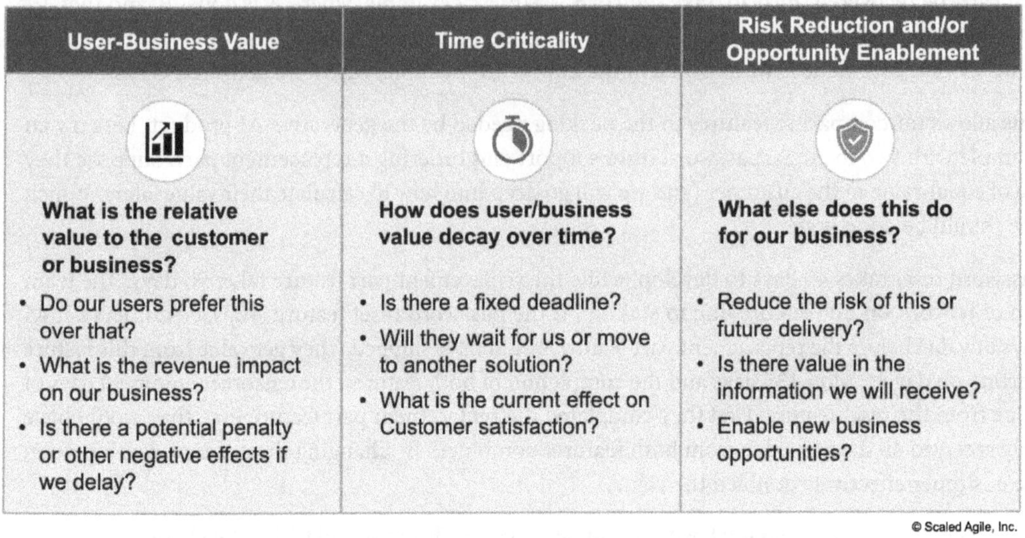

User-Business Value	Time Criticality	Risk Reduction and/or Opportunity Enablement
What is the relative value to the customer or business? • Do our users prefer this over that? • What is the revenue impact on our business? • Is there a potential penalty or other negative effects if we delay?	**How does user/business value decay over time?** • Is there a fixed deadline? • Will they wait for us or move to another solution? • What is the current effect on Customer satisfaction?	**What else does this do for our business?** • Reduce the risk of this or future delivery? • Is there value in the information we will receive? • Enable new business opportunities?

© Scaled Agile, Inc.

Figure 4.2 – CoD

After talking with the Scaled Agile team, we adopted a different approach that can offer similar value. Implementing a repeatable, trainable process that provides consistent results is the most important thing to consider. A **rubric**, a set of criteria, and descriptions create a consistent method for judgment. This makes it easier for people outside of the team to understand and easier to appreciate than the traditional CoD.

For a wealth of details on WSJF, visit the Scaled Agile website:

Article: `Weighted Shortest Job First` (https://scaledagileframework.com/wsjf/)

Reinertsen's book is genius for improving development processes:

Book: `The Principles of Product Development Flow` by Donald Reinertsen (https://amzn.to/3A3u34C)

The book is precious for those who want to understand development processes beyond WSJF. In enterprise development, the SAFe is a well-thought-out, robust, and mature model to support Agile across large organizations. I strongly encourage exploration and deep dives into what is available.

With the basics of WSJF, go away knowing there is value in delivering equivalency in a shorter period. However, not all features are equal. We propose a new method to calculate a user-centric version of the CoD. It is called the **User Needs Score** (**UNS**).

User Needs Scoring

> **The UNS is a user-centric approach to the CoD, accounting for the scope, frequency, and severity of a problem or feature.**
>
> UNS is based on the fine work uncovered in Bruce Tognazzini's book *Tog on Design* (1992), which Phil Haine shared with me while working as a consultant. I then took this work and adapted it into the numerator for the WSJF calculation.
>
> Tog's book referenced a study by Robin Jeffries et al. that evaluated different techniques for finding usability problems, including heuristic evaluation and guidelines, which we will cover in *Chapter 9, Guidelines and Heuristics* usability testing (*Chapter 2, Conducting Effective User Research*), and cognitive walkthrough (a solid technique).
>
> Article: `User Interface Evaluation in the Real World: Comparison of Four Techniques` by Robin Jeffries et. al. (`https://dl.acm.org/doi/pdf/10.1145/108844.108862`)

In Jeffries's study, the participants were asked to analyze issues to "take into account the impact of the problem, the frequency with which it would be encountered, and the relative number of users that would be affected." They used a metric ranging from 1 (trivial) to 9 (critical). We will build on this concept with a more repeatable way to score.

The official CoD from SAFe is user-business value + time criticality + risk reduction and/or opportunity engagement, as shown in *Figure 4.2*. The elements of the CoD make sense; do work benefiting customers (user-business value), have value in delivering sooner rather than later (time criticality), and help change the playing field by reducing risk.

We need a consistent and repeatable metric that non-experts can easily explain and repeat. The approach is to follow three similar metrics and provide rubrics that can be applied consistently.

Here is the User Needs Scoring model. This score will be for the feature/bug fix/enhancement and will be the numerator in our WSJF calculation:

- **Scope**: How many users does it impact? (3 – All, 2 – Some, 1 – A few or a limited role)

- **Frequency**: How often is it used? (3 – Always, 2 – Sometimes, 1 – Infrequently)

- **Severity**: How bad is the problem? (4 – Severe, 3 – Critical, 2 – Important, 1 – Minor Importance)

We then model the CoD as *scope* times *frequency* times *severity*. This means the values range from 1 (1*1*1) to 36 (3*3*4). We will explain each value and how to score with many examples so anyone can apply this model.

Scoring the user value and dividing it by the cost brings both sides of the discussion into play. Scores can be compared between teams, and in SAFe, they can be used to get stories that other teams need to complete into a fair and manageable order (or at least to have the discussion). Whatever is *most important* is scored highest in WSJF and should be done first. We do this ordering by scoring each item. This is more repeatable than just *moving things around* in an Agile tracking tool until it looks right.

$$WSFJ = \frac{Scope\ (1\ to\ 3)^{*}Frequency\ (1\ to\ 3)^{*}Severity\ (1\ to\ 4)}{Job\ Duration}$$

$$WSFJ = \frac{User\ Needs\ Score}{Job\ Duration}$$

We can adopt an approach that Agile teams use to evaluate these UNS estimates, which might vary between team members. In Agile, there is a concept called **Estimating Poker**. The idea is to independently calculate the cost of a feature and then discuss why the values were selected when they are different. This helps uncover unknown elements that a member might not have considered. We can apply the same approach to estimating user needs and continue to do this for development costs.

In Estimating Poker, team members can collaboratively review and discuss the UNS, and we expose gaps in our understanding and expectations. Two to three people must compare estimates to judge the UNS. This will teach the team how to do this scoring. One person can score alone but will miss the value of other people's perspectives. A simple scoring method on a one to three scale makes it more likely to agree. However, the poker method is there to help communicate and clarify differences and come to a consensus when disagreements occur. The score will also help decide whether a bug should be fixed before adding a new feature. Going through many examples will make this method more straightforward.

Estimating Poker (also called Planning Poker) appears two-thirds of the way into this article. Please read it to learn more.

Article: `Estimating Poker within the story article (https://` `scaledagileframework.com/story/)`

Scoring enterprise solutions

The best way to learn how to create a scoring method is to try doing the scoring method. We provide the rules, show examples from typical enterprise applications, and coach through how to score some tricky stories. Then, practice with your own stories.

How to score items

First, the *wrong way* to prioritize a backlog.

Look at a list and see which ones the product owner thinks should be done first; consider the ones where the developers can do it quickly. Another common mistake comes from the strongest-willed team member who speaks the loudest to get their work to rise to the top. This is what we want to avoid. An unbiased approach to scoring stories allows the value of the story to drive the need. The story's value should speak for itself. Stories might match use cases from the previous chapter or be software bugs. Sometimes, use cases are broken down into multiple stories to complete a story in a sprint. For our discussion, the story is the piece of the use case we plan on delivering in a development period.

The *right way* is to use an *unbiased* method to put the items in order using a three-step rating system. Anyone knowledgeable about the stories can generate a score. Hopefully, product owners, designers, engineers, or other related team members can do this and then discuss and resolve things that all parties haven't considered. This is done by answering the following questions. We will back up these criteria and levels of performance with many examples and descriptions for each score:

1. **Scope**: How many users does it impact? Answer with a score from 3 to 1.

2. **Frequency**: How often is it used? Answer with a score from 3 to 1.

3. **Severity**: How bad is the problem? Answer with a score from 4 to 1.

4. Calculate each item's score by multiplying them (e.g., 3*2*2 = 12).

5. Order stories by the score. As new stories come in, score them. They will naturally fall into the correct order of user value.

6. Development estimates the cost of these stories.

7. Divide the score by that estimated cost. Generally, work down from the most significant user needs scoring items.

8. If something costs too much, it is a candidate to break down into smaller stories.

How to score consistently

By learning some simple rules, independent people with a shared understanding of the scores should be able to score the same item similarly. However, differences will exist. Let someone else score the same story, and if they come up with a different score, discuss why. Reaching a joint agreement occurs nine times out of ten. If there is confusion about the scope of the problem, someone is likely underestimating it. Scope rarely shrinks over time, so go conservative. Use the larger value. We will drill down into the rubrics for each score in the UNS. I have seen teams split the difference (they can't decide on a 2 or 3, so they score 2.5). This is an option, but this hedging approach is typically unnecessary. One needs to ensure the story tracking tool supports this. I developed integrations with Jira and Oracle's internal bug system to use drop menus to make it easier to select whole numbers. However, we appreciate not wasting time, so do it when needed. Don't automatically split the difference with a disagreement. Discuss the reasons behind the scores. We can explain the rubric behind each score.

> **Note**
> When scoring the feature, isolating the dimension being scored is critical. Ignore the other two factors. Score each factor independently of the other two.

Scope – how many users does it impact?

3 – All, 2 – Some, 1 – A few or a Limited User Role

When considering who this change is made for, think of that group in the context of the total number of customers for the product. It cannot be based only on the people using the specific feature. *Everyone* signs into a secure portal to access enterprise ChatGPT or other services– so rating problems with sign-in get a score of 3. At the same time, only a few people customize their home page. One could rate that a 1. Don't change the scope and say, "Well, *within* the people who customize, *all* of them will use this feature." It doesn't work that way. Scores need to be measured against other stories and across different teams. The goal is to determine where the most good would come from by applying resources, so cheating doesn't help the product. Focus only on the number of users impacted. Ignore Severity and how bad the problem is at this step.

Indeed, being unable to sign in is different than a visual distraction such as the words "sign in" wrapping oddly on a mobile phone. We will handle that in the last user metric.

The enterprise solution might be in a chat interface; parts might be conversational results in hybrid UIs or data rendered from a ChatGPT backend. No matter the form, each can impact a few or many users.

Here are Scope examples that score as 3s:

- Common questions and answers (because "everyone" does this)
- Sign in
- Issues on the landing page that prevent sign-in or registration
- Setting up an account
- Very slow to respond in almost all cases
- Missing the primary use cases

Here are Scope examples that can score as 2s:

- Frequent questions and answers
- Registration for the service, portal, or website
- Customizing the user's profile
- Slow to respond in some conditions
- Ability to export chat logs (the score depends on the use case)
- Missing secondary use cases

Here are some Scope examples of 1s:

- Uncommon questions and answers
- Adding an avatar to their profile
- Ability to export chat logs (the score depends on the use case)
- Slow to respond in a particular condition
- A missing help link in a feature used by a small audience
- Missing supporting use cases

We want to be flexible on *everyone*, like in *Figure 4.3*. Think of it as almost everyone, or about 80% or more, 2 representing the next 15%, and 1 would be 5% or less of users.

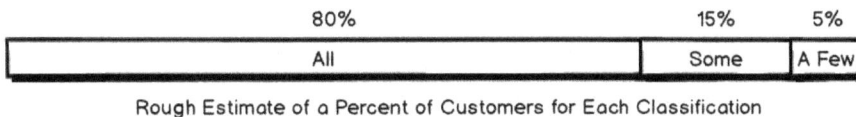

Rough Estimate of a Percent of Customers for Each Classification

Figure 4.3 – All users are 80% or greater, while Some and A Few are in the tail

Discuss and debate, but work to score consistently. Sometimes, the term everyone means very close to everyone, such as customers who sign into online banking. However, we know some older adults might avoid online banking, and a few of those who sign in are high-value customers. If all things are equal, there is more value in doing something for all users rather than for a few. Once the number of impacted users is understood, consider how often the issue, task, or action will occur.

Frequency – how often is it used

3 – All of the Time, 2 – Sometimes, 1 – Infrequently

If every time users come here, the problem exists, it is easy to make this a 3. If it only happens in a specific mode or state, then it is a 2, and those annoying errors that pop up rarely would be a 1. Judge whether issues occur sometimes or infrequently because multiplying by one doesn't do anything. Consider that intermittent errors are challenging to reproduce. If errors are noted in a log analysis or bugs that keep coming up, set this as a 2. There is flexibility, but all the time means all of the time; use a consistent approach to judging between 2s and 1s. If it is less than 5% to 10% of the time, it is a 1. Don't consider the importance of what is being scored. This will be judged in the third score.

Here are a few Frequency examples of 3s:

- The answer to the question is always wrong
- Every time a dialog box is opened, it is empty

- A typo or grammar error that appears all of the time
- A scroll bar always appears even when not needed or wanted
- The tone or style of the conversation needs to be corrected
- The chat doesn't recall the context set in this session

Here are some Frequency examples of 2s:

- The answer to a question sometimes needs to be corrected
- If your saved name is too large, then it truncates
- A time-out error appears after using the product for 10 minutes, and this is too early (if the time-out is usually 4 hours)
- A dialog box appears off-screen some of the time
- In many conditions, the style or tone of the conversation needs to be corrected
- The chat doesn't recall context sometimes

Here are some Frequency examples of 1s:

- In a specific use case, the style or tone of the conversation is wrong
- An error appears rarely, and no understanding of why. It only happened once in that session, and everything is working
- When saving a chat session, it errors out on rare occasions
- Every once in a while, when going back in the chat window, it forgets what step we are on and shows the wrong information that was already affirmed

When looking at logs, it might not be obvious how often something is happening. It would be best to look at a larger cross-section of logs or test instances to learn about frequency. But it is sometimes more of a judgment call for a two or a one because there is so little data. Remember, this is just frequency; don't get freaked out because some are bad issues; we should catch that next.

Severity – how bad is the problem?

4 – Severe, 3 – Critical, 2 – Important, 1 – Not Important

This is the easiest to score, or is it? Severity is sometimes hotly contested. Our values are inverted compared to most organizations' common Bug Severity (Sev). Because all organizations have Severity, it would seem easy to do. But not always. Our score needs to use larger numbers to express more value. We can break down how people think about Sev and the scores we assign to each severity.

An actual **Severity 1** issue only comes up occasionally. Severity 1 means **service is down – unavailable**, and no workaround. We don't often see Sev 1s in development because the code is not in production, and thus, the production system isn't being evaluated. But from a usability perspective, if the task can't be completed, that is a usability Sev 1 and thus is worth the most, so we assign it the largest value, a 4. Likely, these are just bugs. An organization should have a list of definitions to help prioritize bugs; start with that. If not, create one. I had one printed out by my desk to remind me of classes of bugs and their values. At my previous company, we had a category just below Severity 1 called a **Severity 2 Showstopper**. Because they were showstoppers, a significant issue that impedes progress, we would also classify these as 4. From there, we go to Sev 2, worth 3 points. It represents critical issues that cause significant user roadblocks. Sev 3 is worth 2 points. These are problems that cause some issues but likely won't cause the customer to scream and yell. Finally, Sev 4 is assigned 1 point. These are minor issues. Most organizations have the 1 to 4 severity scale, so we have the 4 to 1 values.

> **Note**
> Severity is well understood in most organizations; extending the definition to include LLM issues and inverting the values is a good start. If the organization scoring model extends beyond 4, consider mapping it down. Applying a five—or six-point scale for Severity will dramatically increase its weight in the UNS. It's a choice, but we will discuss the downside of weighting in the UNS calculation later in this chapter.

Severity is well understood for bugs but might not be used for criticality or priority of new features. For our purposes, we use one measure here. Examples consider the issues, stories, features, or interactions in UIs or ChatGPT-enabled solutions. They are just examples to further the conversation.

Examples of Severity for 4 – Severe (the equivalent of Severity 1):

- **American with Disabilities Act (ADA) and Web Content Accessibility Guidelines (WCAG) or a country's equivalent**: Major accessibility issues, including missing labels, non-standard abbreviations, using only color to distinguish **UI** elements and chats, and recommendations that are not navigable or unreadable by a screen reader. Chat or recommendations are not accessible.

  Article: `Web Content Accessibility Guidelines (https://en.wikipedia.org/wiki/Web_Content_Accessibility_Guidelines)`

- **National Language Support (NLS)**: Missing message files are causing pages not to render, garbled error messages, can't translate the string, or the wrong language to appear in a chat or recommendation.

  Article: `Internationalization and Localization (https://en.wikipedia.org/wiki/Internationalization_and_localization)`

- Help doesn't appear or leads the user down the wrong path.

- **Performance issues (beyond the level of service)**: The system or product needs to be more responsive. Its response time is significantly worse than any set goal.

- The wrong model is engaged (i.e., routed to time reporting when the prompt should have sent the user to expense reporting).
- Company-specific language is misunderstood, initiating the wrong flow or task.
- The user is stuck in a loop, and LLM won't let them out.
- The LLM misunderstands channel capabilities, which causes task failures.

Examples of Severity for 3 – Critical (typically a P2 bug):

- Help is wrong but doesn't cause further errors by the user
- A significant percentage of users need assistance to complete the task
- Context lost during workflow
- Typos (that impacts or changes meaning)
- Scalability issues (such as a shuttle component with thousands of elements)
- The user isn't prevented from making a grave mistake (and there is no undo)
- Multiple entities are not understood
- Company-specific language is not understood
- Users have to repeat themselves in a different way to be understood
- The LLM misunderstands channel capabilities, which causes task issues

Examples of Severity for 2 – Important (P3's in the bug system):

- Help is unclear, wordy, or too complex but factually correct
- A large number of users would need assistance to complete the task
- Concepts presented to the customer are complex to understand or too technical
- The layout needs to be more explicit
- Incorrect page header or section label
- Incorrect breadcrumbs (bump up if the user loses context)
- An incorrect time format causes a user error (such as scheduling a meeting)
- Grammatical errors

- Most of a collection of entities is understood, but one is missed

- The solution does not effectively use advantages for a specific channel

- Significant inconsistencies (such as button order) that cause user error or a missing unit that would confuse the user

- The user isn't prevented from making a grave mistake (but undo is available)

Examples of Severity for 1 – Minor (P4s for a bug):

- Minor inconsistencies (button order, using Delete for Remove, OK instead of Continue, or missing units after numbers, but it would still be understood)

- Incorrect usage of blank table cells versus N/A or unavailable

- Lack of correct pagination or segmentation

- Overly wordy or too terse

- A different layout, table, format, or UI element can improve an LLM response.

Not all of these are as cut and dry as we might have led on. Don't fret; decide what is essential to customers. We don't expect each reader to agree with all of the preceding examples. Make decisions on what goes in which bucket, share examples with the team to reduce confusion, create a shared understanding for your rubric, and fix the issues based on their score.

Examples of scoring

Practice to learn the method. To score, we use the UNS values and multiply them together. Then, include development to cost efforts to prioritize and solve these issues.

Dev teams usually start with T-shirt sizing. They estimate the cost of development with **Extra Small (XS)**, **Small (S)**, **Medium (M)**, **Large (L)**, **Extra Large (XL)**, and so on. They must do the math at some point, converting this to story points, a standard unit of measure for Agile development or development days. The Fibonacci sequence is commonly used for this estimation when using numbers. The numbers (1, 2, 3, 5, 8, 13, 21, 34) grow with each gap to convey the complexity of measuring a large project. The larger the estimate, the more significant the gap because more extensive efforts have more variability when estimating. They are commonly used for costing development. Once there are numbers, divide the UNS by the development cost. This is only done once this story is close enough to development that understanding is clear enough to cost it beyond T-shirt sizing. Development can use a proxy like T-shirt sizing (1-XS, 2-S, 5-M, 8-L, 13-XL, etc.) that the team agrees on. However, do an actual costing exercise to understand better what is being built. *Table 4.1* contains some GUI design examples. These examples include conversational issues. Once these stories are understood, development can cost them more accurately.

Samples from a ChatGPT Web Experience	Scope	Frequency	Severity	UNS	Dev Cost
1. The Browser Back does not go "Back"	3	3	4	36	XL
2. Text: "Run" should be called "Search" in the toolbar	3	3	2	18	XS
3. Can't submit a prompt by pressing return in a field	3	3	2	18	L
4. The vertical scroll bar is missing from the Conversational UI	2	3	3	18	M
5. "Enterprise Support Recommendations" is a confusing term	2	3	2	12	XS
6. Columnated answers need to be sorted correctly	2	3	2	12	M
7. The default UI language does not match the website	2	3	2	12	M
8. Text and formatted numbers should align correctly	3	3	1	9	XS
9. Images don't have labels	2	2	2	8	L
10. A task region cannot be dragged or moved onto the screen	1	3	2	6	M

Table 4.1 – Scoring issues on User Needs Scoring metrics

It is challenging to get the team to decide on a costing method; this focus is on a value method. To explore why one uses Fibonacci numbers rather than time in hours or days, learn about story points. After starting with this simple explanation, do some searching. If you are on board with story points, skip the article.

Article: Why do we use Story Points for Estimating? (https://www.scrum.org/resources/blog/why-do-we-use-story-points-estimating)

There is a rationale behind the UNS values for each item in *Table 4.1*. Because you are unfamiliar with the application, context is provided with each item. We will relate these examples to LLM issues typical in generative AI applications:

1. The browser's back button does not go "back":

    - **Background**: When building a web application, a process likely goes from one web page to another; it is natural for a user to want to occasionally return to the previous step. So, they click the browser's back button. Without coding for this, the application fails to navigate correctly.

    - **Reasoning**: Because the user is *lost* and can't return to the main UI, this problem is worth fixing immediately. The tech stack had an issue supporting back, which is why development costs are high. It means rearchitecting how the browser session is created, thus making it a challenge. If the cost is less than twice that of item number two (because its score is half as big), it would still be more valuable to fix item 1 before item 2. When LLM chat is integrated into existing apps, all kinds of issues appear: does the conversation pick up where it left off when opening a new tab or window on the same site? How does the back button in the browser impact the conversation thread? Does the conversation get confused with multiple open windows when the customer interacts in multiple windows?

2. **Text**: "Run" should be called "Search" in the toolbar:

    - **Background**: We typically see a search field when designing a UI to search a database. Terms other than "Search" can be confusing.

    - **Reasoning**: Unless dynamic (text that changes based on the user profile or other factors), text on a UI will always be visible to all users; it always scores a three for the number of users impacted. Everyone searches in this app. If the Run label is generated based on a user profile, it might appear only for a segment of the audience and only in some conditions. In this example, it is always on screen and is used all the time, so the Frequency is also a three. When scoring severity, if it causes the wrong interaction, it scores a three; if the text is just confusing, score it a two.

    - **Alternative UI considerations**: If this were a conversational search from a chat prompt, we would train our model to understand various names for our search function (e.g., `Find me all phone models that have...`, `What are the phone models with...`, `List the phone models...`). The default models should pick up this nuance very quickly, but with enterprise-specific tasks, train the model to understand the tasks and the various ways one can ask for them. This will require prompt engineering and fine-tuning, and we will start to cover that in *Chapter 7, Prompt Engineering*.

3. Can't submit a prompt by pressing return in a field:

    - **Background**: There is a similar issue in the ChatGPT playground. They decided that the *return* key on the keyboard should be used to create a carriage return in the text entry field. To submit the text, press *command + return* (on a Mac), as shown in *Figure 4.4*.

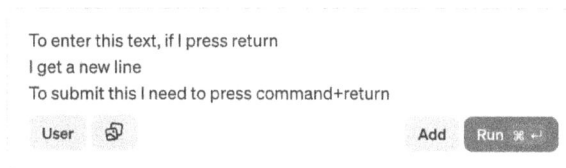

To enter this text, if I press return
I get a new line
To submit this I need to press command+return

User   Add   Run ⌘ ↵

Figure 4.4 – Example of using an unnatural keyboard entry to submit

    - **Reasoning**: We understand how folks naturally press return on the keyboard on mobile or desktop to submit a form. They must stop and deal with the onscreen button press when this doesn't happen. In *Figure 4.4* from ChatGPT, they have to press two keyboard buttons together. That is annoying and not natural. Hence, it is a two on the Severity scale. Users will likely *never* get used to this interaction because it is unlike the 99% of interactions they already do. An OpenAI interaction designer likely lost this battle, but they did get to include the *command + return* command in the button label. A hack at best, but at least it is a visible reminder. This is an example of a visual affordance covered in *Chapter 9, Guidelines and Heuristics.*

4. The vertical scroll bar is missing from the conversational UI:

    *   **Background**: If there is scrolling text and the scroll bar is missing, the user will have trouble getting to their results.

    *   **Reasoning**: In this case, the scroll bar was missing. It wasn't just hidden because of an operating system setting, but it didn't happen to all users (score of 2). It only occurs when there is a lot of text and not all answers are lengthy. However, if the user wants to read the whole answer, the scroll bar not appearing or not being supported is a big issue (scores a 3). This case has some workarounds, but we can still see high scores for this critical issue (Severity score of 3).

5. *Enterprise Support Recommendations* is a term that needs to be clarified:

    *   **Background**: If terms are unfamiliar to the user base, they might miss important information about their task.

    *   **Reasoning**: Customer feedback suggests that this term needs to be clarified. We understand that the results could be written more user-centric and less technical. Some people will understand this term, so it only rates a two. It will always be used; everyone will see it, so it is a three on the *Frequency* scale. But there is help, and the term includes keywords such as *support* and *recommendation*, so the additional word "enterprise" doesn't add any value; it is not critical. As expected, it is not hard to edit a word, so the cost is XS. Conversational style, tone, and language will likely cause issues. Review how knowledge and help is written and work to address how an LLM responds using prompt engineering.

6. Columnated results need to be sorted correctly:

    *   **Background**: The logical order for a list should be the default. If multiple logical orders exist, user control should be allowed, and customer history could be used to decide the default.

    *   **Reasoning**: A challenge with data in documents and tables is preserving table-like features. With this, some information is maintained and clear. For example, showing the country's top 25 college football teams alphabetically isn't helpful. They should be sorted by their current college rank. The table should include the calendar date for the last ranking (so old data is apparent). However, only some users ask for this kind of data, and if the table is sortable, then there is an easy workaround to ask for the data to be sorted. In this example, we don't expect sorting for the top 25 teams in football. In *Chapter 5, Defining the Desired Experience,* we will go into detail about the use of tables in small chat windows. The real estate for a table is sometimes tight at best. A table acceptable on a desktop web experience won't work on a mobile chat experience. In this case, not all users were impacted by the table (a score of

2), but it was always an issue because it was affixed with a poor sort choice (a score of 3). It wasn't a big issue because of the kind of data (score of 2). Critical data, such as deals likely to close this month, demand a good sort order (deal size, likelihood of closing) and the ability to change the order using a GUI or conversationally.

```
Sort the deals by closest diving distance to me

Show the orders by quantity

Display deals for me that require the most attention that will
give me the most commission
```

7. The default UI language does not match the website:

    - **Background**: *Chapter 9, Guidelines and Heuristics* will discuss matching a customer's language and understanding. If there are different terms to mean the same thing (common in an enterprise), the user might only understand terms within the proper context. This applies to spoken and written languages.

    - **Reasoning**: Respect the customers' wishes in an international UI. If they have set their UI to Spanish, they should be given their conversational chat or AI results in Spanish. If they have to ask for it in Spanish, *"Español, por favor,"* this is a fail. If not all the information is in Spanish, the failure is worse. If the LLM supports multiple languages, how will it react to a combination of two languages? Issues around the choice of words, internationalization, or cultural issues should be evaluated for repair or rework. We will talk about some of this in the next chapter. In this example, most customers work in English, so each primary language scores 2 for customers impacted. They would always have this issue (a 3). Depending on the language and user, the severity is typically a 2. If the user can only work in their native language, and the site doesn't support it, score this a 4.

8. Text and formatted numbers should align correctly:

    - **Background**: Currency, table data, and `name:value` pairs must be aligned in specific ways.

    - **Reasoning**: As long as labels and alignment don't impact understanding, these score smaller values even if they always appear. Decide how vital style and a clean experience are to the company. However, as shown, these are low (XS) costs, so with some work, they will be prioritized accordingly and fixed. Formatting, in general, can be an issue for an LLM. In chat experiences, the customer might be able to direct the LLM to change the format (e.g., `show as a table`) or present the findings differently. Improvements to the instructions given to the LLM might be warranted.

9. Images don't have labels:

    - **Background**: When a UI includes images, for example, in search results, they must be labeled with words. It is possible to use an LLM to label these more descriptively than a human would. For example, we might show shoes on a clothing site. "Shoes" is not an uncommon label that a screen reader can read for someone visually impaired. An image analysis LLM might provide the more accurate "Nike tennis shoes with a leather upper, and red, green, and yellow panels on a clear sole with Velcro straps."

- **Reasoning**: Depending on the context, images without labels can be more important than just an accessibility concern. In this case, the photos are more ornamental and are supported by text in the conversation. But watch out for this: make sure the experience is accessible and that labels, if generated by the AI, represent a clear understanding of their purpose. Significant images must be described for the visually impaired. GPT 4o can understand and present details for an image. So the user can hear the description and make purchasing decisions. These descriptions are not necessarily shown on screen. A screen reader will pick up the description from the HTML metadata and speak it. This can be done at a lower cost by sending the request for descriptions to the LLM by batch processing and storing the descriptions in the database. Batch processing in ChatGPT saves money. These robust labels can also be fed into the search to enable robust searching without additional LLM costs.

10. A task region cannot be dragged or moved onto the screen:

- **Background**: Enterprise users have dashboards. Dashboards will get smarter because of LLMs. A little preview: *Chapter 12, Conclusion* will discuss the value of Wisdom, which every dashboard should have. The user tries to add a region that wasn't displayed by default. If they go to customize the page and can't add or move some regions around, they will think this is a bug. *Figure 4.5* shows a Chat region being added to the dashboard by dragging. If it didn't work, that would be this problem.

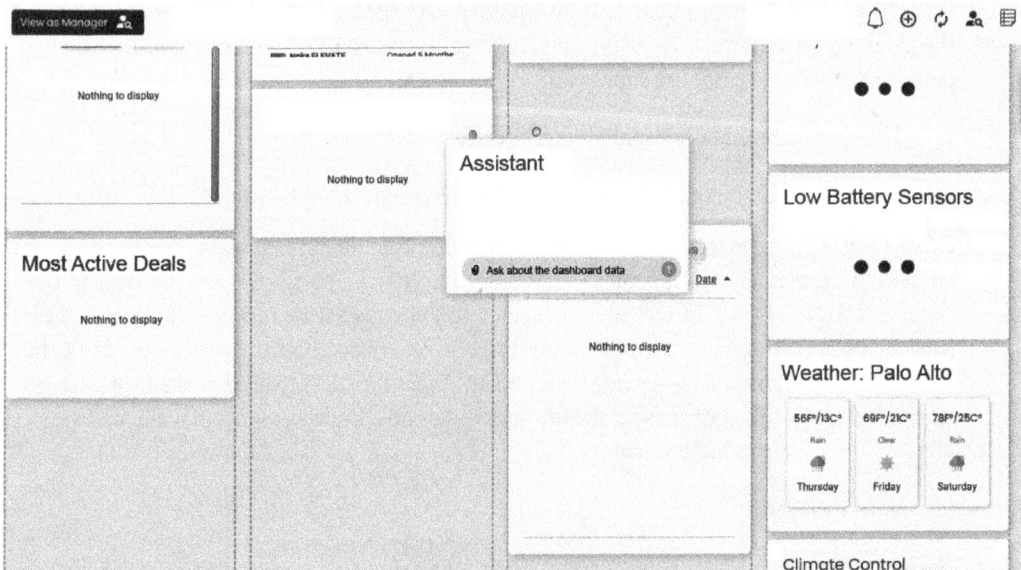

Figure 4.5 – Dragging a region onto a dashboard

- **Reasoning**: This relates to a UI feature that allows the regions and, thus, the ChatGPT experience to be customized. However, only some people do this (score a 1); they leave it where it is on the page. Even though the issue is always there, most people will not notice it.

Even within a ChatGPT chat window, we can present tables, include images in results (such as product images, schematics, or diagrams), use incorrect terms, or even have scrolling issues. ChatGPT is not a stand-alone solution in the enterprise; it is part of the more extensive solution and thus can have not only problems novel to LLMs but can include GUI issues.

Putting a backlog into order

Let's now complete our thought. We have replaced our development cost with story points. We can then calculate and prioritize our backlog. What will rise to the top in *Table 4.2* with the score filled in?

Samples from an Integrated ChatGPT Web Experience	UNSs	Dev Cost	WSJF
2. Text: "Run" should be called "Search" in the toolbar	18	3	6
5. "Enterprise Support Recommendations" is a confusing term	12	4	3
8. Text and formatted numbers should align correctly	9	6	1.5
4. The vertical scroll bar is missing from the conversational UI	18	14	1.23
6. Columnated answers need to be sorted correctly	12	10	1.2
7. The default UI language does not match the website	12	12	1
3. Can't submit a prompt by pressing return in a field	18	20	0.9
1. The browser's back button does not go "back"	36	43	0.84
9. Images don't have labels	8	16	0.5
10. Customization of a task region cannot be dragged onto the screen when empty	6	24	0.25

Table 4.2 – The stories ordered by WSJF (larger WSJF should be developed first)

The row numbers are shown from the previous table to show how the order changed. Some observations from the scoring results are as follows:

- The first three items are all low-cost, also called *low-hanging fruit*. They are easy to pick off and accomplish, but they could be more exciting, couldn't they?

- The eighth item in the table (1. The browser's back button does not go "back") has the most significant cost, but we notice it has the highest UNS. Review this again and evaluate whether a less complex solution can bring most of the value at less cost. If true, that new solution might jump to the third spot on the list. Sometimes, simple solutions can solve a piece of a complex problem.

- Multiple items have the same UNS, but once WSJF is calculated, no two items have the same score. This is really because we only scored ten items. In reality, there will be ties. It just means doing any tied item is acceptable.

Patching case study revisited

In the patching example from *Chapter 3, Identifying Optimal Use Cases for ChatGPT* we covered multiple LLM opportunities. However, we should revisit these interactions now that we have user scoring. To keep it simple, we will take the top three opportunities from *Table 3.4*, reintroduce the steps, and add the WSJF scores, as shown in *Table 4.3*.

Prioritized Steps Identified for ChatGPT Support	UNS (A*B*C)	Dev Cost	WSJF
1) Step 11: Generate tests from existing cases. Identify gaps in testing.	12	8	1.5
2) Step 12: Monitor production instances for abnormalities.	8	16	0.50
3) Step 6: Predict the results of the patch plan, and the implications for missing or conflicting patches.	8	32	0.25

Table 4.3 – ChatGPT use cases prioritized by WSJF

So, we now have the top patch issue to address. Because the UNSs were so close (as discussed in *Chapter 3, Identifying Optimal Use Cases for ChatGPT*), it came down to which item was most straightforward to implement. The cost of doing an LLM solution to *Step 11* is estimated to be less because the model requires fewer moving parts to generate test cases. We need to know meta-information about the software, versions, platform, OS patching level, the list of existing patches, and we can use to tie in with existing bug information to create an LLM advisor. Developing models for *Step 12* to monitor for abnormalities requires additional data sources and, thus, more API work to gather and provide the data. *Step 6* for predicting patch plan success is the most expensive to build and maintain because of the complexities of predicting how *trillions* of possible patch combinations might influence an installation. Each of these steps represents its own LLM model, possibly more than one model, discuss in *Chapter 8, Fine-Tuning*. There is a lot of overlap, but the inputs and outputs will vary. Knowing what patch collections are stable based on customer data from production instances is better understood than knowing what *should* work. The problem gets even more complex when we try to include an understanding of what could make it better or worse, which is why the estimate for *Step 6* is much larger. Your use cases will be easier to understand than our attempt to explain the patching processes for those unfamiliar with the area. There are also examples around the quality of the responses from an LLM as part of a conversation or coming from a recommendation; let me touch on that as well.

We know that the LLM will attempt to hallucinate and will try to respond. *Chapter 7, Prompt Engineering*, will discuss how to control the output by creating guardrails and instructions for the LLM. When LLM generates wrong output this becomes an issue to score. The logs show which customers are impacted by the problem. Monitoring the output will help identify the frequency of the problem. Judge the severity of the issues against the rubric.

There is no one answer for scoring these; it will depend. Hallucinations can be harmful or not so bad. They happen constantly or rarely and could impact all customers or only a select audience. Likewise, this might be a lack of or conflicting knowledge, a missing API, a source not considered by the LLM, poor training, unclear prompts, or just out of the application's scope. With log analysis, reviewed in *Chapter 2, Conducting Effective User Research* , plus scoring and prioritizing effort, the team has what it needs to tackle important work first. Extend software tracking tools to instill this process.

Extending tracking tools with scoring

Most Agile, product, or bug tracking tools can expose additional fields or columns. For example, a development priority drop menu (1 to 4 is typical) might already exist. UNS fields are similar but couched more in user-centric attributes.

Expose all four fields (the three individual and the final score that should be sortable) in the company's tracking tool. This allows for visibility and discussions to validate assumptions. As mentioned, there might be slight disagreements, and exposing scores brings it into the open. Expect to be able to use this to drive the sort order for stories and bugs so that the sprint teams, release management, and customers can see how and why this order exists. Transparency is best here. Score and then sort by scores in the tracking tool. Jira is very popular and supports custom fields and even calculations. WSJF is supported in Rally, and most tools can handle these scores via customization.

Try the User Needs Scoring method

Here is a spreadsheet with examples, coded with the drop menus and calculations. It might help to get started. Bring a few examples, work as a team to understand the method, and fill in stories and scores.

GitHub: `Score Stories Samples Worksheet (https://github.com/ PacktPublishing/UX-for-Enterprise-ChatGPT-Solutions/blob/main/ Chapter4-ScoringStoriesSamples.xlsx)`

Click the download button, highlighted in *Figure 4.6*, to download the file to the desktop. There is no viewer for these files on GitHub.

Figure 4.6 – How to download a file from GitHub

A simple paper and pen worksheet can be shared in a workshop to introduce this concept to the team and coach people through this process.

GitHub: `Scoring Worksheet` (`https://github.com/PacktPublishing/UX-for-Enterprise-ChatGPT-Solutions/blob/main/Chapter4-Scoring_Worksheet_for_Design%20WSJF.pdf`)

Of course, sometimes too much is not a good thing. Resist the need to enhance scoring with more measures and capabilities. Start small. Let me explain why.

Creating more complex scoring methods

We are reminded of the **KISS (Keep it simple, silly) principle**. An organization can *go all the way* and create a score so complex that no one would understand the difference between something with a score of 2,032 and 2,840. I know a system with 17 factors adding and subtracting based on who escalated the issue (a VP escalating the issue is worth more than if I do it), how old the problem is when it was filed, and its severity, among many other factors. Everyone wants to get their factor into a score. Resist that approach.

I find that (and some simple research could confirm) a mortal would have no chance of getting a good feel for working with the output of a score based on 17 factors. Indeed, the proposed simple score method will not differentiate 20 stories, all with a score of 12. However, we suspect there will be more spread when accounting for the cost and work allocated across a few Agile teams. It is OK for items to wait their turn in an Agile backlog. Consider if including rules for ranking tied items would help. For example, a 12 for an issue for all customers should be prioritized higher than a 12 coming from a worse bug for fewer customers. However, the more complexity, the more difficult it is to understand and judge differences. Determine if it is worth the additional complexity, confusion, and overhead. Minimally, we advocate not ranking stories and scoring arbitrarily. Use a method that is repeatable and even can be consistently applied from team to team. So, aggregating rankings on a backlog is an apples-to-apples comparison. Once stories are divided by the costs, the values will spread out. Also, our goal is not spread. If two stories are judged the same, doing either story is just as valuable.

Working top-down from WSJF is not a religion. We don't necessarily follow the WSJF as a dogma. It is a guide. Besides being a guide, different teams sometimes have specific expertise (not all organizations can be perfect about an Agile team doing it all). So, this can influence what teams pick from the backlog. Selecting the third item and leaving the first two for another team might be the most efficient and intelligent approach.

Working with multiple backlogs in Agile

Usually, we have more than one backlog in a large enterprise development environment. So, this discussion is for those familiar with the complexities of multiple sprint team planning. The scrum team for the sprint has a backlog, and so does the project, maybe the program, and even the release. But effectively, these are virtual, and tagging an item for a sprint doesn't mean ignoring the context of all other backlog items. A single large backlog is sometimes helpful to get a few key metrics out of it if done right. In Agile at scale, they look at different roll-ups of stories. At the program level, features that decompose into stories can be viewed. If the feature has a score of 36, it won't mean that all stories in that feature will have the same value. As they are broken down, some stories are more important than others.

The program or product-level backlog might span 30 or more teams, as it did for my old organization. If a few teams represent 50% of the backlog (thinking now about story points or work effort, not in terms of the number of items), then maybe reconsider how teams are allocated. It is hard to compare if teams use different models to cost stories. However, we also want to balance team autonomy. In Scaled Agile, where many Agile teams are developing the same application, there is a need for a well-documented approach to story points. One more attempt to sell story points for those unconvinced.

Article: What is a Story Point (https://agilefaq.wordpress.com/2007/11/13/what-is-a-story-point/)

Don't be trapped by estimating in hours. There is value in abstract story points. However, being accurate with hours is a headache, and because different people work at different paces, there are complications. Mike Cohn writes about this.

Article: Don't Equate Story Points to Hours (https://www.mountaingoatsoftware.com/blog/dont-equate-story-points-to-hours)

If there is one backlog and a consistent story point method, the delivery based on **customer points** can be compared. One team has five stories, each with a score of 12, having (5x12) 60 customer points in that sprint. Meanwhile, another team might do just three stories with scores of 36, 24, and 12, thus delivering roughly the same value (62 points) to the customer. A team tackling 15 stories with a sum of 30 provides 1/2 the value. A reallocation of teams to projects that have more customer points is warranted. This is not a challenge for a single sprint team working one list top down, but when doing Agile at scale, one has to consider when to reallocate sprint resources to be the most effective.

Article: Scaled Agile Framework (https://scaledagileframework.com/)

I can't stress enough the value that SAFe provides. The full version of SAFe, as shown in *Figure 4.7*, includes matured models, concepts, refined processes, and extensive documentation, detail, training, and support. Go to the website, as this overview is constantly evolving.

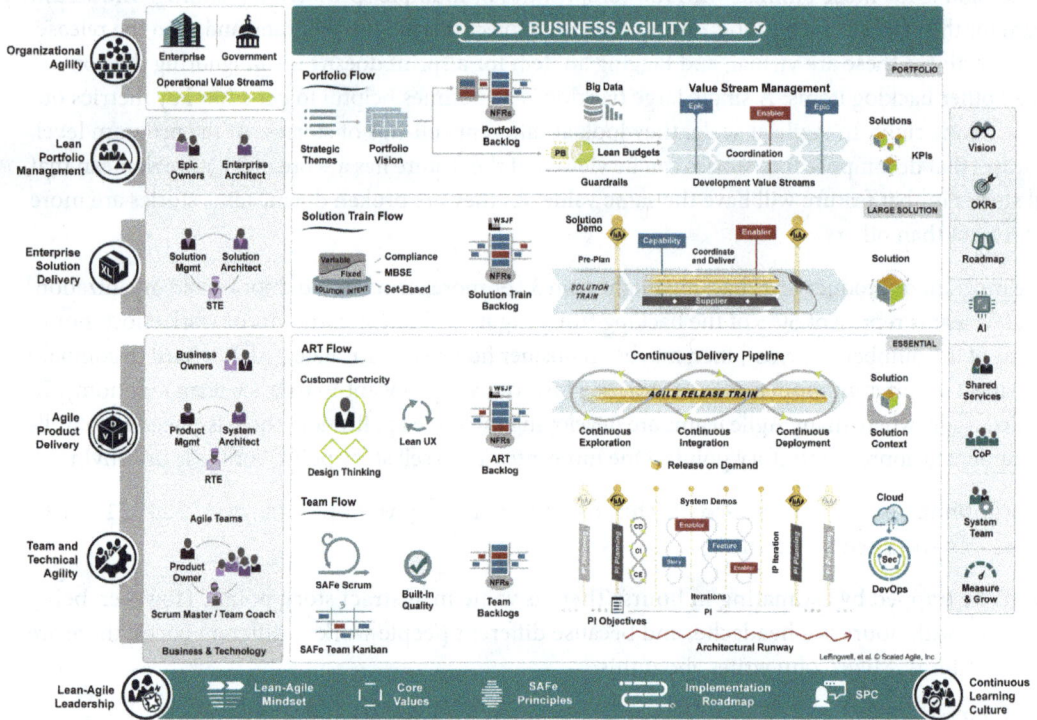

Figure 4.7 – The SAFe map

For example, a typical org might count and report on bugs per software unit. A team with 50 editing and typographical bugs might not be best compared to a team with three performance bugs, and how quickly they fixed those bugs or if it was worth the time and energy. The team with three bugs might have been doing something far more valuable and costly (development cost) than the team with 50 easy bugs to fix. For design, we look at customer points. For development, in Agile, we can look at story points or the level of effort. Then, we can look at this single virtual backlog based on scores and learn something valuable.

An **ordered backlog** is less helpful than one with scoring. Ranking on an ordinal scale (1 is higher than 2) is less potent than scoring (a 15 is 50% more than a 10). We learned this in introductory statistics. An **interval scale** (where the difference between two numbers is an equal interval) is more powerful and carries more information. A user score of 10 is twice a 5. Ranking something 5th and 10th does not communicate this importance. The 10th item is, at best, five behind item 5. That is, items 6 to 9 might not have huge differences in value. An interval (or ratio) scale reveals this.

> **Note**
>
> A **score** is an interval scale, while a **ranking** is not. Scores are more powerful. A ratio scale is similar to an interval scale but has an absolute zero. The difference between them is not essential to this discussion. Just realize the value in knowing that the difference between 2 scores (6 to 12 and 12 to 18) represents the same difference.

This is why they show the points used to rank college football teams. Football fans then know how far *behind* their favorite team is from the next spot up.

Returning to *Table 4.2*, we can see that the first item on the original list, which scored 36, drops way down in priority because of its significant development cost. Look carefully and see the cost of delivering the first five items on the list; fixing all of them is better than fixing the single back button item, ranked first by UNS. This value is only revealed when accounting for development costs and rank by WSJF.

> **Note**
>
> It would seem counterintuitive to not do something so valuable that it scores a 36 even with a long development lead time, which would be strategic and thus prioritized. However, those smaller items give more value. We optimize customer value.

There are scoring issues. Realize this can be challenging. Being aware of possible hiccups means they can be addressed along the way.

Real-world hiccups with scoring

No scoring method is perfectly repeatable when people's opinions come into play. We do our best to try to make a system that avoids debate. I have used this system for many years, and there are gotcha's. Scott McNealy, one of the founders of Sun Microsystems, is often attributed with this variation of a Teddy Roosevelt quote, "The best decision is the right decision, the next best decision is the wrong decision, and the worst decision is no decision." So, let's put a stake in the ground, make the decisions to prioritize our backlog, know some limitations, and move forward.

I know Agile, and this is not WSJF

I defer to the SAFe website or Donald Reinertsen's *Principles of Product Development Flow* book. User Needs Scoring is a variation on the WSJF solution because it can be coached in an hour and repeated easily. The UNS is a proxy for the numerator (it is a combination of the values for user + business value, time criticality, and risk reduction + a score for opportunity enablement value). The actual WSJF uses a Fibonacci-type number, and using that scale has more breadth. Again, this is to help *keep it simple, silly*.

Is WSJF deployed in your organization? If so, excellent work. Is the CoD metric reliable and repeatable? If not, adopt a consistent rubric like the one proposed.

The use of simple numbers one to four

Notice that the SAFe supports the concept of an interval scale for scoring. The Fibonacci sequence of numbers (1, 2, 3, 5, 8, 13, 21, 34) conveys an interval scale. Interval scales are commonly used for costing development. Our straightforward scores are also on an interval scale but don't have the spread of the Fibonacci sequence, as they only go from one to four. We can explore why the three—or four-point scales make sense.

Dean Leffingwell, the lead author of SAFe, suggests that because the larger numbers are less precise, using *"Fibonacci makes consensus easier."*

Article: `Backlogs with WSJF (https://web.archive.org/web/20220919164755/` `https://techbeacon.com/app-dev-testing/prioritize-your-backlog-use-` `weighted-shortest-job-first-wsjf-improved-roi?amp)`

A consensus can be reached quickly with a small collection of numbers, but it is easy to appreciate that larger numbers are rough estimates and, thus, might not be worth debating. Simplicity makes the process easier to digest. If you have strong feelings about using Fibonacci and have good experiences with it for story estimates, *continue using it for cost*. It is an excellent tool for cost estimating. For User Needs Scoring, the spread in our numerator between small and unbounded large numbers won't work. It would overwhelm stories. Scoring from 1 to 10 was evaluated, but the time debating a four versus a six doesn't seem worth it. Having a ten-point rubric is very difficult. Stick with three or four values on the requirements side (the numerator). Development decides the denominator based on story points or the poker estimation we discussed earlier. Another consideration is to weigh each factor in the user score.

Weighting factors

It is always a good idea to consider whether values should be weighted more. We could introduce weighting without changing the rubric or scale (one to three or four). Let's take the Severity scores. We considered doing scores to align with a 100-point scale but didn't find a good balance. 4 points are worth more than 3, but do we need it to be 100, 50, 25, or 10? Does multiplying or maybe not using a 1 for the last value make a difference? Someone should take that up as a research project. What would it look like to weigh each factor? It makes the calculation more complex.

We put the stake in the ground with our numbers. By multiplying the values together, each score is worth 33%. Weighting can be introduced if one score is worth more. I don't know which one that would be. They all seem equally important. In any case, try it as documented, or do better, but don't do less. The numbers allow for easy math.

$$WSJF = 3\frac{(Many\ Score*0.33)*(Often\ Score*0.33)*Bad\ Score*0.33)}{Job\ Duration}$$

What is essential becomes more apparent after a few cycles. It is common to find that scores of 12 or greater are prioritized. Items that score lower sometimes are low-hanging fruit—that is, they are cheap to fix, so they don't take up much time or energy but typically won't drive excitement. They do drive consistency and quality. Severity deserves additional insight.

Severity seems complicated to judge

Every development organization has some form of Bug severity. Typically, technology companies have four levels of bugs, but some companies go to five or six. Here is a GitHub document that provides an extensive collection of examples of severity. With practice, it becomes easier to judge.

> **Deep dive**
>
> GitHub: `Defining severity in a ChatGPT use cases` (`https://github.com/PacktPublishing/UX-for-Enterprise-ChatGPT-Solutions/blob/main/Chapter4-Deep_Dive_into_Severity.pdf`)

What we have seen is that Sev 2s get all the effort. Sev 1s are worked immediately, while Sev 4s, which generally revolve around fit and finish, never get worked. A collection of Sev 4s on the same feature adds up. Would anyone continue to use generative AI with poor grammar, typos, and a confusing layout? Each issue might seem insignificant, but having multiple problems in one element will quickly erode trust. Generally, we see very few of these errors in LLM outputs; it will be more about hallucinations, understanding questions that lack enough context, and lack of the correct enterprise knowledge, which will be a problem. These should be scored in the 1 to 3 range. Practice as a team so the collective expertise improves as the team judges' severity. Don't be scared by those big problems; Agile can help break them down into manageable parts.

The cost is so high that we can't ever get the work done

There is a solution in Agile for this. If the cost is high, consider how to break down the story into smaller parts. We don't want jobs blocked, so we work with agile methods to get the right-sized stories. Delivering a piece of a story can still provide value. Sometimes, backend work is required that won't return user value directly. Judge these stories by what user value they will enable down the road. Judge individual parts independently of the larger story or epic. It is sometimes a creative endeavor to figure out how a large story can be broken up. We can't give a magic solution. Brainstorm, consider where user value is derived from, and look for partial solutions.

Grouping issues into bugs to protect the quality

The grouping of similar bugs makes sense. We don't need a dozen bugs for the same fundamental change.

> **Note**
>
> We are switching gears here. This is about the Severity of the issue as documented by the enterprise bug system. Recall that scores invert these values. Grouping issues allows the team to manage less work in the bug or tracking system. Save time tracking 100s of bugs when tracking 30 would make more sense.

One suggestion is to look at the collection of issues and consider whether they are all related to the same problem. That is, can one fix and resolve three different bugs? Another issue is scope. If a collection of bugs is associated with the same material or within a single ChatGPT instance (when doing RAG), use rules to aggregate bugs:

- A collection of 3 or more Severity 2 bugs for a specific problem, interaction, results, or element is equivalent to a Severity 1 bug (the "three strikes" rule)

- A collection of training synonyms or fine-tuning examples for one concept can be grouped into a single Severity 2 bug

- Five or more Severity 3 issues should be tracked as one Severity 2 item

> **Note**
>
> A missing intent or task in conversational AI can be a Severity 2 "bug" (a bug because the original design/specification defined it) or a Severity 3 ER (if it was not realistic to have anticipated it).

Finally, consider adding a gate to protect the quality of the release. We cannot know the scale of your problem, so adjust as needed:

- No Severity 1s for any instance or model

- No more than 2 Severity 2s open issues per instance or model

A bug and feature tracking system is required. Work WSJF into that software tool.

How to work WSJF into the organization

These are simple calculations. Any spreadsheet can keep track of the results and be used to sort and order a list. However, building this into tracking tools such as Jira, Rally, or any homegrown tool is better. Include custom fields and drop-down menus for the scores, and then use a calculation field to divide the UNS by the development cost. Typically done in story points (recommended), but some organizations use days; any consistent method is acceptable compared to being random. Sort by this each time the backlog is reviewed to select work. This approach ensures the biggest bang for the buck.

Summary

Everything up to this point is to understand use cases and where to apply resources to evaluate and decide what to do with a ChatGPT solution. This chapter is an agile method that applies to ChatGPT interactions, recommender experiences, backend solutions, or any software development project. We needed a method to bridge the gap between defining use cases and deciding what use cases to tackle. Creating a repeatable method applies to any development project.

Do a few things with the learnings in this chapter:

- Estimate the value of existing stories, use cases, bugs, or features
- Work with development to explain the story and get their cost estimate
- Work out issues in scoring stories and prioritize the backlog
- Integrate WSJF into the product life cycle

Once a solution is up and running, we can apply more sophisticated approaches to evaluate LLM performance, but the same scoring concepts can be used there. We are almost done with our pre-development journey – one more stop. The next chapter will focus on designing the right experience. After that, once we go technical into ChatGPT technology, we can reevaluate and verify what was built and how it should improve, again drawing on the research and scorings discussion.

References

The links, book recommendations, and GitHub files in this chapter are posted on the reference page.

Web Page: Chapter 4 References (https://uxdforai.com/references#C4)

5
Defining the Desired Experience

To define the desired experience, the first decision is which experiences demand attention. This is the prioritization discussion started in the last chapter. Knowing where these use cases are to appear helps with scope, planning, and follow-through. Some require building traditional conversational experiences; shockingly, these can be called classic now! Others will use channels (such as Slack, Teams, or a web experience) that can handle interactive elements, such as GUI components. Some will extend features within existing channels to include LLMs in the workflow to solve a limited use case. That last one is a powerful incremental approach. Others will build single-purpose bots or voice-only interfaces. Each of these experiences has unique and general considerations.

This chapter will cover the highest-priority enterprise user experiences one can develop with ChatGPT, along with three overarching design elements: accessibility, internationalization, and trust. Get ready for the following topics:

- Designing chat experiences
- Designing hybrid UI/chat experiences
- Creating voice-only experiences
- Designing a recommender and behind-the-scenes experiences
- Overarching considerations

Designing chat experiences

An LLM-powered chat experience comes in a few flavors: a straight chat-only experience, such as OpenAI's chat UI, or a chat window that is part of a more extensive GUI experience. Each has unique issues, and both have a collection of issues to address. A traditional chat-only experience is an excellent place to start.

Chat-only experiences

Many, if not most, readers will have this use case: a conversational experience backed with a ChatGPT solution. It might support a traditional conversational AI platform (Salesforce, Oracle's Digital Assistant, Alexa, Google Assistant, or one of the dozens of other vendors) or eventually replace them. However, the primary interaction is via a chat window with chat responses. This means the channel SMS (text messaging), WhatsApp, WeChat, Facebook, LinkedIn, a web experience, Slack, Teams, or other services depend on the technology those channels support for integrations, backend services, security, and user experience.

This example chat experience, as shown in *Figure 5.1*, is direct from OpenAI's UI:

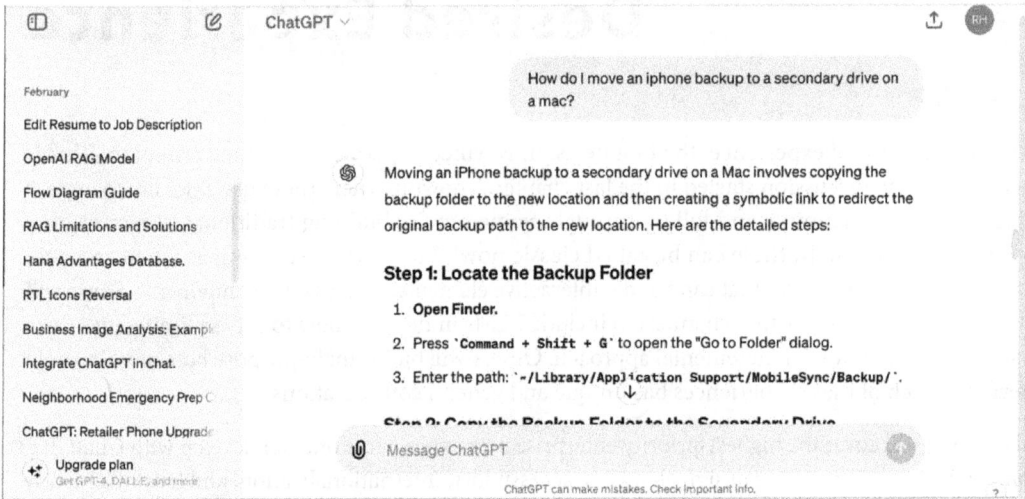

Figure 5.1 – A traditional chat UI from OpenAI

It is typical to have a message input area with a **Submit** button, a single-threaded discussion, the ability to download the discussion, and visual affordance for the customer input and the AI response. It is more robust than most because, as an authenticated user, it keeps track of previous conversations, which is a nice touch. It has room because it is a whole browser window and is mobile-friendly. An enterprise experience will likely embed a chat framework into a website, portal, or app. Some UIs include features such as file attachment, clearing the field, or closing or moving the window.

OpenAI's iPhone client has one nice UX feature similar to what was introduced at Oracle: scrolling only a little for long answers. Working with Kevin Mullet, we coined the problem "over-scrolling." The window should **not** scroll to the end of the response when the AI returns a message longer than the window size. If this happens, the user must scroll backward to find where the message started and then scroll forward to read. This is a headache for accessibility and just annoying for everyone else. As in *Figure 5.2*, the first chunk of a message in the OpenAI mobile UI is returned with the arrow.

The entire message is returned once the user clicks the arrow, so the user still has to return to where they left off. They almost got it right!

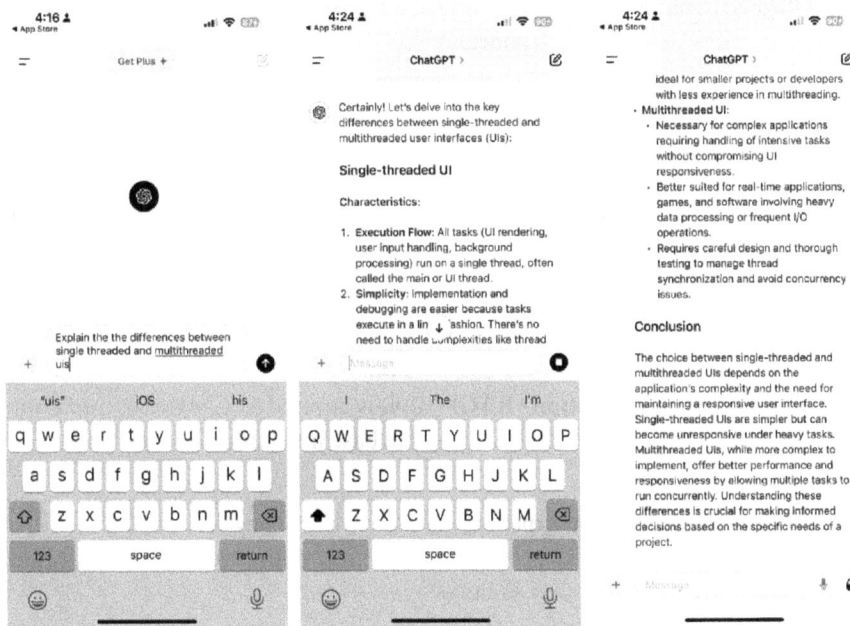

Figure 5.2 – The OpenAI chat message is sent (left), then pauses when the window is full (center). Clicking the continue arrow shows the entire message (right)

See whether your chat client is smart enough with long messages. Does the scroll bar grow, but the screen doesn't scroll as additional details are still not yet exposed? Can the user read messages in one scrolling motion, starting from the beginning of the response? *All chat UIs should have support for over-scrolling.* Slack, for example, shows a **missed messages** button, so it will take the user to where they last left the conversation. This is another good **micro-interaction** pattern for solving this problem and relieving users from the stress of not knowing what they missed.

When designing a chat experience, consider all these micro-interactions and the elements needed to be exposed. Micro-interactions are those small but critical elements that make UI moments seamless and enjoyable. It can be as simple as the button changing color when hovering the mouse over it, giving an affordance that it can be clicked, or the typing indicator to help anticipate the arrival of a message from a friend. They can be used to update the user on system status, encourage behaviors to avoid error, reinforce brand identity, drive the user to a specific task or engagement, or improve the interaction. They are generally very lightweight, following a "less is more" approach. Consider these when creating the UIs, recommenders, or how ChatGPT responds, provides next steps, or interacts with services. A user can enjoy a little delight from a micro-interaction. These are the basics. Here is a little deeper journey.

Article: The Role of Micro-interactions in Modern UX (https://www.
interaction-design.org/literature/article/micro-interactions-ux)

If the plan is to integrate ChatGPT into an existing chat UI, some basics should be in the original chat experience. Covering a few basics next can't hurt.

Integrating ChatGPT into an existing chat experience

Consider whether ChatGPT is the primary mechanism for customer interaction and what other services and data would support it. In the enterprise space, there is something unique and valuable inside the data. The goal is not to repackage the existing ChatGPT service while emphasizing the user experience. What should be considered when integrating ChatGPT into chat? How should it be done? Let's explore the options:

- Use a simple chat window with ChatGPT as the primary actor with knowledge indexed using **Retrieval Augmented Generation (RAG)** (RAG is covered in *Chapter 6, Gathering Data – Content is King*) or other services linked to database integration or interfaces with other company services.

- Supplement an existing Chat "Bot" experience. Enhance the existing deterministic flows (wizards or chat flows) with ChatGPT to improve comprehension, entity recognition, world language support, image recognition, or other abilities.

- Supplement human agent experiences by gathering data or solving problems before being handed off to a human. This is to make the agent more efficient or to provide additional services, marketing, or feedback.

- ChatGPT is used after a human agent has experience handling confidential information, surveys, or feedback.

- Using ChatGPT as a third party in a human conversation (an AI team assistant, supporting an agent, a user, or a team). This can be used as a resource in a group setting. Consider any security implications, where the requestor might have permission to access some data while other members might not.

- Use different LLMs and prompts to support pieces of workflows. Chaining models to improve model output from the previous steps is common and will be covered extensively in the following chapters.

This list can't be exhaustive. The growth rate of creative use for LLMs is exploding. But all of them, if they are going to be chat-centric, have some basic functionality one would expect. The existing chat UI could be ready for some enhanced components.

Enabling components for a chat experience

With a chat window, there is a decision point for what features make sense to support the interactions in the chat window. Most of these elements would be the same if the chat were between two humans or a human and an LLM. However, many developers need to be guided by a suitable set of features:

- Message input window (the typing window for input)

- Threaded response area (the conversation)

- Submit button (to send a message and attachment)

- Visual indicators for the message owner: bubbles, avatars, left/right alignment, and so on (something is required)

- Accessibility compliance: screen reader support, keyboard navigation, alt text for images, high contrast mode, and so on

- Optional components to enhance the overall experience:

  - Download, Print, Export, Clear, and Close (when applicable)

  - Timestamps (as needed and used intelligently; only timestamp some messages)

  - A movable window

  - Previous message history

  - Short-cut suggestions that appear to help finish a thought or action

  - Search in the thread (supported by browsers, consider for applications)

  - Support for not over-scrolling

  - Help, tours, or tutorials

  - An emoji picker

  - Short-cut panels (useful for professional users to quickly insert popular items or fill in common commands and statements)

  - The ability to escalate to an agent (for support use cases, typically done online in the chat or with a button)

  - Customer feedback (optional as a UI element; also supported in chat)

  - Omni-channel integration (so other channels have this context)

  - Typing indicator (recommended if a multi-user UI is used; less critical if the session is only with ChatGPT since it should reply immediately)

A good experience will consider all these options but only implement some. Consider the importance of each requirement. Use the **Weighted Shortest Job First** (**WSJF**) score introduced in *Chapter 4, Scoring Stories,* to decide what part of these requirements are needed. Look at the use case and score these options. A few of these are more valuable than others. And other than links and maybe images, there is very little richness from a *chat-only* experience. Handling links and images is covered in the next section on hybrid UIs.

Designing hybrid UI/chat experiences

Hybrid UIs combine chat with a traditional GUI. This experience can include forms, tables, or interactive elements in a chat stream or merge chat into an existing GUI. Sometimes, chat is in its panel using the graphical UX as content to interact conversationally with the data. Examples like this have been available over the years, but the power of LLMs allows for more robust and accurate interactions. These multi-modal experiences can blend GUIs, voice, speech-to-text, text-to-speech, and LLMs into one solution.

This is the latest version of **Rich Internet Application** (**RIA**), coined around 2002 at Macromedia. I read a paper by Kevin Mullet (written for Macromedia) and found it shockingly consistent with today's experience.

Article: `The Essence of Effective Rich Internet Applications` (https://darmano.typepad.com/for_blog/essence_of_ria.pdf)

> *In software design, the complexities of interactive control remain the most significant impediment to a satisfying user experience. – Kevin Mullet (The Essence of Effective Rich Internet Applications)*

The images in the article are outdated, but the underlying problems are still the same, with a few new concepts (voice and LLMs) added to the mix. While much has changed, much has stayed the same.

For example, the functionality of the Microsoft Word **Format** menu is mostly unchanged, as shown in *Figure 5.3*. The underlying problems of finding and using the breadth of Word's features haven't changed and have barely improved in decades. We bring this up because of the well-understood use cases found in Word.

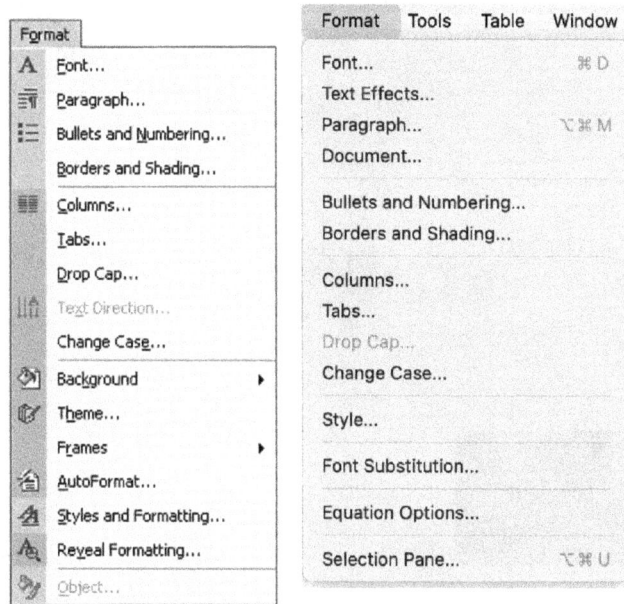

Figure 5.3 – Similarity between the circa 2003 Windows (left) and 2024 Macintosh (right) Format menu

The exciting part about applying AI to Word is that the user's needs should be well known. The most valuable use cases should be easy to determine (but might be hard to realize). It is up to the product people to understand customers' pain points, recognize the LLM's value, and see a path forward.

Getting an order right in a drive-through ordering system is challenging. I am constantly reminded to check the bag before leaving a restaurant. A modern version for solving complex auditory conditions and ordering problems is a multi-modal solution for a drive-through experience from SoundHound. The customer speaks the order; the screen confirms and validates the order with AI audio responses. This demo shows the power of multi-modal interactions. At the time of this video, it did not use an LLM, but it is impressive because of the speed of multi-modal interactions. The display updates at a drive-up kiosk as fast as the user speaks and displays order corrections. *Figure 5.4* is a screenshot from the SoundHound ordering demo.

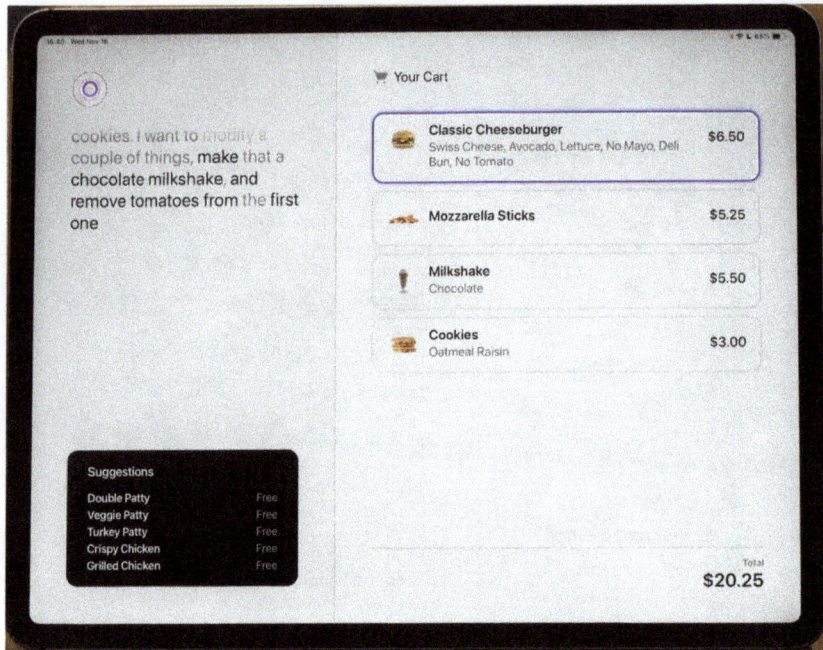

Figure 5.4 – Screenshot from the SoundHound ordering demo

Video: `Restaurant ordering via voice (https://www.soundhound.com/voice-ai-solutions/restaurant-food-ordering/)`

Voice is the input and the display updates with changes. At walk-up kiosks, touch can supplement voice interactions, interchangeably interacting via touch and speech. At the same time, the conversation goes on without turn-taking (you can interject in the speech output and don't have to wait for a pause in conversation). It is a fascinating application because it affords the restaurant more flexibility in language understanding and better feedback for the user as they see their order and can change it. More direct experiences like this are coming, where typing (or voice) directly supports interactions with the interface, including on desktop and mobile apps. As voice interactions become faster for the LLM (in the 10s millisecond response time range), they are fast enough to be natural, pause, or interrupt, just like humans.

Another exciting application of ChatGPT is to supplement and support the interaction experience for creating content. Suppose the use cases include writing emails, generating marketing materials, or writing job descriptions. In that case, you can get a big leg up by gathering and using the material created to develop the next piece of content. For example, an intelligent assistant accessing all the material on a screen can provide many supportive capabilities. With a wealth of quality vetted material, new material can be trained to be in the style and tone of the existing material. Or, with one change to a prompt, an entirely new tone can be developed, but still based on the expertise and content supplied.

The classic example for defining the design space for an application is the integration of generative AI with *deterministic* models to fill in forms or perform tasks. The current crop of chat assistants is based on coding for specific tasks, with specific responses, predictable behavior, and limited flexibility. These straightforward deterministic experiences work well for repetitive tasks with predictable inputs and outputs. However, they have gaps, typically with complex integrations, interactions not accounted for in the coding, and the need for many languages. These platforms are being reworked with LLM integration to address these shortcomings. Many assistants faced some real challenges around natural language processing and recognition. A design can unlock additional value by combining a GUI's visual nature and precise communication abilities with LLM understanding. *Figure 5.5* shows a simulated example of a chat experience with a GUI.

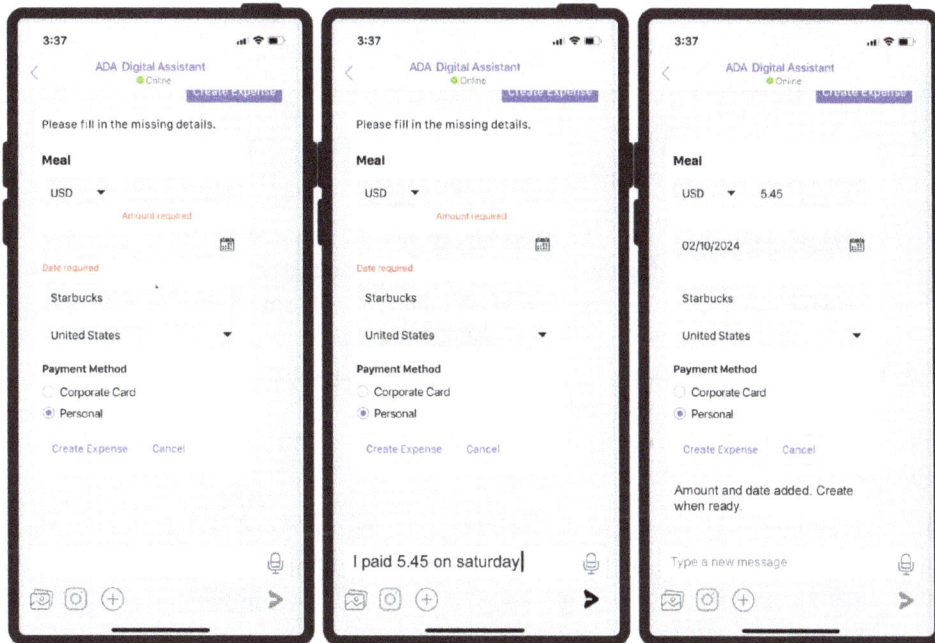

Figure 5.5 – Filling out a hybrid UX with conversational intelligence

This UI appears to be a simple form, but it is in a chat window. A customer interacts with the form or by chatting. In the middle screen, the user typed (or used voice) entering input to manipulate the form. The form was updated on the right, the error was removed, and a message was returned. An LLM can contribute intelligence on how to parse these messages better than traditional chatbot software. They can also be used to quickly adapt the UIs to support more languages than are typically backed by vendors using their internal training models. The current models, such as ChatGPT 4o-mini, continue to improve in these areas.

I want to point out a few micro-interactions that appeared in *Figure 5.5*:

- The AI understands the missing or misunderstood information in the submitted expense or receipt scan (prior to this screen, it would have triggered this response) and presents the errors with a traditional in-line error message. In the future, I could see the LLM asking the user to visually mark something on the receipt that it didn't understand (e.g., circling a number it didn't understand and asking for help). This could drive a workflow for improvement without significantly burdening the user. Don't turn customers into testers, but sometimes you need all the help you can get.

- The form clearly shows what is missing and allows the user to edit with text, voice, or interaction with the form (the keyboard is hidden for this mockup).

- The AI understood "Saturday" as the previous Saturday and entered the information in the correct format, avoiding opening the calendar, viewing, picking the month, and clicking on the date.

- The UI is dynamically updated, acknowledges the added data, removes missing details and errors, and allows users to submit with a button or text.

So, in this hybrid GUI example, ChatGPT can do a lot to interpret the user input format and submit it as the correct fields. Imagine doing this entire interaction without receipt scanning and no form. It would be very tedious and slow. The system and GUI patterns did some heavy lifting to create the complete experience. Similar flows using existing chat platforms might already be in production. Still, they could use assistance in understanding human language, distinguishing entities (specific elements, such as names, businesses, and values), and expanding language support. ChatGPT can enhance and improve these parts of the experience without throwing out the existing intelligence and expertise built up in designing these business-critical tasks.

In addition, scanning a meal receipt in the flow might be an additional area for improvement. Training visual foundation models to understand images of business documents and classify and extract critical information from them seems like a whole world for organizations to explore. Many companies already use AI for expense receipts, invoices, bills, quality control images, counting and determining inventory, disaster scenes, and car insurance. The quality of these processes will go up with more advanced models. This book focuses on ChatGPT's textual and written abilities. However, I encourage exploring image-based use cases when they apply. Models such as ChatGPT 4o-mini can handle images as well (the model can classify images uploaded, for example), and by integrating with other services such as OpenAI's Whisper, transcribed audio can be pulled into a workflow.

Integrating ChatGPT into a workflow is fundamental today. In another example, the product might support writing emails or content for a website. Adding a **Generate Post** or **Write Description** button and having ChatGPT gather the information and inject the copy into an existing window is not too hard. One gotcha: *always support the editing of this material before submitting it*. It might be inaccurate, too specific, or not specific enough based on the content used in generation. Consider also capturing the suggestion and the final submitted (and edited) version that the human changed. Use this as data to help improve the model. The difference between this and the previous marketing content example is that deciding what has ground truth is essential. With the marketing content, what is good content is known, and it can be supplemented with field data on how well customers responded to different campaigns. When writing emails or website content, it might be more challenging to identify high-quality writing to train the model. Otherwise, a generic model will only be provided based on its original training; sometimes, that is fine. You can always ask the model to adapt the language. Even the generic 40-mini can-do tasks like this.

```
                                        Write a thank-you to my
                                   customer in less than 100 words.

Dear Frank Jones,
Thank you for your recent purchase! Your support means a lot to us,
and we are thrilled to have you as a valued customer. If you have any
questions or need further assistance, please don't hesitate to reach
out. We look forward to serving you again soon.
Best regards,
Richard Miller
Alligiance

                                        Re-write it in a more
                                   casual tone, he is family.


Hey Frank,
Thanks a bunch for your recent purchase! It means a lot to us. If you
have any questions or need anything, just give me a shout. Looking
forward to catching up soon.
Cheers,
Richard
Alligiance
```

Suppose the use case requires a sidebar experience where AI can supplement a traditional UI. It could start as a glorified intelligent search, but this can be leveraged to quickly gain previously buried or unavailable insight from current content. It might have an **Analyze** button (or generate some recommendations automatically), know what has happened in the thread, provide a **Generate a Reply** option or do tasks that any human assistant might do to help a team member. Each application will have its use cases, just like any enterprise company in automotive, education, financials, manufacturing, healthcare, government, real estate, sales, service, and a dozen other verticals. Companies that provide enterprise software have a head start because they can work with and adapt complex components.

Oracle's Digital Assistant does an excellent job of advancing the state of the art, as shown in *Figure 5.6*, with a sidebar experience.

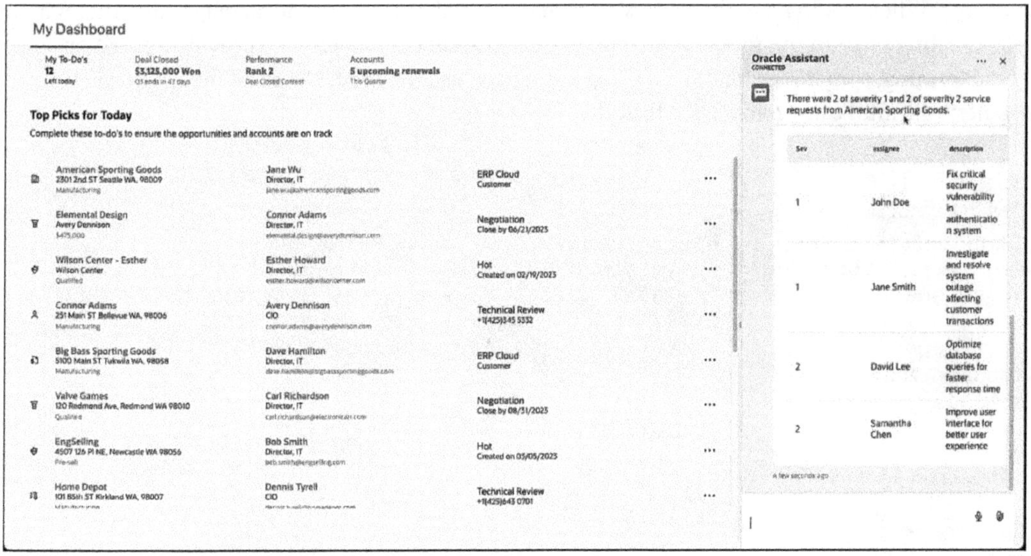

Figure 5.6 – Oracle's Digital Assistant sidebar inside Oracle Sales

This shows how a traditional web enterprise application can be supplemented with a conversational generative AI based on understanding the page's context and all the related data. The demo goes into other details, but the point is that existing services can be supplemented with an LLM. Check out the Oracle demo, which starts at 46 minutes and shows how the co-pilot pattern can work embedded with an existing UI.

Video: A generative AI co-pilot feature in a GUI by Oracle (https://www.youtube.com/watch?v=9CEfru54Oyw)

Besides providing insight, consider interactions with the sidebar that can influence or address problems. In the preceding example, the chat is used to supplement the experience with issues related to the customer, as shown in the dashboard. This page has a wealth of context so that conversational AI could offer proactive suggestions based on the evaluation of the data.

For example, the team is building a sales enablement page that reviews customer accounts and deals. In that case, the sidebar can allow questions about this customer that might be buried in other systems or offer recommendations for the next step in the process. Order status, customer care experience, open trouble tickets, and additional customer history information offer value when on the sales page. However, exposing the generative AI as a resource supplements the common understanding with these details. But don't stop there; use hyperlinks to wrap the text so they can drill down into the enterprise data to further their understanding. Allow the customer to use the generative AI to enhance their editing, posting updates, or drilling down further into insights held by all of this information. Help the user understand the scope of the options the user has at this point. Sometimes, buttons such as **Edit**, **Open Details**, **Send Email,** or **Change Status** are used. Whatever the use cases, consider going deeper into exploiting generative AI models to enhance these steps.

This focus on creating the desired experience by addressing ChatGPT's unique complexities can't cover all UX design guidelines for GUIs. There are dozens of books and websites available. Focus on the fact that UX design within a conversational experience is slightly different. While you might have a traditional web form (with fields and drop menus) in a GUI, what would make that different when placed into a conversational experience? For example, in the **Oracle Digital Assistant** (**ODA**) example shown in *Figure 5.6*, notice a small table in the chat and a very different table on the main page. It is easy to realize that a chat form factor table can't handle everything a big table can do. Let's review some essential tips to consider when designing hybrid conversational experiences.

Chat window size and location

Chat windows are typically vertically oriented and small on web pages, so the main page can still be used. Depending on the target window size, a typical default chat window can take up 11% to 25% of the screen real estate (and less for folks with two monitors). The example in *Figure 5.7* is based on standard desktop screen sizes for August 2024. This data changes very slowly. Changes in the last year varied by less than one percent.

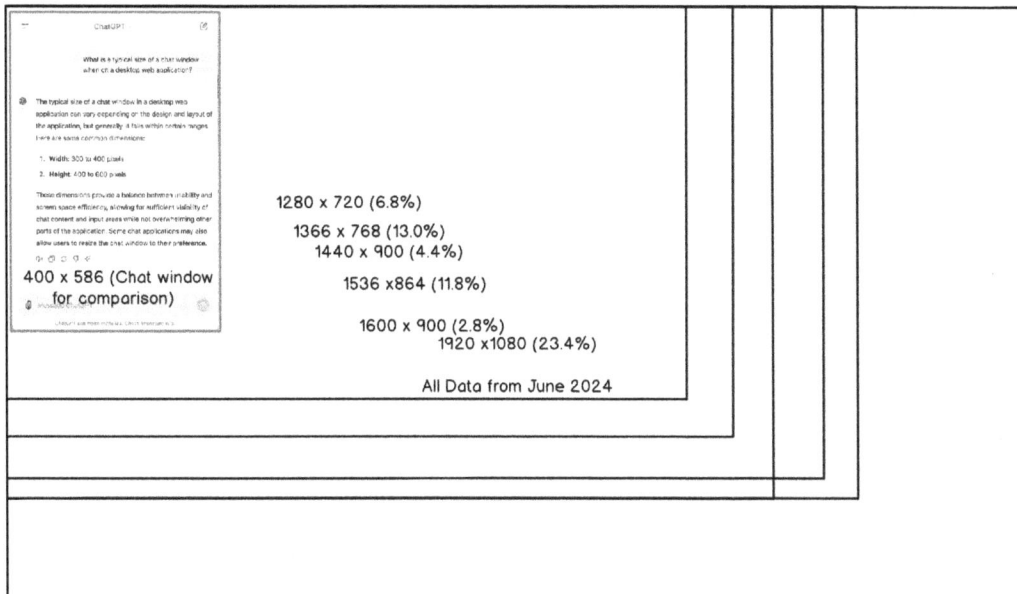

Figure 5.7 – Comparison of desktop screen resolutions to a typical chat window (August 2024)

Article: `Current Desktop addressable sizes (https://gs.statcounter.com/ screen-resolution-stats/desktop/worldwide)`

It would be best to examine customer data to get a more accurate estimate of screen resolutions. I used to run this analysis every year or so to influence all kinds of web application design decisions. Enterprise customers' screens were always significantly larger than in consumer data.

If the primary experience is chat, such as Slack or Teams, size constraints are not an issue on the desktop. There is room for large tables, maps, and good-sized images, forms, and other interactive experiences. For those where chat is a secondary element, provide enough space to support necessary interactions but not so big as to interfere with the primary windows. There is no one-size-fits-all answer. Consider the use cases; the less complex the cases, the less space is needed.

Text-only seems to work fine for the phones of the world, with roughly only 40 characters across (as customers increase font size for accessibility, this number goes down). And the context of use is essential. If the customers are mobile, the width of typical phones or even their specific phone is known (from web analytics). Don't try to do anything else on the screen along with the chat. Horizontal scroll bars should only be used with extreme caution. If it is in a window, consider notifying the user that they have unread messages while away from the chat. Review examples of unread message interactions on any social media platform (LinkedIn, Facebook, Instagram, etc.).

One last example for desktop experiences (web or native). Look at the demographics of users and their browser sizes and desktop sizes (web logging tools such as Adobe Analytics or Google Analytics report this information). Use this to decide how much space the user typically needs for the main application window and how much real estate can be shared with the chat window. Consider the responsiveness of the application window. The ability of the main content window to support good flow and even a mobile-friendly view (even on the desktop) allows support for a sidebar view, knowing that the main content can adapt.

A list of objects is one of the most relevant patterns in enterprise applications that need a lot of space. Sometimes, in tables or card views. We can now address how a chat experience impacts tables.

Tables

Consider the mobile experience's design space when considering what is done in a typical chat window. Tables in desktop windows need to be adapted to a narrower chat width. Either use a different layout or limit the number of columns. For example, email on a desktop is in columns, as in *Figure 5.8*, while on a phone, a card view is displayed, as in *Figure 5.9*. If the model will return cards for results or small tables, realize more room is needed to present comprehensive information directly.

TidBITS	#1693: Vision Pro reviews, secure sharing over the...	3:21 PM	Inbox - Richardhmiller
Aragon High School PTSO	● Aragon A-News - 2/5/2024	2:53 PM	Inbox - Richardhmiller
Peacock	● New week, new streams 📺	2:51 PM	Inbox - Richardhmiller
Senator.Becker@outreach.senate.ca....	● Winter Storm Safety and Recovery Information	2:37 PM	Inbox - Richardhmiller
Wellsfargo Bank Yield (via Google Dri...	Item shared with you: "Yours Access ID Blocked - F...	2:33 PM	Inbox - Richardhmiller
The New York Times	● On Politics: The retribution presidency	2:23 PM	Inbox - Richardhmiller
The Golf Club at Moffett Field	● This Week's Daily Deals on Tee Times	2:19 PM	Inbox - Richardhmiller

Figure 5.8 – A traditional table view won't work well in a small chat window

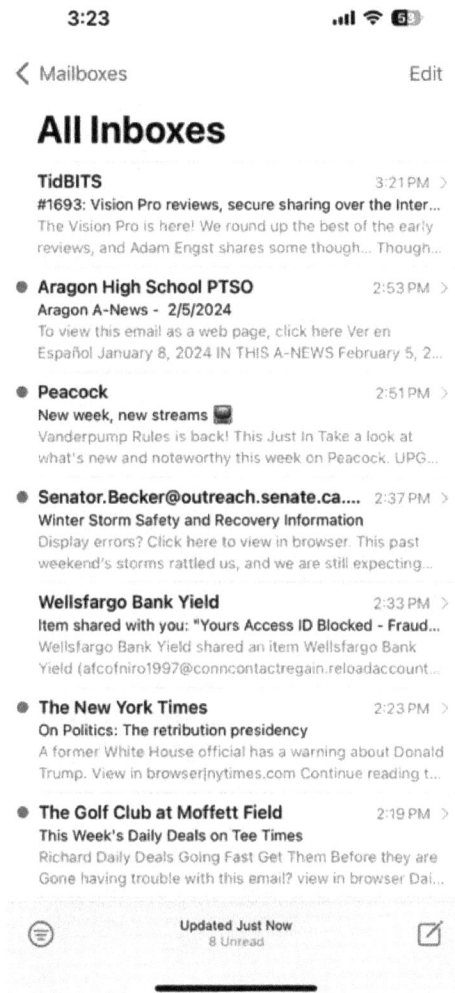

Figure 5.9 – A card view would work in a chat window

Consider format and content when presenting tables. Models such as ChatGPT 4o-mini (the "o" is for Omni, so I guess they think it knows everything) are very good at formatting data. However, a mobile user will still have the same window size issues as those in pre-ChatGPT experiences. And because this has a scrolling window, scroll bars show up. But never have double scroll bars. While a scroll bar would be fine on the desktop, use a next function to show more data since there is already one scroll bar in the scrolling window. Or find another alternative. If an existing table component is mobile-friendly, it might automatically adjust to a card view in the tight confines of the chat panel. It all depends on how the model models support the user experience. If the chat from OpenAI is the UI, there is no option to create richer interactive experiences. If the team owns the client, the world is your oyster, and call on OpenAI models to fill in the pieces to create rich experiences. Even with recommender UIs, there are places to provide feedback for the recommender to make more refined decisions. Prompt the user for information to fill in the blanks in the model's understanding to improve the results. This is on your plate, not the AIs. *Figure 5.10* compares the traditional table format with a poorly implemented horizontal scroll bar. By changing the prompt to provide the results in a more mobile-friendly format, we get more readable (but less comparable) data.

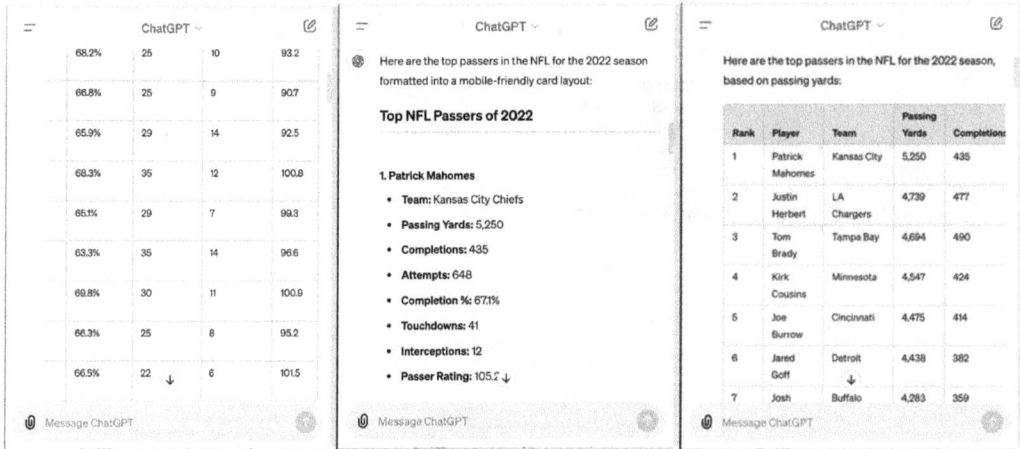

Figure 5.10 – A table view (left and center) compared to a mobile-friendly list view (right)

The horizontal scroll bar is not even in view (and when scrolling, as shown in the middle pane, the player's name will scroll off-screen; it is not frozen or locked like in a spreadsheet), making analysis hard and the right-most columns impossible to associate with the player. Try prompts such as this on ChatGPT, use a phone, or resize the browser window to the smallest size possible, and try to review and use the table data:

```
Show me a list of the top NFL passers in 2022 and their statistics in
a table
Show me a list of the top NFL passers and their statistics, and link
the name of the player to their Wikipedia page
```

Consider the use cases, how the data should be best presented, and the channel in which the user absorbs the results. Links in the table allow the user to dive further into the data. This is typical on a web display of data. Understand that they might already have interactive richness on the traditional website.

> **Note**
>
> If more room is needed and there is more room (e.g., if in a small chat window in a larger desktop window), build the ability to expand the window. Consider opening the table in a new window if a table-heavy use case demands it. It is possible to identify that the data isn't fitting in the space and offer the user a way to see it all, such as sending a file, opening the data in a new window, or other innovative solutions.

As for the table content, please keep it to a minimum. Summarize tables and support drill-downs. To interact with one of the results, consider how to select it and expand the details. Does it make sense to bring that object into a main browser window, where there is plenty of room, or expand the details in the conversation? There is no one correct answer. Let the use case guide you. In the preceding example with American football statistics, a robust experience around table data could have links on data to drill into details, columns that can be sorted, or access to richer analytics on players' names and teams. There are lots of options to explore. Unfortunately, the tools and patterns that the enterprise supports might be limited. Always remember the end state so improvements and new patterns can be prioritized for the chat channel. Editing in a form (and, more recently, in tables) is fundamental to enterprise interactions with data and tasks. A drill down into forms helps us understand some of the complexities when done within a chat experience.

Forms

Fortunately, if created for a mobile-friendly experience, most forms are ready for a chat experience. Only a few customizations should be considered.

First, consider how big the form is now. Can it fit in a chat window? Does it need many pages of fields to be filled out? A crazy long form might not do well in a chat experience. Then, consider what happens when the form is submitted. Since the chat should persist, what does the form look like after submission? I usually suggest converting it to a card. This transition serves as an affordance to the form that was submitted. Also, it is essential to consider when mistakes are made and how the user can re-edit the information. Like including an **Edit** button or link to return to the form, make the change, and then save, returning to the card view. As a little preview of *Chapter 9, Guidelines and Heuristics*, supporting the edit affordance and undo are two items that should be addressed in a heuristic evaluation. That chapter will dive deep into using heuristics to understand and classify user experience issues.

Recall the Starbucks expense in the expense assistant example. After "Submit this expense," as in *Figure 5.11*, should the user continue to see the editable form in the chat's history?

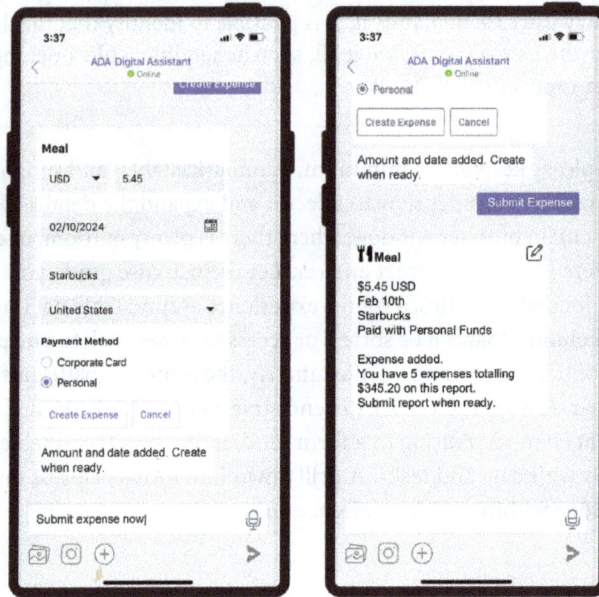

Figure 5.11 – Using visual affordances to turn a submitted expense into a card

A traditional GUI would remove the edited form when submitting the expense item. So, persisting this editing view conversationally also doesn't make sense. Disabling the form is one option, but using the **Cancel** option at the top of the image on the right and updating the existing form to the submitted expense version is better. This ability to revisit and edit existing items on the chat history might be a challenge, depending on the ownership of the chat window technology. With a card display, the content looks finalized; it will feel complete. To edit it further, the AI should allow that conversationally or consider a button. Notice the **Edit** icon on the meal in *Figure 5.11*. Editing conversationally before submitting the entire expense report, including this single expense, is allowed.

Also, consider form elements that are robust or complex, such as a search experience. One component in the form might want to use space allowed by the width or height of the chat window. For example, searching for a product name sometimes product names can take a long time in the enterprise. Consider wrapping the results to a second line instead of truncating. The problem with truncating on a mobile phone or small window is that it is hard to show the full name. And if a few items differ only by the last few characters, how would a customer tell them apart? This isn't specific to LLMs; it is just a good design for small windows, such as **Chat**.

For example, if they are looking for a product demo to download and search for `Agile Product Lifecycle`, which returns two products, it would be impossible to tell them apart if the search results were in a small window or drop menu and truncated:

- **Agile Product Lifecycle Management…**
- **Agile Product Lifecycle Management…**

If the full text of the items is wrapped, they can select the correct one:

- **Agile Product Lifecycle Management
  Integration Pack for Lavasoft E-Business Suite**
- **Agile Product Lifecycle Management
  Integration Pack for Fission Application Suite**

So, consider how this small viewing area impacts components. If components are already mobile-friendly, then most of the problems are solved. However, there is still a difference between components *built* for a small screen and those *designed* for a small screen. The truncation problem I just shared is far too common.

The last consideration is how to interact with the form or card when conversing. Let's look again at the form and card concept shown in *Figure 5.12*.

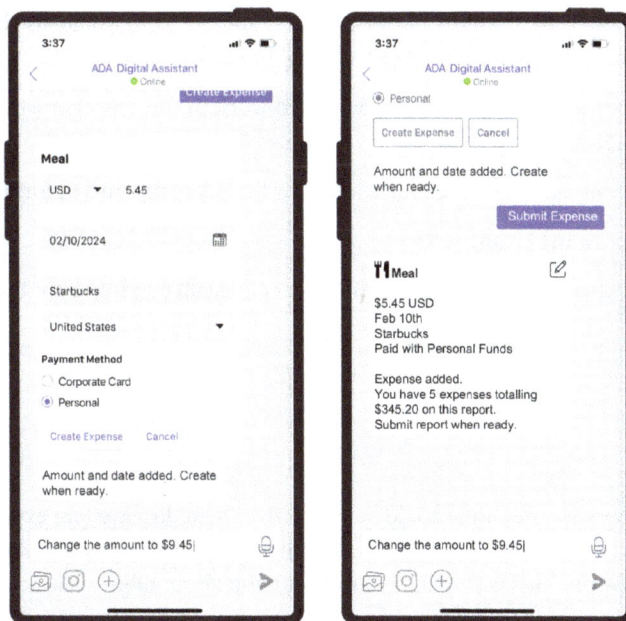

Figure 5.12 – The form and card views when starting to edit conversationally

If the user enters `Change the amount to $9.45`, does the amount element in the existing form update (left example), or do they have to re-render a new form and hopefully remove or hide the now old and incorrect form or card? *Don't show the wrong information to the user*, and updating the card with a visual micro-interaction might be better than scrolling another form onto the screen. The micro-interaction could be a twinkle and an animation showing the amount changing and maybe the message updating. Then, the user saves scrolling through the entire screen to invalidate the old form and create the new form. There is no need to show old or multiple editable forms in the history. Showing errors on an old form when the information is corrected is a terrible look.

With the form submitted and a card appearing, update the card directly with a twinkle and animation and update the text to show the change is accepted. These patterns are not readily available on any platform. It is fair to point out that conversational interaction patterns are still immature. Many kinds of interactions will need to be invented, tested, and matured.

Let's summarize:

- Make forms sensitive to the edit via text or direct edit.
- Use a card to show the completed state of the object.
- It is unnecessary to show all the information from the form on the card. Consider how important it is for the user to recognize (and not recall) that this object has an invisible element.
- Consider whether the use cases require editing everything in the chat AI. Some objects are complex and would be hard to edit conversationally, and the form would also be massive. There are use cases here that a larger viewable area might best serve.
- Create a mechanism for complete components to work in the tiny chat window confines for the parts of use cases that make sense.
- Allow for conversational edits of the form and how to update the card or form with those edits.
- Don't show forms in the wrong state.
- If making direct changes to a card or form, use an affordance to highlight the change.

Sometimes, a form is used to generate a chart. Now, let's cover a few critical ideas around using charts in ChatGPT solution.

Charts

A chart presented in enterprise use cases should be interactive when displayed for exploration. Selecting and drilling into a data point, viewing more details, or exploring data segments to answer a different question is all within reason. This is similar to our discussion about tables. The accessible version of a chart is typically a table, so it should follow the same conventions. Conversational analytics requires charts to be more than static images; not every toolkit can support this. Consider the context for the chart so conversational interactions might tease new insights from what is presented, such as `Show me` the breakdown of the `1st quarter by salesperson`. The chart is a steppingstone to further the conversation with the LLM.

For example, consider the axes. Don't default to a scale or unit the customer doesn't want: `Show me the sales in Euros, not dollars`, or `Set the salary axis starting at zero and change the range to the last five years`. Conversing with data gets complex.

If the user explicitly asks for a chart, give it to them, but a more innovative experience could be providing them with a chart because it makes sense. Suppose the LLM knows the user is visually impaired (because it was told or from profile information). In that case, showing a chart without the ability to show the data in a table is a problem because the screen readers will have an easier time with a table than a chart (but I suspect LLMs will be solving this as well). `Change the table to a chart` would be a logical response if the AI didn't realize the user's abilities. This knowledge might be used to drive subsequent interactions. Accessibility considerations will also be covered later, but it is helpful to keep accessibility concerns in mind.

When presenting these charts, one should consider what charts and design elements should be used. There is a wealth of bad charts, including not starting an axis at zero, rotating labels, using excessive colors, not using redundant coding for colored bars, lack of labels, too many pie slices, misusing line charts, and many other fails. The basic chart should also be designed, or the LLM should craft it from instructions. If this is new to you, ask ChatGPT for advice:

```
What are some UX failures when it comes to designing charts and
graphs?
Explain the top 10 ways to make charts and graphics high-quality.
How does an LLM decide to design a chart when asked for data?
```

Currently, only the paid versions of ChatGPT allow charts to be created, but that won't be an issue for enterprise customers. Interacting with data is more interesting than just presenting static charts, which only tend to drive more questions than answers. And not to mention asking ChatGPT questions about the data that the chart might not reveal. Making data (and thus charts) conversationally interactive will be a massive space for enterprise use cases. Knowing how to design exceptional visuals is critical to effective communication. And being able to generate conversational insights from data is likely more important. The instructions provided to the LLM must instill some of that expertise if the results don't match expectations. To explore the fundamentals of effective chart design, read the collection of Edward Tufte's classic books linked on the reference page.

Books: `The work of Edward Tufte (https://uxdforai.com/references#Tufte)`

 Now that we have dabbled in charts, it also makes sense to talk about static images.

Graphics and images

The enterprise world has two kinds of images: decorative and functional. Showing "pretty" graphics can jazz up the user experience or give it style and character. These decorative images are not required and, thus, don't require any support regarding accessibility and might not even impact usability. However, they can influence the customer's mood, affecting the user's experience. The second, being more critical, are images that provide a function. Avatars, background color coding, and bubbles help communicate the user's content in the thread. Icons grab the eye to make a feature or status stand out. Given the smaller size of most chat windows, this is a valuable application of imagery. Generally, it would be best to have some affordance to the conversational back-and-forth elements, but not every affordance. Sometimes, images are the application's primary purpose, such as with inventory apps, social media, online ordering, or image generators.

Suppose the application is an AI image generator. Consider how much space is available to render images and the resolutions available to download. In this use case, it is better to generate an image in the size that makes sense for the display because this can be 10x smaller than a "full size" image and thus faster to display. Choose the download size with a drop menu or UI element. The major image library sites use this pattern.

> **Legal caution**
>
> Generative images are a new frontier of issues. Legal issues can arise if models are trained on copywritten and trademarked material, resulting in images that are generated too close to the source. This space will mature, but be aware of it. The base model was sourced from the Internet. This will also be an issue with text, voice, and music. Avoid famous people's voices and likenesses. James Earl Jones, Christopher Walken, Scarlett Johansson, or any celebrities' legal counsel will put offenders in their crosshairs.

Suppose the application generates an image to include in a workflow, such as creating an advertising image of a product or a marketing image. If the model is trained on enterprise material and told to use those materials, it might address any legal concerns. To constrain the end-user, the instructions for the model could be set to constrain the image within a corporate standard style. Users should not have to do this every time. Once they approve the image, consider storing it with the resulting workflow and the messages used to generate it; this knowledge can help further model training. If the user needs to edit, they can quickly restart from the same message and tweak it further without retyping and, more importantly, having to remember what they typed.

In a chat-only experience, such as SMS messaging or a chat channel that doesn't yet have component support, images are likely supported to some extent. So, look at the framework to see whether images can be tagged with a meta-data label for accessibility or if an explanation has to be written out in the channel as text. An LLM can be used behind the scenes or even in batch mode to generate descriptions for images and captions for tables and figures. Use LLMs as tools in a business workflow and understand what the channel supports to give customers control and usability around images.

Icons are a unique set of images. They are used in the design of hybrid UI, have some use in pure chat experiences and recommender, and have no place in a back-end UI. I haven't explored using an LLM to generate icons, but from experience, it is excellent for brainstorming but not yet for the fine details needed at the pixel level. This conversation will thus be constrained to using icons in hybrid experiences.

Icons are used to convey an action or state or for visual feedback. Sometimes, they support words (such as a button with an icon and label) and, in some cases, are just icons. The label may be available on a mouseover or in a wider-screen view. In this case, consider the visual style and tone the collection of icons represents and whether there is a chat or conversational capability to interact with the icon. For example, if an icon shows the user is online (like a green dot in the corner by the username), there is presumably a way to go offline or into another state. If this is a support application that thousands of chat and phone agents use, there is typically a multiple-step process to take them offline or online. Is this something that they can do conversationally? It comes down to the use cases. A driver clicks buttons in a car to set the temperature, but it is much easier and less distracting to say to the car, "Turn on the A/C to 68 degrees." GUIs and voice interaction can work hand in hand.

And since interactive elements such as icons and images, their siblings, buttons, menus, and choice lists are worth some explanation.

Buttons, menus, and choice lists

In some traditional chats with autonomous agents, the experiences could be more conversational. They *look* like a chat but only offer clicking buttons; the user can't type. *Avoid this use case at all costs.* It is a red flag and will derail any goodwill with the customer. They become trapped in a poor experience. However, in an actual conversational AI, it is sometimes helpful to offer suggestions or hints. The primary issue with conversational AI comes from one of the primary heuristics of "recognition over recall," covered in *Chapter 9, Guidelines and Heuristics*. This means it is faster for users to pick from choices and recognize something than to pull from their memory and recall items.

A quick primer on recognition over recall

Article: `Recognition over recall (https://www.nngroup.com/articles/recognition-and-recall/)`

This tenet applies to so many places in user experiences, such as menus. Although some applications have many features, finding a function takes time and effort. It is well known that picking from a list is more accessible than recalling a menu command. Now, with ChatGPT, the exact name of a command to activate a function is not needed. Questions can be asked without using the precise words of a command and still find good results. This helps mitigate some of the problems with text entry experiences. This is a beautiful change and makes chat experiences possible in the first place. However, it is still helpful to reduce the user's cognitive overhead. So, if there are just a few choices or you have a good idea of what is next, it is reasonable to offer suggestions. Sometimes, those appear dynamically on a typeahead bar. Like in text messaging, it tries to predict what word will be typed, as shown on the left in *Figure 5.13*. This predictive text is precisely how LLMs work, so it is the perfect place to assist users with LLMs as long as they can keep up with their typing speed. Sometimes, the responses are more robust than word or button choices in the response to submitting the prompt, as in the example on the right of the figure.

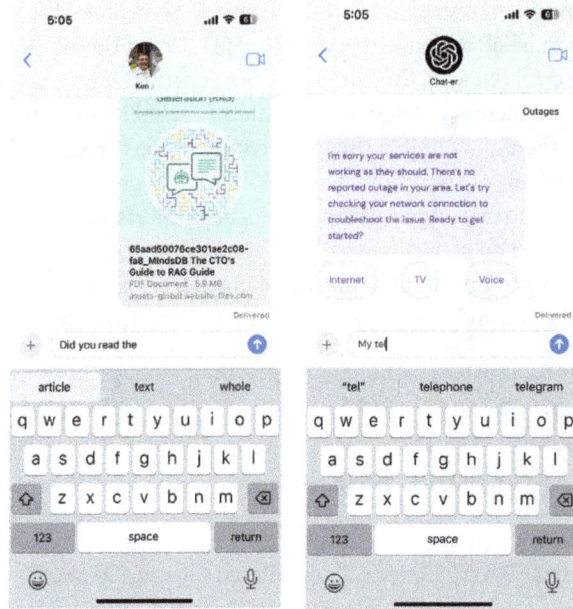

Figure 5.13 – Predictive typeahead buttons in Messages on an iPhone (left)
versus the suggestion buttons in-line after a prompt (right)

Use predictive completion in three cases: when what is next is understood and when the user is lost or needs guidance. All possible paths are too many. Even three or four choices are enough. Consider the data. If the data can predict with 90% certainty that one of four items is correct, then it makes sense to offer suggestions. If four options give a 30% of success, don't offer them. And don't forget to gather this data to help future models and decide how to improve the UI. Generally, it would be best to consider what happens when the menus or buttons are selected. The elements should be removed after clicking them. The same style of any GUI buttons can be used to create consistently, but in most UIs, buttons don't disappear when clicked. However, in a chat UI, the buttons should not appear available, and disabling the buttons is a partial solution, but why waste the space? Generally, buttons are removed after selection. Choice lists or drop-down menus offer more options in less space but with the additional headache of having to open the menu to see the possibilities. If there are a lot of choices, then it is more complex to deal with search or scroll bars. Follow existing guidelines to use these correctly, then remove them from the UI display when done.

In a GUI, there is room for buttons with large labels, maybe to draw the user's attention to something on a large screen, but there isn't room for them in a chat window, so keep menu labels short. Buttons are suitable for actions; use links for navigation.

Links

Links are the basis of web applications, but linking is expected for any app. Tables of contents in documents are the old-school version, and menus are a version of links on a desktop. Links are valuable in hybrid UIs and are among the few UI elements available in any traditional chat experience. Hybrid, chat, and recommender UIs can use the links, and there are a few tips to consider regarding how the links should be handled.

Linking to outside sources

Linking provides URLs to services or content that can't be offered inline. Links can give a reference, provide more details with an article, access a form or service not within the control of the chat, or open a new window or panel for the user. Determine how to present links to the user. If the experience opens a new window, will the user return to the first window to continue the conversation? If the chat is in a side window or panel to support what is happening in a main window, will the experience update if the links resolve in the main window so any subsequent conversations know about the change? Make decisions around these interactions.

Labels on links should be clear and indicative of their purpose. As shown in *Table 5.1,* instead of showing long links in the output (left), use a well-defined label (center).

Don't Do This	Readable & Manageable	Explanation
Hi Richard. A new Explanation of Benefits (EOB) is ready. It may include time-sensitive or other important information. https://click.edelivery.uhc.com/?qs=7b10282356f8afb07d1e82e239f9722d466ca84134771c3c35ee1b60c735d8940b32b169349bc53c30141f6c40140989f75b8589cd2af251	1) Hi Richard. Please review your new Explanation of Benefits (EOB). It may include time-sensitive or other important information. 2) Hi Richard. I can take you to the Explanation of Benefits (EOB) ↗. This link is outside the portal, sign in is required.	1) Readable links are more direct and take up less space. Avoid "Click here". 2) Alternative text when chat supports navigating the main window, but this link opens elsewhere.

Table 5.1 – Use well-defined labels for links in the conversational output

Don't expose URLs if the client can support a label. It is familiar enough for users to right-click and copy the URL. Focus on the use case and what makes the most sense for links. If this is a co-navigate UI (use links in the chat to control the page they are on) *and* has links to outside sources, consider how to communicate the difference between these two types of links, as I show in the second example output in the middle of *Table 5.1.* Will it make sense and be apparent to users that they have co-navigated? They might click multiple times without an affordance, not knowing the page behind the chat window has changed. Also, sometimes, the legal department gets in the way. There are websites where an intermediate screen opens after clicking a link, letting the user know they are leaving the site. This warning seems too complex for a UI, likely written by lawyers or because customers are confused by the link. Clarifying these interactions without the extra page and warning would be nice.

Link color

I strongly encourage following HTML and web norms with links when building a client. Specifically, aging the link color after someone has clicked it. Typically, a blue link means unvisited and purple once visited. It is immensely valuable, even as subtle as it is, to keep track of which links were clicked, especially if a collection of links is being shared in a conversation.

Look carefully at Google search, as shown in *Figure 5.14*; isn't it great to return to the results list and it shows links that were visited? Use this pattern for all links. Customers might not notice it overtly, but it will be more usable. This is accessible for screen readers, as it communicates that the link was visited.

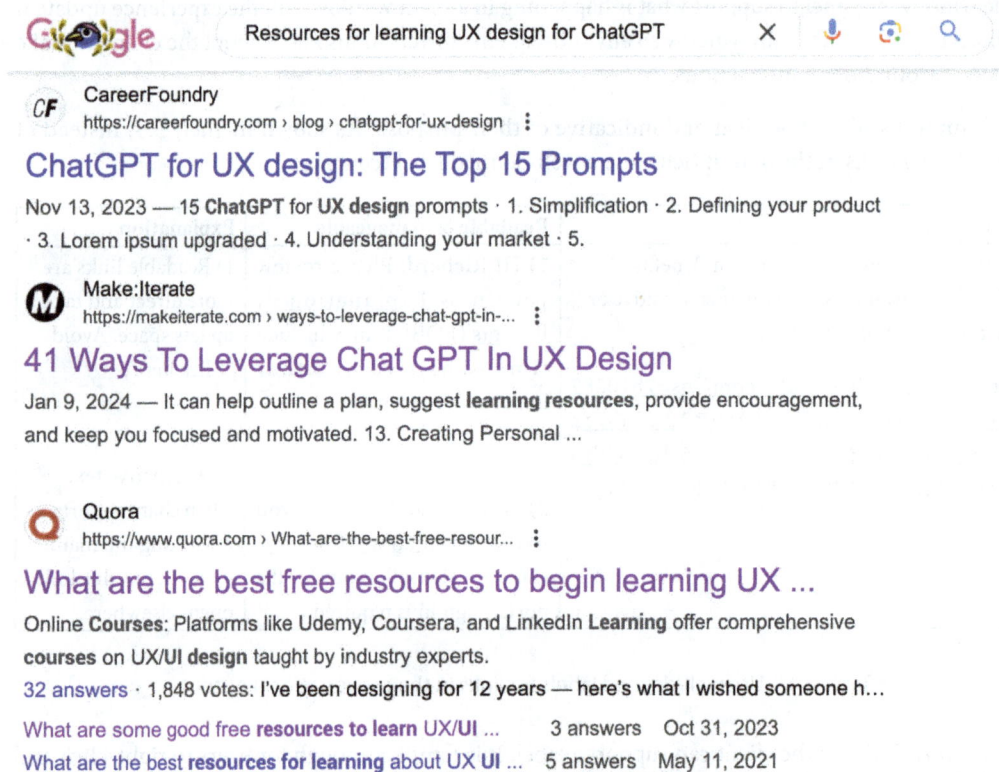

Figure 5.14 – Purple indicates a visited link and is fundamental to providing context

Opening windows

Opening a new window is typical for a web link if it doesn't replace the current page. Opening a new window is required for a chat window because it can't replace the current small window with a web page. But realize that if the user goes off-site to return to the experience, they must figure out the window containing this chat. If they close the newly opened window, they should be returned to where they started, but this doesn't happen as often as expected.

If the link opens a new window on the same site, there will be a different problem to solve. Can the chat window reopen and be reconstructed to pick up where the user left off in the previous window? Or there are now two windows open that have the same chat. And this can continue, so can the site handle 3, 4, or 10 open browser windows, all with the same chat? Can it keep up with the changes across windows? As this depends on the chat platform. Chat vendors should provide the solution. Just make sure to understand the requirements so either the team manages the problem or has the vendor address it.

Since a regular link in a chat window is expected to open in a new window, no additional affordance is needed. However, a sidebar must provide an affordance to help the user understand that it will open in the same window. Don't use a link in a chat window to start a process; this is why buttons exist. Keep links for navigation.

These considerations allow for smoother, more engaging, consistent, and natural ChatGPT hybrid and chat-only experiences. The hybrid experience is the most robust and complex use case because it has to account for all the traditional user experience interactions and then layer in the additional complexity of chat. Stepping away from a graphical user experience, another vertical worth our attention is a voice experience, such as those found when calling a company and using their phone tree. Let's review some tips to create compelling voice-only experiences.

Creating voice-only experiences

Phone calls or smart-speaker devices without a display are a few voice-only experiences. Voice-only was very limiting, and there are still human factors to consider, but technology has come a long way. ChatGPT 4o-mini takes empathy and tone to a new level, and because it is better at following instructions, prompts can control answers to be short to accommodate the capacity of a human to process voice in small pieces. There is nothing else to interact with because it is voice only. We don't need to consider voice-only *within* a chat or hybrid experience; that is really voice-optional. A smartphone allows you to type or use voice and then translate the speech into text. All models behind these services will improve with LLM models behind the scenes. But we are not here to talk about enhancing speech-to-text. Let's focus on a voice-only channel.

> **Tip**
>
> When designing for a smart device, the vendors emphasize that only some devices have displays, but even if they do, the experience should work with voice only. Not all use cases make sense to interact physically with the device, even though it is supported. For example, an Alexa device in the kitchen might allow interaction with the display to go to the next step of a recipe, but the user might want to avoid touching it with messy hands, or it might just be out of reach.

As with chat experiences, decide whether a generative solution replaces or supplements the experience. Everyone has headaches with phone trees, the hierarchical prompts that lead to a leaf on the tree when calling a business, requiring pressing *1*, **2**, or **3** to navigate down the *tree* of options. A positive of a well-designed phone tree is that it focuses on results based on choices. This is also why desktop software has menus. It limits the number of options at each turn: walking down some trees is more straightforward than directly recalling or accessing a specific node.

Replacing a tree by starting with an open-ended "ask me anything" approach can be tricky. If I ask for "help with batteries" in a phone tree at Costco, a big-box retailer, do I want to go to "Automotive," "Hearing Aids," or maybe see whether household batteries such as AAs are in stock or on sale? In a phone tree, the user makes selections focusing on limited areas (Automotive, Hearing Aids, or Shopping), and then the correct choice is clearer.

Consider the choice to introduce text-to-speech listening with generative AI at the root level ("Tell me how I can help") or use it at the next level, "What is your question for the automotive service team?". Generative AI will be more successful with robust content, knowledge, and APIs for a specific part of the problem space. Getting that detail takes time and effort. So, doing it for a particular department might be an excellent first step to limit scope. Draw on the first few chapters to learn which department would have a use case that is best suited. Determine which use case would have the highest WSJF score!

The closer to the top of the call tree ChatGPT is added, the more general and, thus, the more complex the solution. Customers who think it can answer anything will be more likely to ask. For example, a common question for a big-box retailer is a stock check. *"Does my local Costco have any Pampers in size 6?"* Without access to APIs to answer this question, the model can still understand this request, chat about it, and then transfer it to a human agent. Then, solve the problem of requiring the user to ask the human agent the same question. Give the agent the recorded question or transcription with a head start while the user is being transferred. So now the agent can answer the call and say, *"I am checking on the size 6 Pampers; one moment, please."*

Not having to repeat the request would be a breakthrough and an actual omnichannel moment, even if the automation couldn't directly answer the question. Or what about a stock check system that *only* interacted with the LLM while humans only proved the results to the software, thus reserving additional agent time for more pressing and complex interactions? *"I have asked a team member to check on the Pampers. It might take a few minutes. Is there anything else I can help you with while we wait?"* At least until the company gets the necessary APIs to automate stock checks.

The point is to create the best experience within the system while pushing for the necessary capabilities at the same time. If there are no stock check APIs and it is 15% of inbound call volume, then it is probably a problem worth investigating. With a simple phone tree and a team not ready to expand to the questions that will come with an open-ended experience, don't bother adding ChatGPT at the top. The value comes from improving customer service and decreasing hold time for human agents for those common questions that could be answered by combining data sources with ChatGPT. There is a wealth of examples of inadequate implementation of AI. Don't be the company that added ChatGPT, which worsened the experience.

Considering that the user's phone number is known in many cases for existing customers (a good thing to have an API for), a lookup around this customer's profile and history can generally reveal various opportunities for a customized experience:

- Order status and tracking:

  The user is known, so if open orders exist, immediately prompt and assist with details on those orders or route the call correctly to a human agent giving the order as context. *Are you calling about the online order that included the Shipley 4-piece bedroom set? It is scheduled to be delivered tomorrow.*

- Nearest store details:

  Starting with details about their typical store would be reasonable, especially on known holidays. *Happy 4th of July! Your Foster City store is currently open. It will close at 6 p.m.*

- Support for purchases:

  If they just bought a TV, use that information when they ask for support. *I see you just purchased a 65-inch Vizio Smart TV. Is that the product you need assistance with?*

- Technical support:

  An existing open ticket is a likely candidate for further conversation. Of course, they might have dozens of available tickets in the enterprise space, so consider the scale of the issue when deciding how to interact with the customers. *I see your support ticket from yesterday was updated. Are you here for help with your credit card being charged twice?*

Map the customer's phone number to their company history to enable personalized assistance until more is known on the call. You have probably experienced this with messages like, "*I see you have called us, Richard, from this number before. If this is you, press one or stay on the line.*" Still, I suspect every vendor is working to improve. If you use a third-party solution, evaluate its ability to engage users, review logs and usage to identify opportunities for enhancing routing and responses, and use the methods from the prior chapter to prioritize improvements.

Here are a few short bits that can be helpful with voice experiences.

- If disambiguation is needed to understand the situation (as in the example asking what kind of batteries were in stock), keep prompts short and straightforward.

- Offer secondary channels for complex interactions that would be difficult to conduct on a phone call (such as sending an email to complete a task later on a phone or computer).

- Log and track all errors or navigation issues that the user reports.

- Monitor path analytics (Funnel reports) to understand where customers hang up or get lost. This can expose opportunities for an LLM.

- Ensure the voice persona suits the audience. Test and get feedback. Use A/B testing to consider alternatives. Advanced users could use a different persona.

- If using multiple personas, associate the right persona with specific user groups, such as new users, experienced customers, or VIPs.

This covers experiences with interaction. What about cases where there are little or no interactions? Are there things to do to create high-quality experiences? Let's cover bot-only or behind-the-scenes experiences next.

Designing a recommender and behind-the-scenes experiences

Sometimes, ChatGPT is used to analyze data, make recommendations, offer suggestions, or display results. That is an excellent use of this technology, and because the experience is simple, ensure the output is also simple. These experiences are not conversationally interactive or hide the use of the LLM. They might be called "headless" because only a recommendation is provided, and there is no front-end experience, such as a chat window.

The following are some of the examples of bot-only or behind-the-scenes experiences:

- A recommendation flag or marker based on a ChatGPT analysis. This could be an icon showing **Urgent** or **Warning** on the cloud service dashboard or a **Call Now** flag for a sales lead. The user might not even know that an LLM is now behind this notification. Traditionally, this was handled by a simple rule, maybe some machine learning, and now an LLM. This is because the LLM can deliver more accuracy, address more complex situations, or be more timely.

- A call to action: **Call the customer by Thursday; they are 30% more likely to close this deal.**

- A process improvement: **Response time dropped 11% last week. To return to the previous service level, increase call center staff by three agents during business hours.**

- A task recommendation: **Ship this package using DHL for the most cost-effective solution.**

- A virtual non-player character in a video game interacts more naturally and dynamically with players.

- Idea generation. Offer suggestions for new ideas to solve problems.

- Editing: Integrate with built-in authoring tools for marketing or coding that offer better suggestions than traditional deterministic spelling and grammar solutions.

In these cases, there is no direct interaction with the model; at best, you accept or don't accept the options or recommendations. In many cases, UIs already had these experiences, but the sophistication of a generation AI solution did not power them.

Guidelines for bot-only experiences

There is such a wide range of use cases this is just general guidance:

- Use data logging to understand whether recommendations are being followed.

- Create a mechanism to get feedback on use while considering any data from logging. Don't ask questions to which the answer is known. For example, from call logs, the system knows that an agent called the customer when recommended, so don't ask whether they made the follow-up call. Use survey expertise to ask the right questions.

- Don't require feedback.

- Keep suggestions short and to the point.

- Consider secondarily exposing recommendation details, learn more about this recommendation feature, or analyze what factors led to the recommendation.

- Don't overuse recommendations. If every row of data has the same recommendation, then the value of the recommendation will be lost.

- Monitor and revisit recommendations. Creating metrics on their success or failure will help position investment in future improvements. Without monitoring, it is unknown if it was a waste to develop it. I know one client uses ChatGPT to update resumes to be better suited to get through the automated screening tools that review resumes. It was suggested that feedback be incorporated against actual vendor tools, not just relying on whether the resume writer edited the resume further after getting adjusted by the AI. The correct analytic focuses on whether the edited resume passed screeners and resulted in the applicant making it to the next step. *Asking the right question and building the right metric is key.*

Sometimes, revealing more details of what is happening behind the scenes is helpful.

Exposing ChatGPT when it is working behind the scenes

The exciting thing about ChatGPT is it can appear anywhere. The user typically doesn't need to know the details of a behind-the-scenes solution. Usually, details are not exposed, but there can be some exceptions in AI cases. It is generally accepted as good behavior not to pretend to be human. In the case of a chat UI, the LLM doesn't impersonate a human. Customers are told they are talking to a virtual assistant. It can also benefit the user to know this; they might be more forthcoming and will know they will get a quick response. In a hybrid experience that includes a GUI, or in the case of a Chat AI participating in a group conversation, it should be clear when the AI is participating, and it should be clear whether there are any guardrails preventing the AI from listening.

In a bot example, where a simple recommendation is offered, machine learning and algorithmic recommendations in UIs appeared more than a decade ago. The behind-the-scenes algorithms are not explained and might be proprietary (like credit scores). It is unnecessary to clarify whether it uses or includes results from ChatGPT. Decide based on the use case. Get feedback on recommendations, and this is encouraged to happen with some regularity. It might help and can hurt to explain some details if feedback is provided. Even the simple feedback mechanism could be powered by AI. Imagine a purpose-built model whose job is only focused on getting feedback. Having a direct feedback request with possible follow-up can be sufficient. One upside of exposing "**Powered by AI**" is that it might help engage customers to provide feedback. Position this as the user helping themselves because their feedback will be used to improve results. And folks do like being heard. Customers might consider this help-free consulting or be annoyed thinking the AI is in "beta." However, I think this opinion will soften as users recognize the value of a well-trained generative AI solution.

Hopefully, for all the solutions that can be invented, most enterprise ChatGPT solutions will be covered within one of the types of experiences discussed. There are a few overarching areas to consider for all experiences.

Overarching considerations

Some topics apply more universally, regardless of the type of experience. Accessibility, internationalization, and security are three essential items worth discussing.

Accessibility

It should not be surprising that I call out **accessibility**, also called **A11y** (because who wants to type all those letters). Large multinationals tend to have contracts with government entities in the enterprise space, and they require A11y. Each country or region can have its own rules; sometimes, those roles can be broken if well documented, but our goal should be to make an accessible experience. Some forms of accessibility are more challenging than others. All customers can benefit from design thinking applied to A11y. For example, keyboard shortcuts and navigation are required for those with visual impairments, but many computer experts use them extensively. Closed captioning was introduced in 1979 for people who are hard of hearing. Still, bars have them on, and others use closed captioning to learn a new language or to help improve understanding because of a wide range of audio issues from streaming and cable services. Voice recognition was developed for those with mobility impairments. Without that history, Siri, Alexa, and the voice assistants of the world would not exist. The examples go on and on. Consider the use case and ensure a solid A11y solution. Sometimes, the framework, such as an Apple iPhone, has considerable accessibility support. So much is handled by the framework; lucky for your team!

Use the existing standards and learn how they are applied in AI use cases. GUIs, chat, and voice all have A11y concerns. Even if you are in a country where they might not apply, you can learn much from them. We don't want accessible experiences; we want *usable,* accessible experiences. The first three references have the broadest impact:

- **Web Content Accessibility Guidelines (WCAG):**

  Article: `WCAG (https://www.w3.org/WAI/WCAG21/quickref/)`

  Developed by the **World Wide Web Consortium** (**W3C**), WCAG (pronounced "double u-cag") is a guide for web content accessibility. The foundational guidelines are widely adopted and used as a basis for other standards. For example, California mandates WCAG 2.1 Level AA and Section 508 for their public websites. WCAG 2.0 is also referenced as the standard ISO/IEC 40500:2012.

- **Section 508:**

  Article: `Section 508 Government standards (https://www.section508.gov/)`

  Section 508 outlines accessibility requirements for electronic and information technology used by the federal government of the United States. Compliance with 508 is typical for more extensive enterprise solutions sold to the federal government, while sales to state governments might also reference this in their project requirements.

- **Accessible Rich Internet Applications (ARIA):**

  Article: `ARIA guidelines (https://www.w3.org/WAI/standards-guidelines/aria/)`

  ARIA is a set of attributes that define ways to make web content and applications more accessible, especially for dynamic content and advanced user interfaces. This is important when doing intelligent or live updates on a page.

- **Americans with Disabilities Act (ADA):**

  Article: `ADA website (https://www.ada.gov/)`

  The ADA mandates that public entities, including digital services, web applications, and sites, be accessible to people with disabilities. As the ADA is not a standard, compliance often aligns with WCAG guidelines. The guidelines specifically call out algorithms and AI regarding hiring practices.

  Article: `AI Guidance (https://www.ada.gov/resources/ai-guidance/)`

- **European Standard EN 301 549:**

  Article: `European Standards (https://www.etsi.org/deliver/etsi_en/301500_301599/301549/03.02.01_60/en_301549v030201p.pdf)`

This European standard is based on WCAG and provides additional requirements to meet public procurement needs for **Information and Communication Technology (ICT)** products and services. Some of their standards, such as this one for "videotelephony" (video calls), are very technical:

Article:`Human Factors in Videotelephony (https://www.etsi.org/ deliver/etsi_etr/200_299/297/01_60/etr_297e01p.pdf)`

- **ITU-T F.922:**

    Article:`ITU-T F.922 Guidelines (https://www.itu.int/rec/T-REC- F.922-202008-I/en)`

    The **International Telecommunication Union (ITU)** standard provides guidelines for accessible user interfaces on telecommunication devices. This specific one is for visually impaired persons, but other standards in the same section might also be valuable.

It would be best to understand and, when needed, refer to the specific standards relevant to the target audience. If this is your first chat UI, then consider how a screen reader reads the conversation, how to deal with the LLM coming back with a response that might be delayed (how to notify the user if they are in another window, for example), and to make sure that any GUI elements have the same support as they would if they were on a web page.

If you have been building existing UIs to A11y standards, do test suites already exist for the models? Compliance and even exceeding mandates and standards help create a more inclusive experience for all users, and speaking of all users leads us to think about a worldwide audience.

Internationalization

This is another one of those long words written as **I18n** (skipping those annoying 18 characters between the *I* and the *n*). The world of ChatGPT has opened the doors to communicating with the audience in their native language. That is fantastic news. Can the enterprise handle working in their language? It isn't just a matter of turning on translation. That might lead to a worse situation than if the language wasn't enabled because it is only one piece of supporting the customer in that language, at least to some extent. Define what level of support makes sense. If there is a ChatGPT conversational solution for asking knowledge questions, the next step would be to support the customer's follow-up questions or escalate to a live agent in that language. Many companies are not good at that.

Internationalization, the overarching development effort for language support, includes localizing text translation and considering locale-specific elements.

Example of using in-conversation translation

Even the largest support companies sometimes have minimal multi-language support outside their native language. In one large multinational, even when another language is supported (such as the Asian languages, where English is typically not an option for this company), a team translates each service request. Then, the translated request is sent to the engineering team, and each response is translated back into the customer's language. It is a game of telephone, and these intermediate steps cause noise and confusion.

Could this process be improved by putting a real-time machine or LLM translation service into the loop? Would the technical nature of the requests and conversation confuse the participants? This technology is coming, but getting there will be hard work. It is one thing to ask for directions to the train station. It is quite another to get into the details when asking what specific SGA parameters and configurations need to be adjusted or optimized to enhance the performance and throughput of a highly transactional Oracle database system, considering factors such as concurrent transactions, response time, and resource utilization. Will the translation keep up with the differences between `BUFFER_POOL_KEEP`, `BUFFER_POOL_RECYCLE`, `JAVA_POOL_SIZE`, and the eight other properties? A far cry from, "When is the next train to the city?"

If conversations are less technical than this crazy database example, that is good news. But this is why enterprise solutions are needed. We aim to build on the knowledge within the company and presumably use this to expand communication capabilities in customers' native languages. It makes it evident that monitoring, debugging, and improving the machine or human-in-the-loop translation services are critical to success.

Translating knowledge

Translating knowledge into other languages is a start. The cost was traditionally expensive. Each article goes through an editorial process as part of any automation; this allows for a high-quality experience. Technically savvy translators must build up experience to translate complex or technical material. Guidelines and translation dictionaries are needed to maintain consistency and quality. Even with all the advances in automation, deciding the quality and cost trade-off threshold is still required. Machine translation might be cheap (or free) and fast. Determine if it is good.

Note

There is an adage in the tech world, "Cheap, fast, and good, choose two." If it is cheap and fast, it is likely not good. As shown in *Figure 5.15*, it is probably not cheap if it is fast and good. And if it is cheap and good, it is probably not fast. There is an argument for design that some of these are untenable options. There are use cases for all of these, but they might not be best for the business.

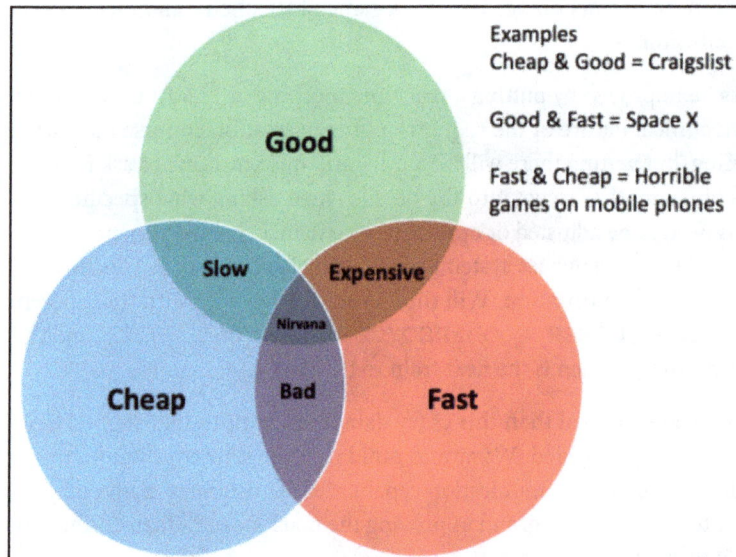

Figure 5.15 – Cheap, fast, and good, choose two

Everyone wants cheap, fast, and good. It seems to be a unicorn, magical and non-existent. Please prove me wrong, gladly. I have always considered this adage a truth for product development. Let me share Benek Lisefski's counterpoint for some additional reading on this.

Article: `The Big Lie of Good, Fast, and Cheap` (`https://solowork.co/story/the-big-lie-of-good-fast-cheap`)

Pick your battles. There are fast improvements in LLM-based translation, and the cost of quality translation is falling rapidly. Automate or allow some of this on the fly, but always check or spot-check the results. At least for popular articles, pay for professional review and editing even if the LLM did the primary translation. However, for less popular articles or lesser-used languages, could it be enough to let the user know this article has been offered in their language but still might have errors? And don't translate an article more than once. If the system translates on a pay-as-you-go approach for older articles, cache the results so you don't have to hit the model again the next time the article is requested. Automation can help batch translate and reduce costs, but because of the complex nature of the company languages, this process requires checks and balances. Here is an approach to using LLMs in translation situations.

How to decide what languages to support

Sometimes, the decision is made for you. A salesperson pre-sells the product in a specific language. I have seen this happen; the customer is still waiting years later. It is easier said than done. This is not as common in large companies because no one deal is typically big enough to force this hand. However, this can be a deal breaker in smaller companies, or it is mandated to enter a new territory because of a strategic partnership or country-specific laws. If you are involved in deciding what languages to support and how much to help them, here is an outline to get started. Adjust as needed.

There are four buckets to define: primary, fully, partially, and not supported. The company has a primary language, the base languages for the company; English is used for this example. A process is needed to define each resource the customers will touch, the cost of support, and the quality goals for that language. It is tough to achieve close to native quality. The cost to professionally translate technical knowledge might be prohibitive using traditional methods, but this would ensure that complex descriptions are explained correctly. Having native-speaker-level language quality in more than a few languages is challenging. ChatGPT's translation cost is multiple orders of magnitude less expensive, maybe achieving 80–90% of the value with the proper monitoring and fine-tuning. Is that worth being able to expand markets?

Some languages would have partial support; some things won't be done in those languages. And there will always be languages that are not supported.

All of this depends on the languages the LLM supports. This is still a maturing space. Although many models talk about supporting dozens of languages, training will improve their expertise in those languages. The quality of each language varies considerably.

Many pieces of the solution can include design elements for supporting languages. What happens when they have a ChatGPT conversational experience in German and then escalate to a support center that doesn't speak German? What is the plan? What is the customer's expectation? Consider how ChatGPT can support language needs in the customer's journey. In *Table 5.2*, I have shown some examples (**in bold**) where an LLM can enhance or extend language support.

	Primary Language	Fully Supported Languages	Partially Supported Languages	Not Supported Languages
Use Case for User	Expects high-quality responses in their native language with the most up-to-date language, details, and tone		Expects good quality experiences with localization	The least popular language should get the best effort from machine translation
Percent of User Base	Primary	Greater than 5% of I18n users	From 1% to 5% of I18n users	Less than 1% of I18n users
Language	English	Japanese, French, German, Spanish, Italian, Brazilian-Portuguese	Finnish, Dutch, Hebrew, Swedish, Arabic, Chinese, Korean	Thai, Icelandic, Slovak, Romanian
User Interface & Help	Complete	Complete, Professionally Translated	**Complete, LLM translated with a fine-tuned model**	**Auto-Translated with LLMs, cached after translated**
Knowledge	Complete, Active Updates	Key Knowledge, Quickly Updated	**LLM-based batch translation, with fine-tuning**	**Auto-Translated with LLMs, cached after translated**
Conversational Assistant	Full Support	Full Support	LLM Based on Translated Sources	**Real-time LLM translation (if available)**
Tier 1 Agents	Native Speakers	Native Speakers	Supported with Translation	Not Available
Tier 2 Agents	Native Speakers	Limited, **LLM Support**	Very Limited, **LLM Support**	Not Available
Milestones	Timely	Quickly, by Q2	Slower, by EOY	Not Supported
Cost	N/A	The incremental backlog cost is 1.25M, 100K/Qtr. ongoing	The incremental KnowledgeBase cost is 125K and 15K/Qtr for new material	UI language cost is new and is about 15K per language
Extenuating Circumstances and Notes		Contractually obligated to Arabic Tier 1 in Dec 2026	Dutch customers mostly speak English, so defer	Romanian agents speak supported European or English languages

Table 5.2 – Hypothetical comparison of support levels and strategic plan for languages

This is just an example of how to consider the impact of language support. This table is hypothetical. Clone it, expand it, and consider the implications for each box on the business language strategy. When designing for I18n, there are also details to work out. Let's explore a few considerations.

Accounting for I18n when designing

One of the most significant issues for internationalization likely comes from the knowledge base. Is the content ready to be translated? Is it organized in a way that lends itself to translation? Will culturally specific idiomatic phrases fail automated translation? Consider *garbage in, garbage out* (even that phrase is idiomatic). It is incredible how well the current models understand idioms. Chat GPT continues to impress me with its results.

```
                             What happens if I ask you to translate
                             an idiom like "Your back is up against
                                    the wall" into German? Will it be
                                      a literal translation, or can you
                                     translate the idiomatic meaning?
```

```
If you ask me to translate an idiom
like "Es geht gleich los runter"
into German, I will provide a translation
that conveys the idiomatic meaning rather
than a literal translation. The
idiomatic meaning of "Your back is up
against the wall" is that someone is
in a difficult or desperate situation
with few or no options left.
In German, a common equivalent idiom is:
"Du bist mit dem Rücken zur Wand."
This phrase captures the same sense
of being in a difficult, constrained situation.
```

ChatGPT would take something like, "It's raining cats and dogs." Instead of translating it literally, it understands that it is raining heavily. It translates it more accurately using the German word for streams: "It's raining in streams (heavily)" (Es regnet in Strömen). In Knowledge bases, we hope not to see these problems, but it is challenging to avoid idioms. I have used dozens in this book, and I feel it makes it more fun. But I have done my readers a disservice if they don't translate well. Related to idioms is jargon.

Jargon doesn't translate well

Remember that some terms in your world might be well understood in other languages because they use the same word and don't translate it. But don't count on that as a strategy. Consider how technical jargon, shortcuts, acronyms, and initialisms play out in other languages. For example, *PIN* is short for *Personal Identification Number* for an ATM. In Germany, the Geldautomat still uses a *Persönliche Identifikationsnummer*, and some form of PIN and ATM are typically understood. However, if the company's vocabulary is not universal, account for it in the models, knowledge, or dictionaries. Related to this is how to speak to customers. It would be best to learn how to address them. Handling language jargon is slightly different than handling technical jargon. It might be well-known in *Database* circles to talk about **DB** or **RDBMS** (which stands for **Relational Database Management System**), so those might translate just as the letters, but generic language jargon should be avoided. *Chapter 7, Prompt Engineering,* helps address this to control the LLMs output. It is hard to catch these issues because we are so used to language. *Not seeing the forest for the trees* can be a challenge. Oops! There I go, using an idiomatic expression without even noticing.

UI issues in LLM solutions can be addressed just like traditional GUIs. Topics such as right-to-left language and icon direction, using colors not to offend certain groups, and even removing culturally offensive imagery can be addressed. Still, thousands or millions of articles could generate a profound amount of rework to get them ready for translation. This is a cost hog.

As designers and product leaders, we control the user experience, but we know those upstream knowledge sources will likely significantly impact it. So, spend the time making those ready for an international audience. Let's review a few more language-related areas, starting with punctuation.

Punctuation and grammar

A funny story can make a point.

A panda walks into a bar. He orders a sandwich, eats it, then draws a gun and fires two shots in the air.

"Why? Why are you behaving in this strange, un-panda-like fashion?" asks the confused waiter as the panda walks towards the exit. The panda produces a badly punctuated wildlife manual and tosses it over his shoulder.

I'm a panda," he says at the door. "Look it up." The waiter turns to the relevant entry and, sure enough, finds an explanation.

"Panda. Large black-and-white bear-like mammal native to China. Eats, shoots and leaves." – Lynne Truss (*Eats, Shoots & Leaves*)

We do not control ChatGPT. So how does this help? Of course, each response from ChatGPT is not controlled by humans. Quality can be expected by ensuring the resources given are well written and monitoring its output to verify that it works as expected. We can't expect ChatGPT to return a quality answer if the content shared is not written well. If it wasn't clear, the statement should have been punctuated to explain that the panda "eats (bamboo) shoots and (tree) leaves."

Note

If writing is not your strong suit, go beyond the required manuals of style reading. The book is also included in the online library.

Book: Eats, Shoots & Leaves: The Zero Tolerance Approach to Punctuation by Lynne Truss (https://amzn.to/3M7x8DH)

Grammar also applies to recommendations UIs. An LLM can be used behind the scenes to generate the data or text to insert into recommendations. Grammar and other issues, such as plurality, come into play. Consider the length of the messages and communications when writing these statements or messaging prompts to write well-written prose.

Length of labels and strings

Test UIs for text expansion. For example, if ChatGPT writes a notification, account for the length of the message. It is not unreasonable to see more than a 200% increase in word length when coming from English. The button *Edit,* in German, uses the word Bearbeiten. It is 150% longer. In Hungarian, Szerkesztése, clocks in at 200% longer. With ChatGPT powering a recommendation engine, ensure the UI has enough room to write the English and translated versions. Prompt engineering can be used to create instructions that limit the size of responses. In *Table 5.3*, the Hungarian translation is up to 53% longer. That is not too bad.

String	Language Details
Tip: Edit the bills for this Employee ID by the end of the day, or the state penalties will accrue.	English 100 Characters
Tipp: A nap végéig szerkeszd meg ennek az alkalmazott azonosítójához tartozó számlákat, különben az állami büntetések felhalmozódnak.	Hungarian 133 Characters
Tipp: Szerkeszd meg ennek az alkalmazottnak az azonosítójához tartozó számlákat a nap végéig, különben állami büntetések keletkeznek.	Hungarian (Google) 136 Characters
Tipp: Szerkessze meg a számlákat ezzel a Munkavállalói Azonosítóval a nap végéig, különben az állami büntetések felhalmozódnak.	Hungarian (ChatGPT 3.5) 147 Characters
Tipp: Szerkessze meg a számlákat az alábbi Munkavállalói azonosítóhoz a nap végéig, különben az állami bírságok emelkedni fognak.	Hungarian (ChatGPT 4o) 153 Characters
请注意：请在今天结束之前编辑此雇员ID的账单，否则将会累积州的罚款。	Chinese 34 Characters
Hinweis: Bearbeiten Sie die Rechnungen für diese Mitarbeiter-ID bis zum Ende des Tages, sonst werden staatliche Strafen anfallen.	German (ChatGPT 4o) 142 Characters

Table 5.3 – Comparison of English to Hungarian, Chinese, and German translations

Sometimes, the opposite is true, and another language can be shorter. In Chinese, as shown, the character count is 34, but the double-byte character set takes up more space. Still, the issue with wrapping and truncation typically occurs with longer text. For messages that are sentence length or longer, account for at least 50% growth; for shorter items such as labels, consider 100–200%, depending on the languages supported. A reasonable guideline is a minimum of 30% plus two extra characters. However, if the design can accommodate larger text, there is no harm in most cases. If there are tables where excessive wrapping would result, consider creating layouts for the other languages to accommodate their unique situation. For A11y, consider customers zooming in and scaling text, which can also make wrapping and truncation an issue. This can also impact how fields appear with their text. In some languages, such as German, compound words have unique rules for where to wrap. For example, in a human resource solution, there are labels for absence batch parameters, so when an absence job (vacation, holidays, or sick time) was run on a schedule, it knew the properties to use. The translation from an AI can vary dramatically:

- Absence batch parameters (English)
- Abwesenheits-Sammelverarbeitungsparameter (literal in German)
- Fehlzeitenerfassung im Stapel-Modus Einstellungen (descriptive version)
- Einstellungen für die Stapelverarbeitung von Abwesenheitszeiten (descriptive focusing on the batch process)
- Stapelverarbeitung von Abwesenheitszeiträumen (common in HR software)
- Mitarbeiter-Abwesenheiten Sammelbearbeitung (emphasis on the employee)

A human translator might settle on Abwesenheitsbatchparameter; notice there are no spaces, so wrapping in a UI might be an issue. If words are not wrapped correctly for the language, the customer will view it as a bug.

Tip

Consider whether abbreviations can cause displays to be off. Not all languages have the concept of abbreviations. Understand how to translate these as whole words and have room for them in Chinese or Arabic. Some languages might not recognize a character as a word to translate ("#" for number, "Q1" for quarter one, "ft" (feet) might be better calculated in meters outside of the USA, "info" might not translate as "information," etc.). FYI, GPT 4o-mini understands # and *Q1* in translations and converts them correctly to *number* and *first quarter*.

Leave ample space to avoid using ellipses with truncation. Ellipses must be used if a word is truncated, but try to prevent truncation. Be sure the user can read the entire label or text. I have seen many UIs that show truncation with ellipses (…), but there is no way to read what is there; not all components have a tooltip. And worse, when translated, the order of words can be different. "**Customer Number**" might be truncated in English as "**Customer N..**" leaving helpful information. In Spanish, the truncated version of "**Número de cliente**" is "**Número de ..**", leaving nothing of value in the visible label. That is a failure for the user experience and frustrating for the customer.

I created a simple Google sheet, as shown in *Figure 5.16*, to test length issues. Enter a string into the spreadsheet, and the tool will highlight the languages with much larger lengths. This is good for understanding how small words or phrases can grow in length. This is not perfect, as this is Google's translation, but it is fast and cheap. Clone this for your purposes.

Tool: `Translation length tester (https://docs.google.com/spreadsheets/ d/1P-FLn8Kc4wcOgFbUfGuCamAL7O6IXvF_qkDAhvS8QdA/edit#gid=0)`

If you are using an LLM for UI text, make sure there is plenty of room in the UI for these long translations.

A	B	C	D	E	F	G
Enter String to Translate	Translation	Language	Key	Length	Growth	Reverse to English
Undo	Undo	English	en	4		
	Deshacer	Worldwide Spanish	es	+4	100%	Undo
	Desfazer	Brazilian-Portugese	pt-BR	+4	100%	Undo
	annuler	French	fr	+3	75%	Cancel
	Rückgängig machen	German	de	+13	325%	Undo
	Disfare	Italian	it	+3	75%	Disfrociate
	撤消	Simplified Chinese	zh-CN	-2	-50%	Dismiss
	元に戻します	Japanese	ja	+2	50%	Return
	Kumota	Finnish	fi	+2	50%	Repeal
	Ongedaan maken	Dutch	nl	+10	250%	Undo

Figure 5.16 – Example of the Google sheet you can use for quick length checking

Concatenation

One great trick in ChatGPT could be to combine a defined statement template with the output from ChatGPT. In examples such as `"It is recommended to " + <ChatGPT response>`, or `"Refill the shipping container by " + <ChatGPT response of a time> + " today."`, this will experience translation problems. The structure of a sentence in English is not the same for other languages. This static part of the text might be at the end of the sentence. Break up the answer into two distinct pieces. Create the recommendation title with static text and then generate the *entire* recommendation from the model. In these examples, put that recommendation in a box labeled `Recommended Actions`. Or label the time `Today's Refill Container Deadline` so the output "`3:45 PM`" stands alone.

> Tip
> Ask ChatGPT to translate the entire string. That is better than writing code to create the formatting for each language.

Let me explore the complexities of translation a little more. Plurals are one of the most well-known language problems.

Plurals

Plurals are one of the more complex issues with translations. With a chat experience, ChatGPT handles this, and this topic can be skipped. However, if the bot is recommended with *templates* to fill in pieces of text coming from ChatGPT, this situation can get complex. In English, there are simple rules; we can use a notification alerting the user to their vacation balance as an example:

- Zero: You have no days off remaining.
- One: You have 1 day off remaining.
  - You have one day off remaining. (alternative)
- Many: You have 23 days off remaining.

English has three cases for plurality. A special rule for zero is useful (so the cleaner textual *no* can be used instead of a zero). The singular word *day* is used for one, and all other numbers use plural *days*.

It is easy to write a rule in software to account for this:

```
if days == 0:
    print("You have no days off remaining.")
elif days == 1:
    print("You have one day off remaining.")
else:
    print("You have " +str(days)+ " days off remaining.")
```

This could work for negative balances, but a good UX person would get one more condition added. This gets more complex when doing this logic in other languages because many languages have different rules. To accommodate languages, there are software and standards to account for zero, one, two, a few, many, and other use cases. In addition to one noun form for the singular and another noun form for the plural, as in English, different languages have unique noun forms when the count of the object is two ("dual") or a few ("paucal"), or when the number ends in one or ends in zero… There are many special rules! As shown in the preceding example, spell words for small numbers (1 to 9) in English. This can make every solution grammatically correct but more effort to code. To go deeper into the complexity of using numbers, see this reference:

Article: `When to spell out numbers (https://www.masterclass.com/articles/ when-to-spell-out-numbers-explained)`

As mentioned, if ChatGPT outputs a complete thought, it should follow the correct rules. Work is needed to create a hybrid experience where the output includes numbers in static messages, such as recommendation text. The resources listed later in this section provide some more detailed explanations. Here is an example of English and Czech, which is not even that complex.

```
You have {COUNT, plural, one {# mail} other {# mails}} in your inbox.
You have {COUNT, plural, one {# mail} few {# mails} many {# mails}
other {# mails}} in your inbox.
V doručené poště máte {COUNT, plural, one {# zprávu} few {# zprávy}
many {# zprávy} other {# zpráv}}.
```

This format shows how the translated strings support the corresponding plural forms. Don't expect to inject numbers directly into ChatGPT responses. Either let it do it correctly or write the code. *Table 5.4* shows one end-to-end example, including what not to do, what to do, and how it will look when done.

Don't Do This	Readable & Translatable	Example Output
I'm ready to submit your invoice with {0} and {1}. What should we name the invoice?	I'm ready to submit your report with {NUM_ITEMS, plural, =1 {1 item for} one {# items totaling} other {# items totaling}} {AMOUNT} {CURRENCY} and {NUM_ITEMS1, plural, one {# foreign currency items} other {# foreign currency items}}. What should we name the invoice?	I'm ready to submit your invoice with 5 items totaling 140 USD and 1 foreign currency item. What should we name the invoice? I'm ready to submit your invoice with five items totaling 140 USD and one foreign currency item. What should we name the invoice?

Table 5.4 – Example of a string to be written in ICU format for translation

Notice that the variables are readable, not just 0s and 1s. Do this to validate that the statements are readable quickly. Good design thinking extends into the code sometimes. Whether these variables are numbers or words must be known to translate or validate. The specific number needs to be known if they are numbers so the code can generate the correct format. Notice I provided two example outputs, one using numbers and one using number words (the written form). I prefer numerals for numbers below 10 in UIs, even though they are not grammatically correct in English. It helps these values stand out and prevents the user from having to read the entire text. People don't read!

> **Additional references for plural resources**
>
> Article: `Pluralization (https://lingohub.com/blog/pluralization)`
>
> Article: `Plural Rules for Internationalization (https://cldr.unicode.org/index/cldr-spec/plural-rules)`

How to handle caps in design may be less exciting and less well-known. This is not specific to conversational AI, but it is still helpful to realize when someone wants to use them in recommendations.

ALL CAPS

Don't ask ChatGPT to return output in ALL CAPS. To format for emphasis, use **Bold**, for example. Not only are all caps harder to read in English, but there are also languages where this is not acceptable (French and Greek) and has no equivalent in some languages (Chinese and Arabic). Avoid this problem by not using all caps and telling the LLM to avoid it. STRINGS IN UPPERCASE ARE SIGNIFICANTLY HARDER TO READ. It is not suggested even for titles, tables, or labels. Use Title Case, as used in newspaper headlines. For database fields that are stored in uppercase, change it, edit the strings, or, as a last resort, force them to be lowercase. The 5–10% improvement in readability is worth it.

Work and letter recognition

Learn more about some of the research on word and letter recognition:

Article: `The Science of Word Recognition (https://learn.microsoft.com/en-us/typography/develop/word-recognition)`

Accounting for locales

Designing for locales involves considering how a specific region or country handles something. For example, if building a scheduling service and someone asks, *"Book a hotel for this weekend,"* what nights would that booking be for? In the US, it would be for Friday and Saturday nights. Makes sense? However, different countries have different definitions of the weekend. In some Muslim countries in the Middle East, it would be for Thursday and Friday, while in Iran, the weekend is only Friday night, as they observe a six-day work week. ChatGPT understands these concepts to some extent, but make sure integrations also have this understanding. Many ideas and words can be locale-specific. For example, when working on the design of an expense assistant, it understands the expense of a car, taxi, Uber, Lyft, Limo, etc. But there are Didi, BlaBlaCar, Cabify, and many other services in various countries. Ensuring models support the locale-specific concepts will make the experience more robust. Sometimes, it is not just a country-specific product or service; it could be jargon covered earlier in the chapter. If the UI will handle understanding the customer's work week, it should likely learn how to address the customer properly.

Addressing a customer and vocative case

Many UIs start with "Hello, Max!" or "Welcome, Madison." A direct translation will not work because, in some cultures, addressing someone by their first name is inappropriate.

My friend shared an excellent example of how formal or complex it is to speak to the customer in their name. I have shared this in *Table 5.5*. So be sure to have enough information to accommodate their first name, last name, or various forms of their names (the fancy term is vocative declension), depending on the language.

Language	Problem	Basic Name	Example
English	None	Mark Stevens	Hello, Mark
Japanese	Requires last name + honorific	Yoko Ono

Akira Kurosawa | Hello, **Ono-san**

Welcome, **Kurosawa-san** |
| **Arabic**

(Latvian,

Lithuanian, Vietnamese) | Prefers polite address | Mohamed Hassan | Hello Mr. Mohamed (suitable for a welcome screen)

Dear **Mr.** Mohamed,

Alt: Dear **Mr. Hassan** (for email) |
| **Czech, Polish**

(Greek,

Latvian, Lithuanian) | Requires vocative declension | Pet**ra** Lukáš

Mar**ek** Martin | Vitejte, Pet**ro**

Dobrý den, Lukáš**i**

Milý Mar**ku**

Martin**e**, jak vám mohu pomoci?

(notice o, i, u, and e at name ends) |

Table 5.5 – Direct address in various languages that can cause issues

If all else fails, *don't use it at all, or don't use it in those languages*. One workaround is to use an email address. I don't recommend it. It is better to say "Hello" than "Hello, jill@miller.com." Software can be used to build declensions (changes to the ending of words, in this case, for the person's name) for specific languages. I asked ChatGPT to do it for Hungarian, and it generated Python code. It doesn't have to be a design problem; it is something to be aware of when using people's names in the experience:

```python
def decline_name(name, case):
    declensions = {
        'nominative': '',
        'accusative': 't',
        'dative': 'nak/nek',
        'genitive': 'é',
        # Add other cases as needed
    }
    # Handle special cases or irregular declensions
    special_cases = {
        'János': {
            'accusative': 'Jánost',
            'dative': 'Jánosnak'
        }
        # Add other special cases
    }
```

```
    if name in special_cases and case in special_cases[name]:
        return special_cases[name][case]
    suffix = declensions.get(case, '')

    # Apply vowel harmony rules or other specific rules
    if case == 'dative':
        if name[-1] in 'aeiou':
            suffix = 'nak'
        else:
            suffix = 'nek'
    return name + suffix
# Example usage
name = "János"
case = "dative"
print(decline_name(name, case)) # Output: Jánosnak
```

One nice thing about a ChatGPT solution is that it can ask how they want to be addressed. Even if ChatGPT can't remember this from one session to the next, store it in a setup variable and then remind ChatGPT how to address this customer. ChatGPT will see enterprise data with people's names when helping to compose customer emails. If working in an international arena, ensure this is handled correctly. Addressing the customers is not the only cultural-specific issue.

Cultural and bi-directional hiccups

If customers use Arabic or Hebrew, be reminded that responses that include images or icons might not work as well or as intended without a flipped version. These languages are read from right to left. If an arrow is a pointer to a list of items (or any icon representing forward motion), it will look odd, with the arrow pointing the wrong way. Icons such as text bubbles, arrows, toggle icons, audio controls, pagination, progress indicators, and drop-down menu icons must be orientated for reading from right to left. There are issues with some icons in some cultures. These issues still apply in conversational AI. Thumbs up in some cultures is offensive. I learned that the hard way!

Article: `Cultural design (https://material.io/blog/localization-principles-techniques)`

A deeper dive into the bi-directional issues can be found here:

Article: `Bidirectionality issues (https://m2.material.io/design/usability/bidirectionality.html#mirroring-layout)`

Be aware of cultural differences when generating solutions with ChatGPT. Use prompts and verify culturally appropriate answers with native speakers or a trusted model. The more the design supports these small things, the more customer trust is built.

Trust

We have all heard about trust issues. If the AI provides bad or wrong data, this can quickly derail a conversation and cause the customer to lose trust in the AI. This is the same as when working with a human agent. As soon as a human agent says something so silly and wrong, it derails the conversation, and the tone of the interaction will immediately change. The goal is to mitigate these errors and issues. This next part might feel like a TV show, but here are the top 10 ways to improve trust in a ChatGPT experience, in reverse order.

10. Regular updates

When appropriate, let folks know the staleness of results. For example, if they ask a time-dependent question that can be easily changed based on recent information, tell them when the model was last updated. "*I see that the answer to this question has changed repeatedly, but as of July 5, 2024, it is...*", and "*My information is up to date as of last week...*". Do not let customers think they have the most current answer; they will act on that. It would be better not to answer at all. Even recommendations might need some supporting info to be clear that the recommendation is based on the latest information. This also applies to backend services. Sometimes, it is worth showing what data was last ingested or the last update time and name of the file. *Chapter 7, Prompt Engineering*, can address this. These little things can help build trust.

9. Empower users

With tools to customize the experience, make it apparent this is an option and how to adjust. They will trust the system differently if they feel in control regarding the UI, privacy, data storage, and other essential factors. This is about understanding and then implementing use cases that empower the user. Speaking in their language using terms they are comfortable with is an empowering feeling.

8. Ethics

As mentioned, most enterprise use cases don't delve into ethical concerns. However, in problematic areas around medical decisions, legal ramifications, lending practices, driving decisions, government practices, personal privacy, mental health, or other answer spaces where ethics is essential, know that the answers might not be the right choice. Avoid these topics, when possible, implement extensive safeguards, and consistently monitor this. Lawyers will likely get involved in providing lengthy disclaimers. That is the way of the world. Prompt chaining, by using one ChatGPT to check the work of another one's output, can help to verify the output.

7. Bias

Regularly monitor and address biases in responses. Implement measures to minimize biased outputs and ensure fair treatment of all users, regardless of demographics. Consider whether the knowledge base contains these biases. *Chapter 6, Gathering Data – Content is King* is next and will go deeper into this subject. Use emerging tools to check and monitor output for bias.

6. Privacy and security

Recall how easy it is for someone to share data with a general public model without realizing the implications. The more significant consideration in enterprise models might be at the company level. Does company A's model impact other customers' models? Does the business run on a model affected by all customer engagements? Product owners must ensure that company-sensitive data is not making its way to other companies. By siloing customer data and with independent model updates, then a good chance of protection at the company level. This also means either a lot more work to help every customer's model improve, outsourcing this improvement to the customer or third parties, automation, or ignoring the problem. The reason why so many first-generation chatbots failed is that they ignored the problem. Don't fail. Put checks in place so systems can't be manipulated to share the wrong data. Some platforms, such as Salesforce, cover this extensively.

Ensuring that only the right people in the company have the correct access is still a challenge. Just because someone works for the same company doesn't mean they deserve access. Sales, acquisitions, financials, and human resources all demand security. Salesforce goes out of its way to remind people that "Your data is your data. Your data is not our product." This would seem a pretty clear line. The downside can mean less competent models without sharing data across companies. It is possible that improvements learned in one data silo might not work in another. So, there is more to monitoring and improving. A real challenge to scale for the enterprise.

The bigger the company, the more it has to lose and the less risk it is willing to take. Security is a big deal, and I suggest spending the extra time and effort on security audits, **penetration (PEN)** testing, performance tuning, and usability testing.

5. User feedback

The general feeling of user feedback is that it goes into a black hole when given. Sometimes, feedback is anonymous, so it is hard to communicate directly with the user. Occasionally, customers are asked for follow-up or to provide their information if further questions are needed. One way to build trust is to listen to and communicate feedback. If they provided their email, do more than send automated emails that their feedback was sent. That doesn't build much trust. Even this step could be automated with an LLM to provide a more thoughtful version of a "we are looking at your issue" response. A model can respond in a few ways:

- That a feature request is in the works by comparing the request to existing published plans

- Sharing existing bugs filed, comparing against existing bugs

- Provide a link to monitor an existing bug or community threads

- Provide alternatives based on a knowledge search

All ideas support the idea that there is a loop in the feedback process. Let the humans do the more complex work and let the LLM assistant handle minor interactions.

For anonymous feedback, point users to documentation or help sites that collect and communicate about the feedback. If they see that people like them are being listened to, they may gain some trust in the solution.

It is wonderful to ask users for feedback proactively. Very few customers will respond, and it can be unpleasant to ask, especially if the user is in the middle of something important. Although covered in *Chapter 2, Conducting Effective User Research*, it is an excellent time to remind folks with some tips for gathering feedback:

- Don't require feedback. Don't request feedback at every turn or for every customer (use sampling).

- Keep feedback simple, such as *"Did we help solve your problem today?"* Based on the answer, probe why not.

- Someone (the LLM?) must follow up if they were asked for contact information.

- Log the context of feedback. If they are prompted while *creating* an urgent service request , "Is your problem solved?", the answer is no. Consider *when* to request feedback so relevant questions are asked.

4. Consistent style and tone

This book spends a lot of time on style and tone, which impact trust. Maintaining a consistent communication style lets customers know what is expected. However, this doesn't mean the style and tone can't be changed. If the situation calls for it, then the tone should be understandable for that context. Again, it becomes part of the instructions shared with the LLM, covered starting in *Chapter 7, Prompt Engineering*.

3. Explainability

In a conversational UI, this is the way, while in a GUI or bot use case, they might have an **Info** button, more details, or some element to *please explain*. I used the **Info** (i) icon for a decade next to simple recommendations to link to more details or provide it in a window. Sometimes, people want to know more details and the reasoning behind a decision. This can build trust to expose the reasoning; if the reasoning is sound, it can enhance that trust factor. The UI can use this to help guide the customer, building trust again. This is covered in *Chapter 9, Guidelines and Heuristics*.

2. Transparency

This is more important for a Chat AI than a bot or recommendation. Be clear communication is with an AI when in an actual conversation. When a UI pops up a simple recommendation or suggestion, it is a less compelling need to say precisely *how* this was generated. And it is likely already a combination of AI and other tools. An earlier section in this chapter, *Designing a recommender and behind-the-scenes experiences*, details communicating with an AI behind the scenes. See the section earlier in this chapter.

1. Accuracy and relevance

There is a reason why this is number one. And because this is only an ordered list, it is not apparent how much more important this is than the nine previous items. If you read the previous chapter's discussion on ranking versus scoring, now this might make more sense. Accuracy and relevance could be as important as *all the other issues combined*. Without accurate and relevant answers, the solution is doomed. There can be some flexibility with other trust issues, but not so much with wrong answers. Sometimes, the problem can be much worse if they trust it is the correct answer and do not understand enough to recognize it is wrong. That is the worst problem to have. An incorrect answer allows the customer to dismiss the answer, ignore the answer, and possibly leave disgruntled. However, a wrong answer that looks right can cause downstream repercussions based on customer decisions. Include the right resources to formulate answers and create test cases to detect issues (more on testing in *Chapter 10, Monitoring and Evaluation*). LLM solutions require care and feeding to uncover these concerns rapidly. User feedback can only catch known errors; it won't catch unknown errors. Those are hard. Build expertise to tune your radar to find unknown errors. The following chapters on prompt engineering and fine-tuning provide techniques to be accurate. Given that the goal is to build trust, security can go a long way to help with trust. It is logical to finish this chapter on this important topic.

Security

Security takes many forms. At a minimum, conversations should be end-to-end encrypted; the user already gets point-to-point encryption with the typical web HTTPS interaction. Custom applications should not do less, while some channels like SMS are already inherently insecure. Vendors make a point of promoting security in their apps. Consider what happens to the data within the enterprise network after the HTTPS connection has landed and information is shared with backend services or a database.

Is this data safe from prying eyes? For example, some companies don't allow staff to accept credit card information over the phone. Shield this information from insider exploitation using a virtual agent to receive and confirm payment information. This might even feel safer to customers; the virtual agent can be used to market trust.

Consider how models are being updated and with what data. Customer data must be isolated in the enterprise and can't leak into other customers' models. This book is for applying UX-centric design methods to enterprise solutions; our advice in this space is limited to making sure the customer's data is secure and communicating this detail to the customer. This is the same for any prompt engineering efforts where instructions must include details so the user's prompt has context. *Chapter 7, Prompt Engineering,* covers this. The design also has to account for when this data is unavailable because the user is not permitted to access it. For example, create context around sales deals so that users can ask more detailed questions about the deal. Some salespeople might not have full access to the details, so consider this when framing the prompts.

The takeaway concerning security is to be the customer's advocate and ensure that the conversation is safe and will not fall into the wrong hands. Not all data and content will be stored in a secure channel. Integrations with other services, or even access to some knowledge, might be limited to specific audiences, and LLM integration needs to account for this. Integrations become even more critical in the next section, which discusses the hybrid UI/chat experience issues.

Summary

Learn the capabilities of the tools in your framework, or better yet, help find the right tools to move the user experience bar forward. Every tool has capabilities and limitations. For any experience, the UX implications of accessibility, trust security, language, and internationalization support are likely a given. We hope the guidance, tricks, and tips help create a **Functional, Usable, Necessary, and Engaging (FUN-E)** experience. Strive for all four attributes (remember it as *funny*).

It is time to build a ChatGPT solution. Use cases should be ready to develop, and the context of use should be known. The context of most experiences is easy to figure out. In many cases, the context of use is not a choice. If the company provides phone support, a voice-only experience is a given. This is where we change gears and go from UI-centric discussions about generative AI solutions to model-centric discussions with an eye toward UI practices. Integrating enterprise data, like knowledge, databases, and other services, is next.

References

The links, book recommendations, and GitHub files in this chapter are posted on the reference page.

Web Page: Chapter 5 References (https://uxdforai.com/references#C5)

Part 2: Designing

This part focuses on the most technical part of LLM design, the work done in the LLM, or when using other tools and techniques to make an end-to-end LLM solution. We will start by introducing practices such as **Retrieval Augmented Generation** (**RAG**) to integrate enterprise datasets with the generative ability of an LLM. You'll then learn about the fundamentals of prompt engineering for enterprise applications. The examples follow a non-technical route, so you can get the most out of learning about the steps involved without coding. We'll then explore fine-tuning to make models think and act based on examples provided that will follow the style and tone needed for any business or enterprise. We'll also explore a few case studies and give some hands-on experiences to get a feel for the process.

This part includes the following chapters:

- *Chapter 6, Gathering Data – Content Is King*
- *Chapter 7, Prompt Engineering*
- *Chapter 8, Fine-Tuning*

6
Gathering Data – Content is King

There is an assumption in this book: enterprise ChatGPT solutions are needed in almost all cases because a company has something unique to offer its customers, and it possesses an exceptional understanding of its products, services, and content. This content is private or unique and thus not part of **large language models (LLMs)** built from scraping the internet. Models are built on crawling the 2+ billion pages of web content to teach the model. A third party, Commoncrawl.org, is commonly cited as a primary source of this material for major models (GPT-3, Llama). These models, which are massive collections of text, learn the statistical relationships of words and concepts and can be used to predict and respond to questions. Creating a model can take months; most have billions of connections and words. When customers come to the enterprise for answers, the models must include enterprise content that is not part of this crawl to make them unique, secure, and more accurate. This is done with the expectation that the solutions will be up to date, optimized to be cost-effective, and less prone to hallucinations or lying, as some call it.

This chapter addresses gathering data for the LLM and how to include enterprise data sources in LLM solutions using a method called **Retrieval Augmented Generation (RAG)**. We'll discuss the following topics:

- What is in a ChatGPT foundational model
- Incorporating enterprise data using RAG
- Resources for RAG

This chapter and the next few will be more technical for those who have reached this point and are focused on user-centered design concepts. The chapter covers all the ideas and provides access to additional videos and online resources. The book does not require most of these external resources; they are meant to give more details.

What is in a ChatGPT foundational model

When an LLM is built, it is trained on sources of data from the internet. It knows publicly available information about companies and products. If asked typical enterprise-like questions, it can get robust answers – sometimes better than what is available from some vendors' websites. For example:

```
What are the advantages of Hana for a database?
What is a good value for SGA for an Oracle 12.2 transactional
database?
Can you easily replace the battery in an iPhone?
How do I return a product to Costco?
```

Try these questions out and notice a trend. Each answer is slightly more generic than the previous one, and that generic nature is part of the problem.

The following applies to most foundational models such as ChatGPT 3.5 or 4o, Anthropic's Claude, Meta's Llama, or Mistral7B:

- Don't understand specific business or use context or complex products

- Don't have customer history or context to consider

- Can't access proprietary knowledge sources

- Are not trained on service requests or other service data and won't know correct assumptions from incorrect assumptions and inaccurate solutions

- Can't integrate with databases or APIs for retrieval and task performance

- Can't be scaled or tuned for multi-tenancy

Now imagine those questions if they were in the context of rich business knowledge:

```
What are the advantages of Hana for a database connecting to my
service application running ServiceNow on the Washington DC release?
What is a good value for SGA for an Oracle 12.2 transactional database
when connecting to EBS 13.2 with 1200 concurrent users?
How do I replace the battery in the iPhone 15 Pro?
How do I return a product from my last order to Costco that was
oversized and delivered?
```

Integrating data sources with ChatGPT to contextualize the solution to the business can address these richer questions. No matter the design pattern used, such as a chat UI, a hybrid experience, or a standalone recommender, enterprise data will make the solution powerful. Foundational models gain access to knowledge with Retrieval Augmented Generation or **RAG**.

Incorporating enterprise data using RAG

There are other ways of taking data and making it part of an LLM. It is possible to build a foundation model, but as mentioned, the training time and effort are extreme. Even with RAG, there are different approaches. Some technical resources are shared, but this chapter will focus on and teach RAG understanding and how product people can contribute to the development process. First, a RAG explanation.

Understanding RAG

RAG supplements LLMs with enterprise data. RAG is a technique for retrieving information, such as from a knowledge base, and it can generate responses from authoritative knowledge collection with coherent and contextually accurate answers.

This methodology allows us to overcome some generic model problems:

- Material is always up to date since it is evaluated when prompted.

- Tools can reference the document source and, thus, are more trustworthy.

- The foundational model is already trained, so supplementing it is inexpensive compared to building a model from scratch.

- It allows for a robust set of resources (APIs, SQL databases, and various document and presentation file formats) to continue to be managed independently (and still available to other solutions).

- It will *NOT* be used to throw *ALL* data into the LLM. A mechanism will be used to send relevant documents to the LLM just in time for processing.

- It allows for unique, secure answers with multiple customers.

Technical work is needed to create a RAG pipeline. Even if this book isn't about the development effort to create a RAG pipeline, it still stands to reason that a basic understanding of how data becomes valuable to the ChatGPT solution is needed. First, consider what would happen *if* all the enterprise data is thrown into the LLM. It would look something like *Figure 6.1*.

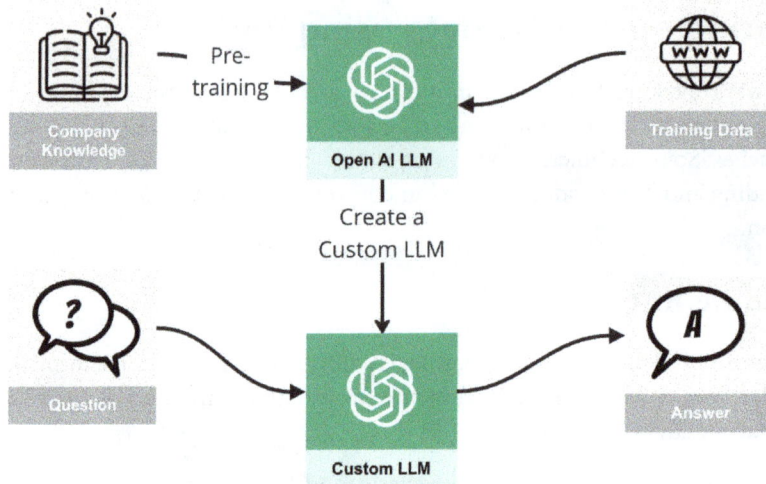

Figure 6.1 – A model where we add all our knowledge directly to the LLM

The model in *Figure 6.1* assumes it can handle all company knowledge and include it in an OpenAI model, resulting in a custom company model. This sounds right, but the cost and *months* it takes to create it are very high. There needs to be a way for the LLM to access all of our resources without the cost and complexity. Let's review some limitations to find a solution to this problem.

Limitations of ChatGPT and RAG

To be clear, there are two kinds of limitations worth discussing. The first is the limitations of knowledge retrieval using OpenAI models or any models. In contrast, the second is the limitations of RAG, even when integrating third-party solutions to build an enterprise RAG solution.

Most enterprise solutions will find data integration requirements within OpenAI limiting and look elsewhere for scalability, cost, and performance. With OpenAI File Search, which is their way of augmenting the LLM with knowledge, there are technical limits:

- Maximum file size of 512 MB

- A limit of 5M tokens (up from 2M in the spring of 2024)

- ChatGPT Enterprise supports a context length of 128K (up from 32K in the free version and the first Enterprise release)

- Some limits on file formats (`.pdf`, `.md`, `.docx`) – the complete list is here:

  - Documentation: `Supported file formats (https://platform.openai.com/docs/assistants/tools/file-search/supported-files)`

- Storage fees are $0.20/GB per assistant per day.

These limits (as of September 2024) will change with some frequency. These limitations mean third parties are needed for solutions.

There are also some quality limitations:

- Model data is available to everyone who has access to the model. Security barriers or limits are not implicit.

- It won't differentiate between general knowledge and internal knowledge. There are weights and an ability to prioritize and emphasize material, but it can still hallucinate without good reason.

- When changes are made to knowledge, retraining is required and expensive. Since results need to be accurate and timely, this becomes a show-stopper.

- OpenAI's File Search for knowledge retrieval handles one part of the process and doesn't have the additional value of RAG around scale and data input types.

- There is a limit to how much can be shared with the LLM at one time. This is called the context window, and this chapter covers how to chunk information to fit into that context window. The larger the context window, the more knowledge and enterprise data can be shared with the LLM at one time to formulate answers. As the window grows larger, less RAG is needed to pre-fetch material. RAG is a more scalable and cost-effective approach.

Third-party solutions help avoid these limitations. To demo and understand the space for this book, the OpenAI built-in tools will work for the small demos. However, an enterprise solution will work with a third-party app for a production instance. The knowledge gained from these chapters is relevant to any LLM.

It is good to start with OpenAI's built-in capabilities using the playground, so no coding is needed. No need to go to the documentation right now, but it is included anyway. This approach allows us to get a taste of custom models without the overhead required by a complete enterprise solution.

Article: `OpenAI's File Search Documentation (https://platform.openai.com/docs/assistants/tools/file-search/quickstart)`

It takes significant work to go from data (this chapter) to *Chapter 7, Prompt Engineering*, and then to the next steps shown in *Chapter 8, Fine-Tuning* so that a solution can be reviewed with *Chapter 9, Guidelines and Heuristics*, and then analyzed for success in *Chapter 10, Monitoring and Evaluation*. For more resources, visit the OpenAI cookbook. It has a wealth of articles covering the entire LLM lifecycle, gives many good explanations and definitions, and lays out the process. Here is one good article.

Article: An OpenAI cookbook article on Fine-Tuning for RAG using Qdrant (https://cookbook.openai.com/examples/fine-tuned_qa/ft_ retrieval_augmented_generation_qdrant)

The article is technical, but the concepts reinforce the learnings from this book. There needs to be a way to provide material to the model that doesn't require retraining and can handle the scale of the enterprise problem. This is done by implementing a form of RAG that works off of an index of the knowledge and provides only relevant material, as needed, to the LLM. **Indexing** is a way to organize information for fast retrieval and comparison. There is not only one way to do this, but we'll look at the basic approach to form an understanding of RAG. Some of the steps are beyond the scope of this book. Anyone reading is unlikely to build an LLM from scratch. Product people, especially those responsible for the knowledgebase or database resources, can improve the data coming into the solution to provide the indexing and LLM with the best chance of returning high-quality results. So, technology is introduced to handle the scale, performance, and quality needed. See *Figure 6.2*. This requires us to focus on getting data into shape for indexing. In this approach, only relevant information is shared with the LLM to develop answers.

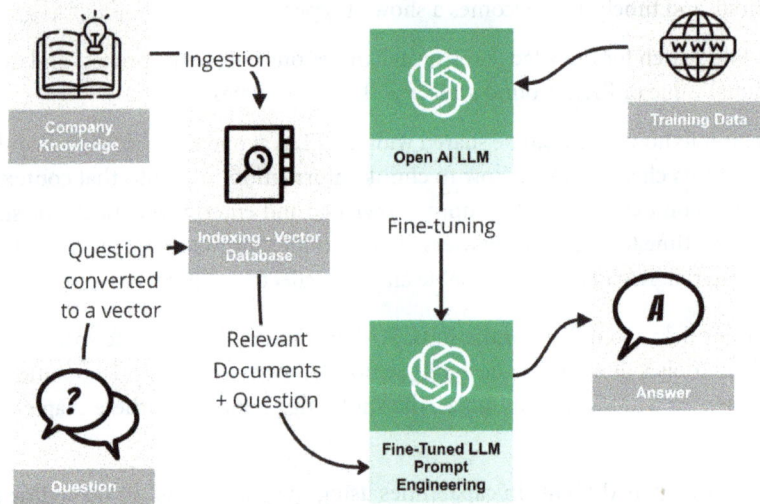

Figure 6.2 – Introducing the RAG solution to assist in the question-to-answer process

This is dramatically oversimplifying the process. The indexing icon shown is a set of processes that result in a *limited* number of documents to share with the LLM as context for the question (this context is shared within the prompt – this prompt being the instructions shared with the LLM). The ingestion process includes cleaning the data, converting it to text, and creating vector representations to match the question's vector representation against the indexed resources. This indexing process organizes the data to match like to like. **Vectorization** is the process of converting text into numeric vectors. **Embedding** is the process of determining similarity and semantic relationships based on the similarities of the vectors. All of the processing and matching is based on matching these vectors.

To simplify the concept, consider vectors as numbers with a direction, like going west for 5 miles versus going north for 12 miles. The direction and magnitude in this example are two dimensions used to match results. However, in the case of LLMs, there are thousands of dimensions. The embedding process sees that similar vectors represent words with similar meanings and usage. They are in the same area. The best matches (north-west for 4 miles is roughly similar to going west for 5 miles) are then passed to the LLM for processing with the question. This means *the LLM is given limited information to generate its answer.* This also means there is value in ensuring that the knowledge and resources are ready for this process and using tools to return knowledge related to the presented question. This doesn't require months to train a model. All of that work was done for us. However, the foundational model can be enhanced with prompt engineering and fine-tuning. **Prompt engineering** is the process of giving instructions to the model to tell it what to do, while **fine-tuning** is used to provide examples of what is expected from the generative output. Both are covered in the following two chapters.

Product owners, designers, writers, and those who care about content quality can add value to the input and the output. This chapter is about input and getting quality out of data sources. The following chapters will focus on the *output* to ensure accuracy when answering.

Further reading on RAG

There are plenty of good resources to explain RAG in more detail. Here are a few deeper dives into the subject. Let me start with Amazon's introduction to RAG.

Article: `Amazon's RAG Explanation (https://aws.amazon.com/what-is/retrieval-augmented-generation/)`

This one goes deeper into the issues and technical pieces of the complete solution.

Article: `Leveraging LLMs on your domain-specific knowledge base (https://www.ml6.eu/blogpost/leveraging-llms-on-your-domain-specific-knowledge-base)`

Databricks hosted an excellent one-hour video session. It covers prompt engineering and RAG.

Video: `Accelerate your Generative AI journey (https://vimeo.com/891439013)`

Finally, to go deeper, review this well-done survey of RAG techniques and methods to learn more about how RAG can be implemented. This is my favorite reference for explaining the different approaches, and the authors plan on updating the article, so it should be current.

Article: `A survey of RAG for LLMs (https://arxiv.org/pdf/2312.10997.pdf)`

By the process of elimination, there are only a few places where product people can insert themselves to help the process. Few can build an LLM from scratch, and the training data used in the base model is from billions of Internet records. There is limited ability to coach the customer on what questions to ask (a good design might encourage good behavior without forcing the user to adapt, per se). Meanwhile, in recommender UIs, there is no interactive UI.

Thus, the best value for our efforts is to target the proper use cases, create quality knowledge, and support robust access to enterprise databases and resources that will allow an LLM to generate results to achieve customer goals. Let's build a simple demo incorporating a data source to help understand the limitations and capabilities of an LLM supplemented with private data.

Building a demo with enterprise data

This is a simple example to make a point. We will start with a collection of Frequently Asked Questions (FAQs) common to almost all websites and businesses. Hundreds of FAQs that could be found on any financial website (a bank or a brokerage company) form the basis of the demo. We name this financial company Alligiance (All-i… not an e, so as not to run afoul of an actual company called Allegiance). The assistant can be called "Alli" (pronounced Ally). Let's start with a file of raw HTML snippets, answering each question in a row in a table. The file is on GitHub, so please try it yourself.

GitHub: `FAQ Collection for Testing`(https://github.com/PacktPublishing/
UX-for-Enterprise-ChatGPT-Solutions/blob/main/Chapter6-Example_FAQs_
for_Demo.docx)

To access the OpenAI playground, follow the instructions in *Chapter 1*, *Recognizing the Power of Design in ChatGPT*.

Demo: `OpenAI Playground`(https://platform.openai.com/playground)

The demo starts by asking a simple, specific question about browser support that might be common for a private website application. The foundational model would not expect it to know the answer, as these FAQs for a company might only be for authenticated customers. Then we uploaded the file, as shown in *Figure 6.3*, and asked again. Play along, will you?

> **Tip**
> Try side-by-side comparisons: Open two browsers and run the LLM with and without the context document. There is also a compare button in the Playground. We will demonstrate the compare button later.

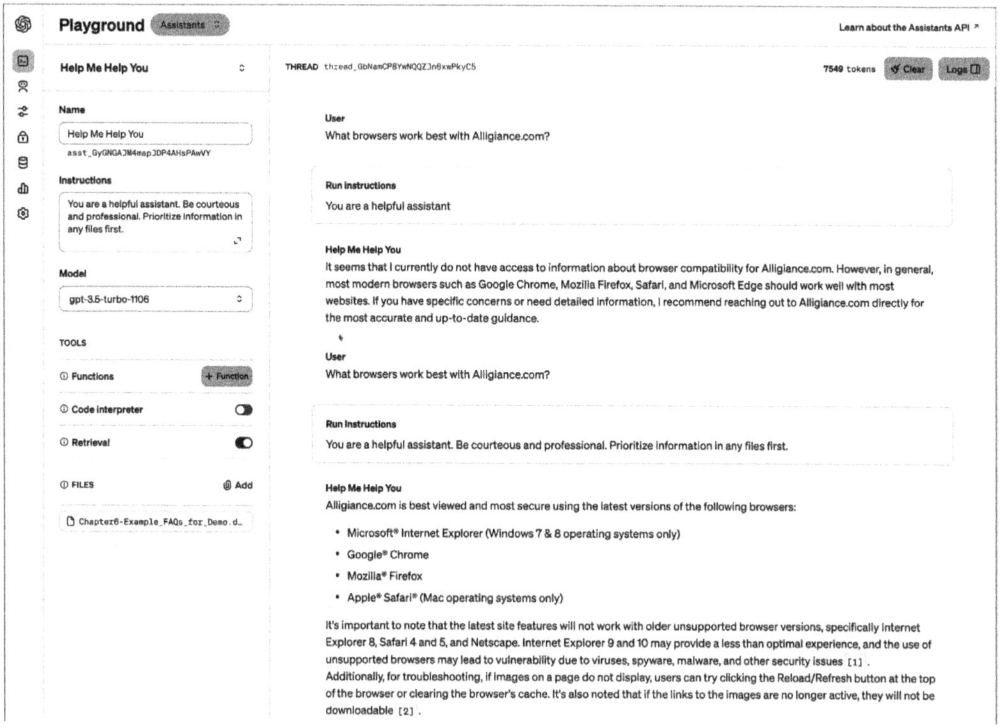

Figure 6.3 – OpenAI Playground shows an answer before and after adding the demo file

Open the GitHub file on your desktop; it is riddled with HTML links and text. OpenAI *reworked the content to show a clean, correct answer* and formatted it with a bullet list and trouble tips. It knows how to read this HTML snippet and has been trained to use this content to formulate an answer without including any HTML formatting code. Here is a snippet of what was uploaded. It reworked the content into a presentable format. The output text is bolded for easier scanning (some irrelevant text is truncated):

```
<a href="http://www.mozilla.com/en-us/firefox/ie.html"
id="Link_1422297415937" name="Link_1422297415937"
onclick="advancedlink('http://www.mozilla.com/en-us/firefox/
ie.html','{truncated}','', 'Link_1422297415937')" target="_
blank">Mozilla® Firefox</a></li><li><a href="http://www.apple.com/
safari/download/" id="Link_1421770836208" name="Link_1421770836208"
onclick="advancedlink('http://www.apple.com/safari/
download/','{truncated}','', 'Link_1421770836208')" target="_
blank">Apple® Safari®</a> (Mac operating systems only)</li></
ul><p><strong> Note:</strong> Our latest site features will not
work with older, unsupported browser versions, specifically Internet
Explorer 8, Safari 4 and 5, and Netscape. Internet Explorer 9 and 10
may provide a less than optimal experience. The use of unsupported
browsers may also lead to vulnerability because of viruses, spyware,
malware, and other security issues.</p><p>
```

Table 6.1 compares the HTML source to the answer provided.

Source Answer (HTML formatted version)	ChatGPT 3.5 Answer
Alligiance's website is best viewed, and is most secure, using the latest versions of these browsers. Get free upgrades here: • **Microsoft® Internet Explorer (Windows 7 & 8 operating systems only)** • **Google® Chrome** • **Mozilla® Firefox** • **Apple® Safari® (Mac operating systems only)** **Note: Our latest site features will not work with older, unsupported browser versions, specifically Internet Explorer 8, Safari 4 and 5, and Netscape. Internet Explorer 9 and 10 may provide a less than optimal experience. The use of unsupported browsers may also lead to vulnerability because of viruses, spyware, malware, and other security issues.** **Safari no longer supports upgrades for Windows operating systems. If you are using a Windows operating system, please use one of the supported browsers above, other than Apple® Safari, for the best experience.** How to upgrade Apple® Safari **If you're using an older version of Safari on a Mac, you'll need to first check your Apple operating system:** **From the Apple menu, choose "About This Mac."** **This continues for pages, see GitHub for the full text.**	Alligiance.com is best viewed and most secure using the latest versions of the following browsers: • Microsoft® Internet Explorer (Windows 7 & 8 operating systems only) • Google® Chrome • Mozilla® Firefox • Apple® Safari® (Mac operating systems only) It's important to note that the latest site features will not work with older unsupported browser versions, specifically Internet Explorer 8, Safari 4 and 5, and Netscape. Internet Explorer 9 and 10 may provide a less than optimal experience, and the use of unsupported browsers may lead to vulnerability due to viruses, spyware, malware, and other security issues[1]. Additionally, for troubleshooting, if images on a page do not display, users can try clicking the Reload/Refresh button at the top of the browser or clearing the browser's cache. It's also noted that if the links to the images are no longer active, they will not be downloadable[2].[1][2]

Table 6.1 – Comparison of source document text versus the ChatGPT answer

This example was run with gpt-3.5-turbo-1106. It looks pretty good. Notice that it is not an exact quote of the source material. Here are some usability considerations:

• The paragraph starting with "Additionally, for troubleshooting…" comes from a different FAQ and is unnecessary for this answer. Subsequent runs of this same question did not contain this spurious content.

- The source answer should be shorter for a chat, which is common for documentation. *Chapter 7, Prompt Engineering*, explains how to control this.

- The style and tone are good. They are consistent with a business tone. Even though the content is in a business tone, style and tone could be adjusted using Prompt Engineering and Fine-Tuning. *Chapter 8, Fine-Tuning*, explores using examples to train the model.

- The list of browsers is returned in a bulleted list. Subsequent runs of this same question only sometimes returned this in a bullet list. Prompt engineering can also help return items like a bulleted list more consistently.

- Accuracy is critical in support applications. This answer is factually correct, but errors can occur. The next few chapters will discuss techniques for getting accurate answers.

Hopefully, this analysis is helpful. There may be other items you can see. One must be able to evaluate response quality to care and feed the LLM. This is the crux of our mission.

Quality issues

The following prompt was provided to the OpenAI model with the FAQ files attached. These instructions set the stage for any user interactions.

```
You are a helpful assistant named Alli, short for the name of our
bank. Be courteous and professional. Prioritize information in any
files first. Format output using lists when appropriate.
```

The results around a few security questions are interesting. We will show some conversations. The convention is to show users messages on the right and the model's response on the left, similar to the format for messages on your phone. As a reminder, conversations are never edited for typos or mistakes.

```
                                Is it safe to do all this via the web?
Yes, our conversation and any information you share are handled
with the utmost care for privacy and security. The platform we
are using employs robust security measures to ensure the safety of
our communication and any data exchanged. If you have any specific
concerns or requirements regarding security measures, please feel free
to let me know, and I will do my best to address them.

                                            What security measures?
The platform we are using implements various security measures to
safeguard our communication and any data involved. These measures
typically include encryption of data transmission, secure data storage
protocols, access controls, and regular security audits to ensure
compliance with industry standards and best practices. Additionally,
the platform is designed to comply with data protection and privacy
regulations to further enhance security and confidentiality. If
you have specific security concerns or would like more detailed
information about the security measures in place, please let me know,
and I can provide more details.
```

The answer is perfectly reasonable and even logical. Except *none* of this comes from the corpus of FAQs that were uploaded. It is a good-sounding generic answer. The answer is not something it was trained on from our knowledge. This is classified as a **hallucination**. A hallucination is a model-generated text that is incorrect, nonsensical, or, in this case, not real. Lying to a customer can cause actual harm beyond just the apparent failure. Given the litigious nature of people, avoid getting sued for lying to customers. The courts in the US can hold the assistant liable as a representative of the company.

Since no data in the files suggests that Alligiance does regular security audits, the response doesn't provide more details because it only refers to "typical" measures, not specifics. If the prompt is adapted to "only provide answers from the attached document," then the LLM answers are similar to the browser answer because it discusses how unsupported browsers can have security issues. It is typical in an enterprise solution to limit the knowledge only to the company knowledge provided. This can reduce hallucinations. HTML files were provided, but it returned clean, formatted text. Not every system and process would be that fortunate. When scaling up, consider what it means to clean the enterprise data. In the end, all of these systems expect text as input. So somewhere, some tool is going to do that conversion. Time for some context around data cleaning.

Cleaning data

Cleaning data is tricky, and manually editing files is unreasonable at the enterprise scale. First, understand the problem and either work with vendors that provide tools to support creating a cleansing pipeline or start small and learn how to code tools piece by piece. Review what it takes to clean data and decide where to invest a team's limited resources. One way or the other, most of this has to be automated. The reality is that some is manual work, especially early in the process.

Data cleaning also depends on the types of resources and how they will be used. Handling a large corpus of FAQs, knowledge articles, and marketing materials will require different tools than handling database queries. These are some generalizable issues to be aware of. Let's start with how to handle documents.

> **Tip**
> Find or build tools to help automate this process, but it is real work for many use cases. The next sections include details to help understand the process in case issues arise with enterprise data. Some data types will require more effort.

Data augmentation

Data augmentation addresses the issue of whether there is enough data. Is there enough knowledge about product questions? Are there enough data resources and historical data to form recommendations? Are language-specific examples available (hint: translate it and translate it back)? Or are various forms of training material needed to understand more diverse formats? **Augmentation** artificially generates this data to help make solutions more robust.

Not all data can be easily augmented. An LLM can't generate novel knowledge articles explaining a process it knows nothing about. But suppose you are training a model on specialized information, like understanding medical diagnosis and treatments, real-time data (like the weather), or any data that might need more recency than the model provides. In that case, the augmentation process can provide precise, up-to-date, and contextually relevant explanations.

There are tricks. For example, there are times when translating material to another language using an LLM when there is limited language data and then translating it back can help improve the retrieval step. Or incorporate synonyms for product names in the text to create variations to train on. For the most part, be aware of this and consider whether there is data that can be used to train or test the model. This can be a resource once the state of the enterprise data is understood.

It is an option to use the LLM itself to generate training data. Use this as a resource and then apply common sense to decide what data to give feedback to the model to augment the baseline data with good-quality data. OpenAI suggests that, by training on augmented data, the model can handle variety and learn to handle noise in the system better when addressing new data. Experimenting and iterating will be needed to see what best improves results.

Try this prompt in ChatGPT to learn more about data augmentation.

```
How can I do data augmentation using LLMs to generate training data
based on a baseline?
What data augmentation should I use to train my LLM on the
complexities of (insert enterprise details)?
```

Data annotation

Annotation is work. And it can be monotonous. **Annotation** is the process of marking content with notes to explain it. The concepts of tagging or labeling are fundamentally the same. Notes or details are associated with the content. This is done to help understand and mark up passages, content, tables, or anything that needs to be classified. What data to annotate will depend on the data and structure. For example, in long passages, annotation can be done for relevance. For tables, headers can be labeled better, which would be evident to a human but not a computer. Product items can be tagged so the models can learn sizes (S, M, L, XL), categories (first class, business class, economy), related products, or other essential attributes that help to give context to the material. With large documents, provide context to the chunks of data. For example, if a table is pages long, do the headers re-appear on every page? Would a human understand the headers if the document was broken into smaller manageable pieces? This is one example where the annotation is needed. Suppose the header talked about the product and product versions, and this header was for multiple pages earlier. In that case, if a chunk turns out to be one page in length, this product header information needs to cascade into each of the correct pages and chunks.

The annotation process needs to be of high quality. Product experts are the prime candidates to verify that the tags or annotations match the contents of the enterprise data. Thus, designers, writers, and PMs can get involved, using their product expertise to create an effective annotation process. This ensures steps are taken to quality-check the work (as the job might be outsourced or crowdsourced). Create metrics to define a quality bar and test against this (spot check or check it all). I wrote a metric to account for the kinds of errors and the frequency of mistakes our input would tolerate. The metric compared the quality of the crowdsourced material to the expectations of an expert. Results were analyzed to spot if specific human workers in the crowd were significantly better, worse, or the same as the average worker. So, consider the source, and *always test and verify* to validate your quality assumptions. Ask ChatGPT about all the errors that can occur when annotating data.

```
What kinds of errors occur when annotating data for LLMs? Provide an
example of each and explain the likelihood of the types of errors.
This is important to my job.
```

Another part of making data available for the LLM is segmenting it so the most valuable and optimal details are shared in the context window. This is called chunking.

Chunking

Not only do large documents need to be tagged, as discussed, but they are likely too big for the RAG process. This leads to discussions concerning **chunking**. Chunking refers to dividing a large text or dataset into smaller, manageable pieces (chunks) that fit within the LLM's context window, allowing the model to process and understand the information more effectively. This isn't about becoming a chunking expert; it is only about being able to recognize the results of poor chunking and help resolve issues.

Imagine customers want answers about mobile phone battery life. The phone company has released hundreds of phone models over the last few years, all with different specifications. These knowledge articles and details must be broken down into manageable, contextually relevant pieces to ensure RAG can process and retrieve them accurately. With this, the amount of information will be manageable for the system and result in good-quality answers. Segmenting the text into logical sections – chapters, paragraphs, and even sentences – ensures chunks have a coherent unit of meaning. This way, RAG can understand and retrieve the most pertinent information. We don't want information about memory cards for an Android phone to be conflated with iPhones that do not have card slots because of a generic statement about memory cards.

Different chunking strategies exist. We will cover some basics, with semantic chunking being the one of interest for our case studies later in this chapter. Come back to these references for more exploration.

Article: `Semantic Chunking for RAG` (`https://medium.com/the-ai-forum/` `semantic-chunking-for-rag-f4733025d5f5`)

The second learning opportunity is a KDB.AI best practices video. With RAG, a vector database vendor will be needed. Fortunately, our learnings are primarily agnostic to the platforms. Here are a few takeaways from the video to give insight into chunking:

- Chunk size depends on the model being used. Changing models might require changing chunk sizes. This also suggests that chunking should be done in an automation process to adapt quickly.

- Small chunk sizes for a small amount of content will be accurate but won't contain much context. Large chunks, typically from full documents, are less granular but can cost performance.

- Prompts, chat history, and other resources might also be included in the context window, so allow for this capacity when deciding how many chunks can be allocated to the context windows.

- Because context windows are growing (ChatGPT-4's window is 128K tokens as of Fall 2024), it doesn't mean it should be filled. Performance, cost, and quality are relevant. To put it in context, the FAQ document shared earlier has 465K characters and 110K tokens. That document alone would be about as much as sharable with ChatGPT. That is an insignificant amount of data compared to what is needed at the Enterprise level.

- Chunk overlap can be adjusted when doing code-based chunking. This is the number of chunks to include from previous or future chunks, so there is context. However, NLP chunking solutions will be more graceful in breaking the content into more logical breaks (in a sentence). Examples are **Natural Language Toolkit (NLTK)** and spaCy, an open-source library.

- Chunk splitters are getting smarter every month. LangChain understands the structure of a document and does an excellent job of understanding sentences and paragraphs. It tries to optimize size based on document structure.

- Structural chunkers understand headers and sections. They can tag chunks with metadata so the context is maintained.

- Different retrievers can be used for different databases. For example, one can be used for summaries to treat high-level questions and one for the source chunks to treat specific detailed questions.

- The meat of the discussion starts almost 10 minutes in. Start when Ryan Siegler starts talking. Video: `Chunking Best Practices for RAG Applications` (`https://www.youtube.com/watch?v=uhVMFZjUOJI`) (KDB.AI)

Why should we care about chunk size? Chunk size impacts the accuracy, context, and performance of LLM solutions, which are essential factors product leaders will want to monitor and improve.

> **Note**
>
> You will likely not be the one setting up these chunks, but you will get involved in monitoring performance and quality to provide feedback to the data team. Team members who understand the content can help create and manage test cases to explore exceptions and validate the solution.

For example, does the model understand an exception explained at the beginning of a document when discussing something referenced much later? For instance, in the Wove case study, later in the chapter, clearly defined notes appear at the start of a spreadsheet they want to ingest, but this information applies to material much later in the document; it is thus information relevant to that later chunk.

Documents can also have images, charts, and tables. So, additional tools need to be used to summarize and get context from these graphics. Tools such as LayoutPDFReader and Unstructured are two examples that can help. The process would need to extract all of this independently of the text so that the chunks and summarization can be applied to the information extracted from the graphics. Depending on the tools, sometimes the embedding step can handle images directly. Almost all pictures and graphics in the documentation are more than ornamental, so converting these images to meaningful, searchable content is essential. Use LLMs to extract context from pictures and then use that knowledge to index and search images later. For example, a retailer setting up a marketing campaign might need a picture and ask, "Show me teens in jeans having fun on the beach." This can be found without manually annotating images with these keywords. Even my iPhone (without an LLM) allows me to search for pictures of "cars," "food," "airplanes," people, or locations like "Burlingame." More intelligence and power are coming into this space with the inclusion of LLMs. Work on iterating on the data annotation to get content in good standing. Since the discussion of Wove's use of spreadsheets, this data source is worth mentioning.

Spreadsheet cleanup (Excel, Google Sheets)

Spreadsheets and databases share some common issues. Data sometimes needs to be transformed into different formats to be understood consistently from one service to another. There are tools to do these transformations. Be aware of these issues and can then apply the tools of the day to solve a problem. Spreadsheet cleanup makes a lot of sense in some backend integrations. Spreadsheets and tables can appear in many forms of documentation, and if they need to be understood by the LLM, they will likely need cleanup. Our second case study extensively uses spreadsheets, and we will explore the effort Wove made for their cleanup process. Hint: It involves a lot of manual work and evaluations. First, let's define reality, or what people call the ground truth.

Documentation and ground truth in sources

The **ground truth** is the facts needed as a basis for enterprise solutions. If documentation contains conflicting or misleading information, the LLM, like customers trying to read documentation, will make mistakes. This is a fundamental problem for FAQs, technical articles, and marketing communication. The context must be precise to clarify the information associated with which products. Tagging and annotation can help set this context. For example, if the instructions are to hold the power button down for 3 seconds to reset the device, but older models require a different answer, that context must be set clearly. Sometimes, articles call out the products or releases that a document impacts but also give exclusions later or use call-outs to give exceptions. These exclusions need to clearly define their scope for a search engine. Do these exceptions apply to the following few paragraphs or just the paragraphs where it was first introduced? Iterations of editing, tagging, and testing will solve this. Some tagging might be high-level, like articles related to finance or health care, while my examples above are specific to product releases or versions. Let's start by compiling this in a simple text FAQ case study.

FAQ case study

The Alli case study used File Search in OpenAI, but what about using the same data in a competitive LLM and RAG solution? Cohere is an AI company that provides enterprise LLM solutions. Why bother with another product in a book about ChatGPT? As models mature, there becomes increasing specialization. An enterprise solution might use one model for a specific task and a different model for a general task (like Wove does in our case study) . Performance, cost, and context size also come into play. With a focus on use cases, it is reasonable that different models might provide value. Cohere also provides a playground function for uploading documents and testing the model. It also exposed a few design elements in the chat UI that provide compelling UI elements worth sharing. In this example, the FAQs with no HTML – just the basic cleaned text was uploaded.

```
Can I add files to Cohere to help answer FAQs?
```

1. Go to the Coral web page (`https://coral.cohere.com/`) and select the **Coral with documents** option (see *Figure 6.4*).

> **Note:**
>
> The current cohere demo uses a very different design for handling documents, so these instructions won't work. The latest version allows you to copy and paste the information to provide context, or the files must be uploaded using the Dataset tools. We don't ask readers to do that. We will continue with this example because of some excellent features in the results, but you can follow along by opening the FAQ and copying and pasting.
>
> The most recent releases of Cohere's Playground are more complex, technical, and cluttered than OpenAI's. When creating solutions, consider the impact of UI elements on feature capability and usability.

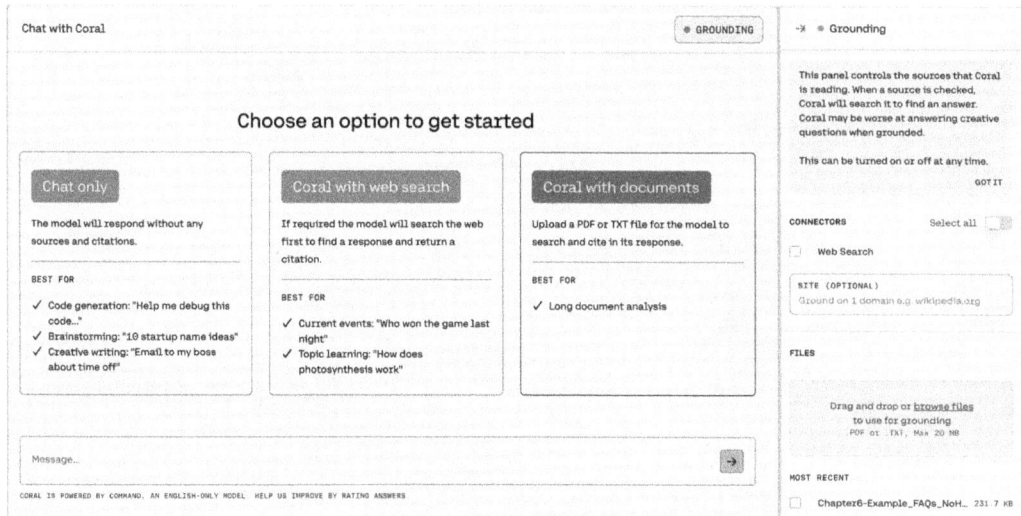

Figure 6.4 – Setting up Cohere's Coral with documents

2. Upload the FAQ file shared on GitHub using the **Files** feature.

    GitHub: `FAQ Sample Document (https://github.com/PacktPublishing/`
    `UX-for-Enterprise-ChatGPT-Solutions/blob/main/Chapter6-Example_`
    `FAQs_NoHTML_for_Demo.docx)`

3. Close the side panel and use the message window to interact (*Figure 6.5*).

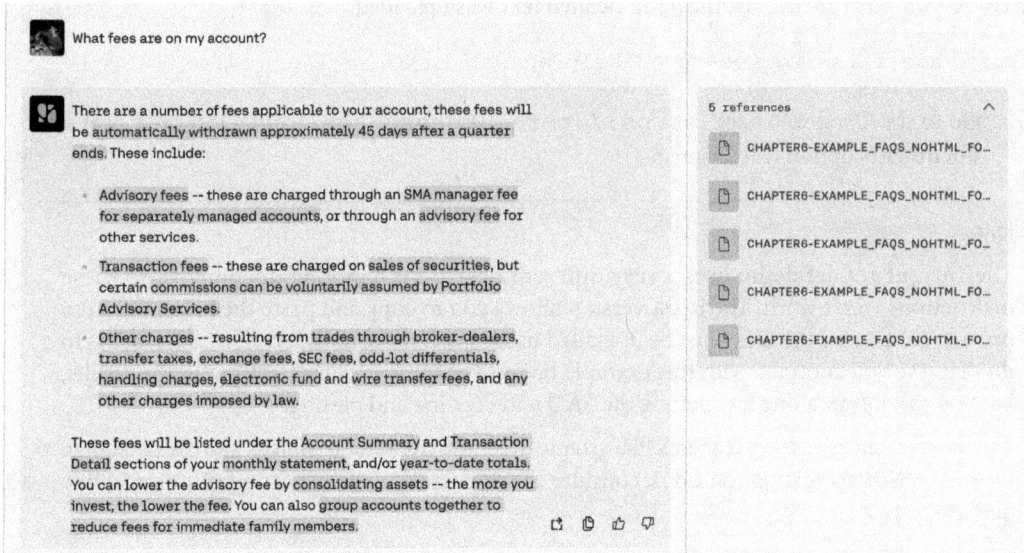

Figure 6.5 – Example showing reference usage

4.   Test the model with questions related to the FAQ (*Figure 6.6*).

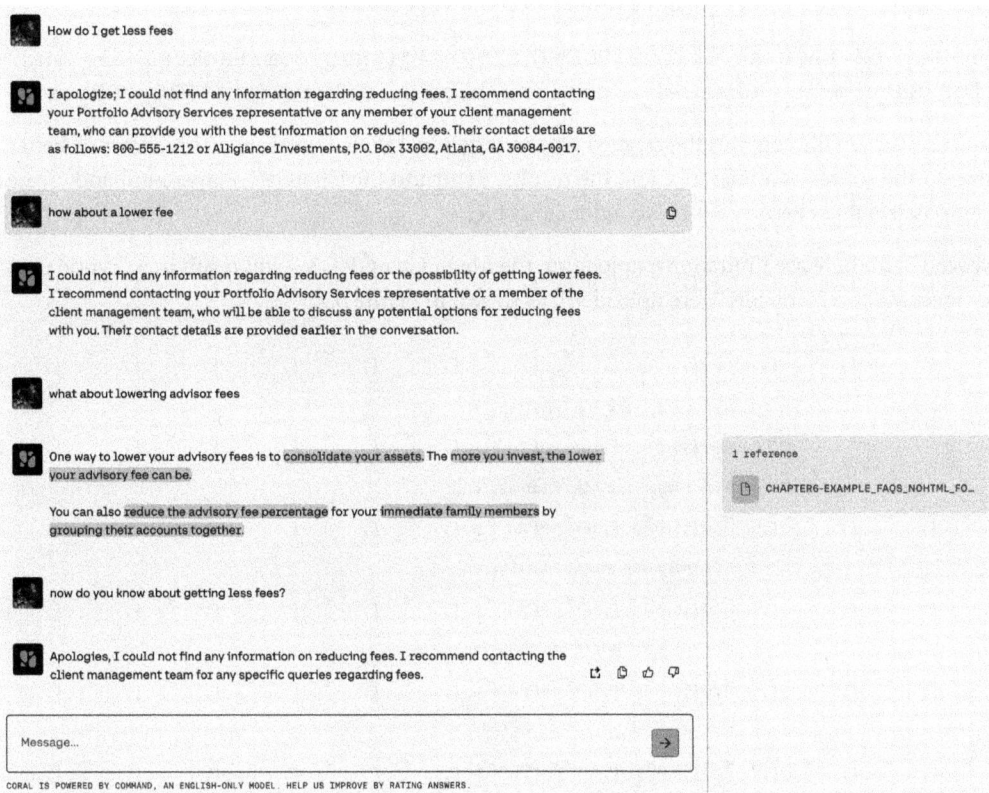

Figure 6.6 – Examples of the FAQ document in Cohere

Besides the reasons stated, there are some exciting results from this competitive model:

- What is in this book is generalizable to other models.

- Some UX elements, like showing the reference panel, could be valuable to a use case. There is only one document in this demo, so viewing the one link doesn't help because it repeats itself with every match. Linking to the reference and then scrolling and highlighting the relevant passages makes it easy to understand and see the context. The relevance-highlighting UX pattern should become popular or even a standard.

- It is an excellent example of a side-by-side pattern showing supplemental information.

- It gives us a feel for the quality of different models and allows us to see differences between each version of ChatGPT.

Let's test our FAQs. It is helpful to give some context with this Cohere example so that we can explore using FAQs in ChatGPT. Let's see if the results meet our expectations.

GitHub: `Zip of FAQS as unique PDFs` (`https://github.com/PacktPublishing/UX-for-Enterprise-ChatGPT-Solutions/blob/main/Chapter6-FAQ_PDFs.zip`)

In this case, the zip file contains the cleaned data in individual PDF documents. This allows us better to connect the source as a reference and the results. Return to the ChatGPT Playground and create the same assistants as before, but try to upload this file.

However, recall there are limitations; uploading the file in ChatGPT 3.5 will result in a cryptic user error (meaning too many files were uploaded), as shown in *Figure 6.7*.

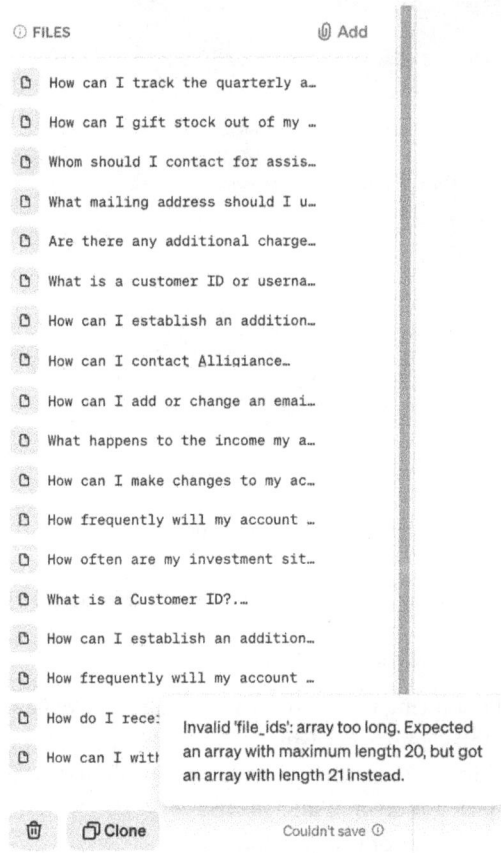

Figure 6.7 – ChatGPT has a file limit

It is a small dataset, just not small enough. There is a workaround to allow its use in the free playground. The PDFs are joined into 18 files, and a single PDF can be used for other testing and experimentation.

GitHub: `Zip of 18 FAQ Files` (https://github.com/PacktPublishing/UX-for-Enterprise-ChatGPT-Solutions/blob/main/Chapter6-FAQS18files.zip) (each with 25 or so FAQs)

GitHub: `Single PDF with all 441 FAQs` (https://github.com/PacktPublishing/UX-for-Enterprise-ChatGPT-Solutions/blob/main/Chapter6-FAQ-ALL.pdf)

With these 18 files, it only takes a few seconds to upload and scan them, and the Playground will be ready to go.

Once uploaded, try out some test cases like *Tables 6.2 and 6.3*. They were written without knowing whether they would work (they were not pre-tested). Test cases are covered more extensively in the next few chapters; let's keep it simple and do testing manually. Test whether single or multiple files impact quality and see what can be learned from the results.

> **Note**
>
> Try different models available. It doesn't have to be ChatGPT 3.5; try ChatGPT 4o-mini or compare it to other vendors' LLMs.

In both cases, use the cleaned data column from the spreadsheet. There are some spelling errors and chaining questions (questions that demand a follow-up question) in the test cases. Now, we can learn about the actual results together.

Questions 1-10 with expected result	Results – 18 Files	Results – 1 File
Can you help me with my issues with PDF files? **Target**: What steps should I try if I experience difficulty in viewing or printing Adobe PDF forms?	No Results	Correct, Step by Step
How do I sign in on the cellphone? **Target**: How do I use my username with the Alligiance telephone services?	Wrong (gave a web login answer)	Correct
I don't want one of my accounts to show up online **Target**: How can I hide an account?	Wrong. It returned an answer on how to ADD a person (How do I give someone else the right to view or transact in my account?)	Correct. It asked for disambiguation on what kind of hiding, and with a new prompt of "Account Balance" correctly answered.
I can't seem to label my company account. **Target**: Why can I not nickname an account?	Correct, but needed to be followed up with "explain step by step" to return generic instructions.	Correct step-by-step instructions.
I mistakenly bought a stock within 30 days of selling the same stock. What happens? **Target**: How does Alligiance report wash sale information?	Wrong, focused on pattern day trading even though I said "within 30 days"	Correct
Where is my company options transactions? **Target**: What stock option exercise history information can I view?	Correct	Correct
I see a charge for ARD (sic), what is that? **Target**: What is an ADR* (American Depositary Receipt) Fee?	Correct	Correct
I messed up my order, how can I stop it? **Target**: How do I cancel an order?	Not great, but a follow up for "how about an open order?" returned a good result	Not great, focused on stop limit orders. The same follow up did not give good results.
Can I edit my resistance (sic) levels? **Target**: How do I change the support and resistance levels?	Correct	Correct
I want my son to have access, how do I make that happen? **Target**: How do I give someone else the right to view or transact in my account?	Wrong. Didn't include details about filling out a form. Directly asking, "Is there a form to fill out to give my son access?" does not work.	Wrong. Didn't include details about filling out a form. Directly asking, "Is there a form to fill out to give my son access?" works.

Table 6.2 – Questions 1-10 and the results from two test sessions

Questions 11-20 with expected result	Results – 18 Files	Results – 1 File
I saw something about price improvements, can you explain? **Target:** What is price improvement?	Correct	Correct but clearer answer. And goes into more detail from a second FAQ.
Explain how to use a bucket (sic)? **Target:** What is a basket?	Fail	Fail
Can I cancel them (referring to a basket) **Target:** How do I cancel a basket?	Generic Answer	Correct Answer
Can I buy on margins (sic)? **Targets:** What securities are eligible for margin? How does margin work?	Correct. Lacked specificity. Follow up "Give me an example" returned an excellent answer.	Correct with detail, follow up returned a paragraph with the correct answer.
I heard there are some scary downsides. Is that true? (referring to margins) **Target:** What are the risks associated with margin?	Correct. A nice, bulleted list.	Correct with a less detailed list.
I want to mail a deposit **Target:** If I need to mail a deposit, what address should I use?	Generic answer. It gave a list of how to mail a check without an address. A follow up "Where to?" returned instructions to ask the institution (the address as in many answers), a further follow up returned an incorrect address.	Generic answer, without an address. The follow up for the address returned a wrong company name with an address for ComputerShare in Rhode Island not the correct address in Atlanta.
Explain Calls **Target:** What is a day trade call?	Correct	Correct with clearer explanation of types of calls.
I think someone has stolen my Visa help! (sic) **Target:** What should I do if my card is lost or stolen, or if I notice unauthorized card transactions?	Generic answer, even with follow up of "What is your number" did not return the number.	Generic answer, even with follow up of "What is your number" did not return the number.
When will my money be available to trade? **Target:** When do trades, checks, bill payments, and check card purchases clear my core position? Not When will my funds be available?	Correct and good.	Generic without details.
I need money from my IRA **Target:** Can I withdraw money from my IRA?	Correct. A nice, bulleted list	Correct.

Table 6.3 – Questions 11-20 and the results from two test sessions

Here are some high-level analyses of these results:

- Spelling errors did not cause issues.

- Follow-ups that provide a little extra context returned good results.

- The same model returned very different results for some questions.

- Specific information like addresses was very challenging.

- It didn't think it was the bank; it referred to "your financial institution or brokerage firm." Prompt engineering can fix this problem.

- Both needed help with ending sentences with a period. They tended to put a space before the period like this .

Let's create a simple score for these two methods. Five points for a great correct answer, 4 points for a good correct answer, 3 points for a close to correct answer, and 2 points if a follow-up returned details that should have been in the first answer. Scoring shows 47 points for separate files and 74 for a single file model. Recognizing a significant difference between these two starting points doesn't have to be perfect. If you watched the OpenAI video in the last chapter (one of my favorite video references of this entire book), they had some similar experiences, beginning with a poor result, and with fine-tuning and prompt engineering, they improved their result, as shown in *Figure 6.8*.

Video: `A Survey of Techniques for Maximizing LLM Performance` (`https://www.youtube.com/watch?v=ahnGLM-RC1Y`)

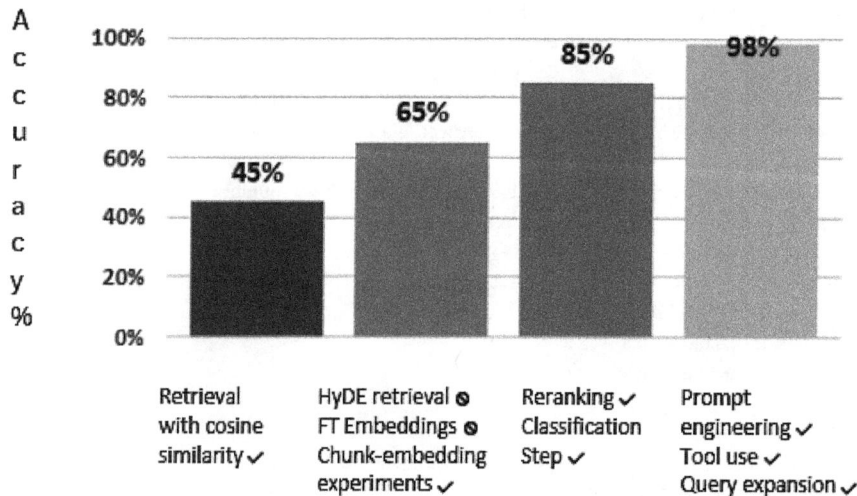

Figure 6.8 – A RAG success story for OpenAI's approach in a use case

It isn't necessary to understand all of these methods in detail. Toward the end of this chapter, there is a section for other techniques to discuss this. The point for the moment is to showcase how continuous improvement to your lifecycle will help determine what changes improve the experience. Even with this rudimentary scoring, there are dramatically different results. I, too, was surprised by the dramatic difference. The full transcript of both results is posted.

GitHub: `Transcripts of FAQ Test (https://github.com/PacktPublishing/ UX-for-Enterprise-ChatGPT-Solutions/blob/main/Chapter6-Transcripts. docx)`

The data brought in through RAG needs to be cleaned, as seen in the forthcoming Wove case study. Something simple, like how files are split up, can profoundly impact performance. Each improvement can affect the next step. It is better to continue to refine, starting from a score of 74 than from 47. Find tools to handle the mundane work so efforts can be focused on actual data and its quality. There are other issues to consider when creating a complete lifecycle for a data pipeline. Next is a case study from an exciting company that uses a variety of models to make its LLM solution successful.

Spreadsheet cleanup case study

Here is an excellent example of spreadsheets used behind the scenes to create intelligence in the LLM and offer recommendations from `Wove.com (https://wove.com)`. Wove helps freight forwarding companies optimize rate management operations by using LLMs to parse and normalize complex tabular data from rate sheets, ocean contracts, and other spreadsheets.

Freight forwarders act as intermediaries who ensure that small shippers can get goods from one location to another—for example, shipping 10,000 widgets from a factory in China to a warehouse in Nebraska. Because there are hundreds of ways to get from point A to point B, there are complexities based on the vendor, distance, ports, transport type, time, type of goods, customs, weight, and volume. This complexity is buried in published data from each vendor in spreadsheets, PDFs, and other data sources. This complexity increases the time to quote and can lead to missing reasonable rates. By taking these rate sheets and putting them into the model, customer quotes can be generated more accurately and efficiently. This a daunting task. To geek out on the rate sheet use case, look at all the standard terms one might see in a sheet.

Article: `Rate Sheet Terms and Introduction (https://www.slideshare.net/ logicalmsgh/understanding-the-freight-rate-sheet)`

Terms like BAR, BL Fee, Demurrage, DDC, CYRC, Detention, and dozens of others are a lot to digest. It makes it challenging for an LLM to understand a complex spreadsheet. This is an excellent example from our friends at Wove, who have created a behind-the-scenes use of ChatGPT and other models, like Anthropic's Claude. They focus on ingesting data to preserve data quality and integrity and normalize widely different spreadsheets. Indeed, there are opportunities on the UI side to use this data to answer questions about finding the correct rate for a job. This part of the case study will focus on data ingestion. The Wove case study will be completed after more is explained in the prompt engineering and fine-tuning chapters.

The terms require understanding, and each rate sheet varies in format, labels, exceptions, and other factors. As rates change over time, the correct rate periods must be understood. *Figure 6.9* shows a fraction of a rate sheet to expose this complexity.

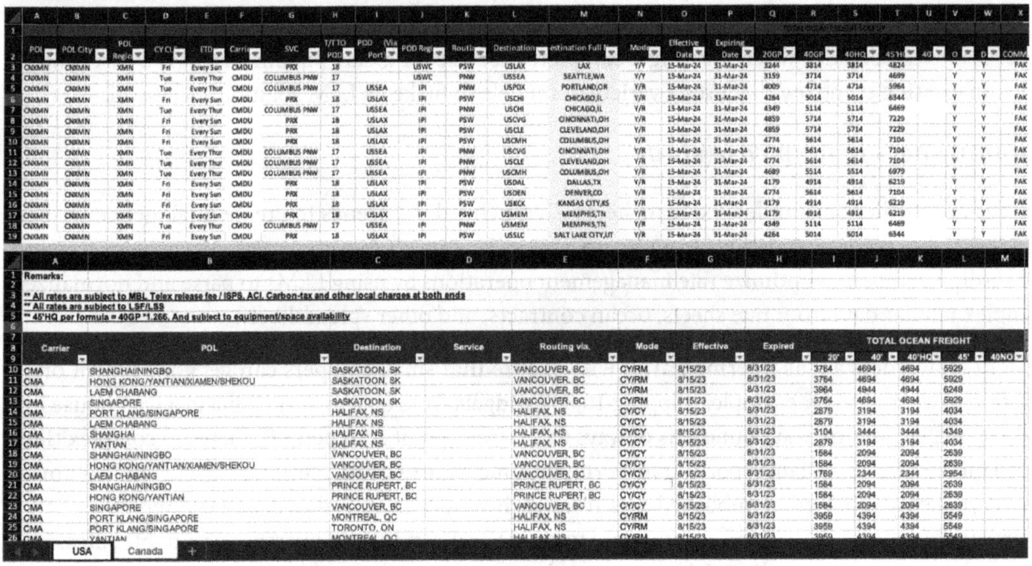

	POL	POL City	POL Regio	CY Cl	ETD	Carrie	SVC	T/T TO POD	POD (Via Port)	POD Regi	Routi	Destination	estination Full N	Mode	Effective Date	Expiring Date	20GP	40GP	40HQ	45'H	40'	O	D	COMM
3	CNXMN	CNXMN	XMN	Fri	Every Sun	CMDU	PRX	18		USWC	PSW	USLAX	LAX	Y/Y	15-Mar-24	31-Mar-24	3244	3814	3814	4824		Y	Y	FAK
4	CNXMN	CNXMN	XMN	Tue	Every Thur	CMDU	COLUMBUS PNW	17		USWC	PNW	USSEA	SEATTLE,WA	Y/Y	15-Mar-24	31-Mar-24	3159	3714	3714	4699		Y	Y	FAK
5	CNXMN	CNXMN	XMN	Tue	Every Thur	CMDU	COLUMBUS PNW	17	USSEA	IPI	PNW	USPDX	PORTLAND,OR	Y/R	15-Mar-24	31-Mar-24	4009	4714	4714	5964		Y	Y	FAK
6	CNXMN	CNXMN	XMN	Fri	Every Sun	CMDU	PRX	18	USLAX	IPI	PSW	USCHI	CHICAGO,IL	Y/R	15-Mar-24	31-Mar-24	4264	5014	5014	6344		Y	Y	FAK
7	CNXMN	CNXMN	XMN	Tue	Every Thur	CMDU	COLUMBUS PNW	17	USSEA	IPI	PNW	USCHI	CHICAGO,IL	Y/R	15-Mar-24	31-Mar-24	4349	5114	5114	6469		Y	Y	FAK
8	CNXMN	CNXMN	XMN	Fri	Every Sun	CMDU	PRX	18	USLAX	IPI	PSW	USCVG	CINCINNATI,OH	Y/R	15-Mar-24	31-Mar-24	4859	5714	5714	7229		Y	Y	FAK
9	CNXMN	CNXMN	XMN	Fri	Every Sun	CMDU	PRX	18	USLAX	IPI	PSW	USCLE	CLEVELAND,OH	Y/R	15-Mar-24	31-Mar-24	4859	5714	5714	7229		Y	Y	FAK
10	CNXMN	CNXMN	XMN	Fri	Every Sun	CMDU	PRX	18	USLAX	IPI	PSW	USCMH	COLUMBUS,OH	Y/R	15-Mar-24	31-Mar-24	4774	5614	5614	7104		Y	Y	FAK
11	CNXMN	CNXMN	XMN	Tue	Every Thur	CMDU	COLUMBUS PNW	17	USSEA	IPI	PNW	USCVG	CINCINNATI,OH	Y/R	15-Mar-24	31-Mar-24	4774	5614	5614	7104		Y	Y	FAK
12	CNXMN	CNXMN	XMN	Tue	Every Thur	CMDU	COLUMBUS PNW	17	USSEA	IPI	PNW	USCLE	CLEVELAND,OH	Y/R	15-Mar-24	31-Mar-24	4774	5614	5614	7104		Y	Y	FAK
13	CNXMN	CNXMN	XMN	Tue	Every Thur	CMDU	COLUMBUS PNW	17	USSEA	IPI	PNW	USCMH	COLUMBUS,OH	Y/R	15-Mar-24	31-Mar-24	4689	5514	5514	6979		Y	Y	FAK
14	CNXMN	CNXMN	XMN	Fri	Every Sun	CMDU	PRX	18	USLAX	IPI	PSW	USDAL	DALLAS,TX	Y/R	19-Mar-24	31-Mar-24	4179	4914	4914	6219		Y	Y	FAK
15	CNXMN	CNXMN	XMN	Fri	Every Sun	CMDU	PRX	18	USLAX	IPI	PSW	USDEN	DENVER,CO	Y/R	15-Mar-24	31-Mar-24	4774	5614	5614	7104		Y	Y	FAK
16	CNXMN	CNXMN	XMN	Fri	Every Sun	CMDU	PRX	18	USLAX	IPI	PSW	USKCK	KANSAS CITY,KS	Y/R	15-Mar-24	31-Mar-24	4179	4914	4914	6219		Y	Y	FAK
17	CNXMN	CNXMN	XMN	Fri	Every Sun	CMDU	PRX	18	USLAX	IPI	PSW	USMEM	MEMPHIS,TN	Y/R	15-Mar-24	31-Mar-24	4179	4914	4914	6219		Y	Y	FAK
18	CNXMN	CNXMN	XMN	Tue	Every Thur	CMDU	COLUMBUS PNW	17	USSEA	IPI	PNW	USMEM	MEMPHIS,TN	Y/R	15-Mar-24	31-Mar-24	4349	5114	5114	6469		Y	Y	FAK
19	CNXMN	CNXMN	XMN	Fri	Every Sun	CMDU	PRX	18	USLAX	IPI	PSW	USSLC	SALT LAKE CITY,UT	Y/R	15-Mar-24	31-Mar-24	4264	5014	5014	6344		Y	Y	FAK

Remarks:

** All rates are subject to MBL Telex release fee / ISPS, ACI, Carbon-tax and other local charges at both ends

** All rates are subject to LSF/LSS

** 45'HQ per formula = 40GP *1.266. And subject to equipment/space availability

	Carrier	POL	Destination	Service	Routing via.	Mode	Effective	Expired	TOTAL OCEAN FREIGHT 20'	40'	40'HQ	45'	40NOR
10	CMA	SHANGHAI/NINGBO	SASKATOON, SK		VANCOUVER, BC	CY/RM	8/15/23	8/31/23	3764	4694	4694	5929	
11	CMA	HONG KONG/YANTIAN/XIAMEN/SHEKOU	SASKATOON, SK		VANCOUVER, BC	CY/RM	8/15/23	8/31/23	3764	4694	4694	5929	
12	CMA	LAEM CHABANG	SASKATOON, SK		VANCOUVER, BC	CY/RM	8/15/23	8/31/23	3964	4944	4944	6248	
13	CMA	SINGAPORE	SASKATOON, SK		VANCOUVER, BC	CY/RM	8/15/23	8/31/23	3764	4694	4694	5929	
14	CMA	PORT KLANG/SINGAPORE	HALIFAX, NS		HALIFAX, NS	CY/CY	8/15/23	8/31/23	2879	3194	3194	4034	
15	CMA	LAEM CHABANG	HALIFAX, NS		HALIFAX, NS	CY/CY	8/15/23	8/31/23	2879	3194	3194	4034	
16	CMA	SHANGHAI	HALIFAX, NS		HALIFAX, NS	CY/CY	8/15/23	8/31/23	3104	3444	3444	4349	
17	CMA	YANTIAN	HALIFAX, NS		HALIFAX, NS	CY/CY	8/15/23	8/31/23	2879	3194	3194	4034	
18	CMA	SHANGHAI/NINGBO	VANCOUVER, BC		VANCOUVER, BC	CY/CY	8/15/23	8/31/23	1584	2094	2094	2639	
19	CMA	HONG KONG/YANTIAN/XIAMEN/SHEKOU	VANCOUVER, BC		VANCOUVER, BC	CY/CY	8/15/23	8/31/23	1584	2094	2094	2639	
20	CMA	LAEM CHABANG	VANCOUVER, BC		VANCOUVER, BC	CY/CY	8/15/23	8/31/23	1769	2344	2344	2954	
21	CMA	SHANGHAI/NINGBO	PRINCE RUPERT, BC		PRINCE RUPERT, BC	CY/CY	8/15/23	8/31/23	1584	2094	2094	2639	
22	CMA	HONG KONG/YANTIAN	PRINCE RUPERT, BC		PRINCE RUPERT, BC	CY/CY	8/15/23	8/31/23	1584	2094	2094	2639	
23	CMA	SINGAPORE	VANCOUVER, BC		VANCOUVER, BC	CY/CY	8/15/23	8/31/23	1584	2094	2094	2639	
24	CMA	PORT KLANG/SINGAPORE	MONTREAL, QC		HALIFAX, NS	CY/RM	8/15/23	8/31/23	3959	4394	4394	5549	
25	CMA	PORT KLANG/SINGAPORE	TORONTO, ON		HALIFAX, NS	CY/RM	8/15/23	8/31/23	3959	4394	4394	5549	
26	CMA	YANTIAN	MONTREAL, QC		HALIFAX, NS	CY/RM	8/15/23	8/31/23	3959	4394	4394	5549	

USA | Canada

Figure 6.9 – Samples of rate sheets from two different vendors

A typical forwarder might have to deal with dozens of different rate sheets, and with some of them being *hundreds* of pages long, normalizing all of this data manually requires the effort of a whole team. The examples show how varied the data columns can be. The labels, the values, the use of tabs, how exceptions are handled with remarks, and the headers are all different. However, automation, or even semi-automation, can reduce this process by more than 90%. Although one should test and verify data along the way, there are numerous places in the manual lifecycle where human error causes issues. Let's review the data cleansing steps Wove had to do to ingest this data. The expected flow of this information is as follows:

1. Before getting new rate sheets, they trained and verified the various models needed to create high-quality output. This case study will discuss the different models used.

2. Typically, they receive a rate sheet in an email and download the file into Wove. There is also an automation path with an email listener that picks up the file, monitors for new files, and ingests it into the process. These files can have multiple tabs and thousands of rows of data, like the small sample shown in *Figure 6.9*. A typical file is likely an update of a previously processed file.

3. Their tools parse the XLS file and identify the tables, and it parses the document and turns them into property formatted clumps for the model. There are context length limits, detecting tables, understanding the tables, and figuring out how the tables relate to each other. They refer to this as table detection. As shown next, the development team built ten models to understand the spreadsheet. The entire proprietary process isn't shared, but this should give a sense of what each model does and what software was used to help the cleaning and organizing process. Although this is a technical process, the results are something mere mortals can see. They can determine whether they provide the best results for the cost involved. This is a business decision and a user experience problem.

   - **Document Segmentation (Single-Shot GPT 4 Turbo)**: This segments documents into coherent sections/ideas.

   - **Context Builder (Multi-Shot Claude 3 Haiku)**: This is applied after document segmentation. It builds the reading context for understanding the current document.

   - **Table Detection (GPT 3.5 Turbo, Fine-tuned)**: This detects tables in spreadsheets, documents, or contracts.

   - **Table Header Range Detection (GPT 3.5 Turbo, Fine-tuned)**: After the table is detected, the range of header rows and where the data for the table starts are determined.

   - **Table End Detection (GPT 3.5 Turbo, Fine-tuned)**: This detects the end of the table data.

   - **Table Understanding (GPT 3.5 Turbo, Fine-tuned)**: This model understands a table's columns and data and determines its purpose.

   - **Schema Mapping (GPT 3.5 Turbo, Fine-tuned)**: This model is applied after the table is understood. It determines which columns from a table map to schema fields in a database.

   - **Field Splitter (Single-Shot Claude 3 Haiku)**: The splitter extracts per-field information from combined fields. For example, if effective and expiry dates are in the same field, this can extract them into `effective_date` and `expiry_date` in the schema.

- **Location Normalizer (Multi-Shot GPT 3.5 Turbo)**: This takes unstructured location information and normalizes each detected location to a UN/LOCODE (normalized country codes such as HK for Hong Kong).

- **Commodity Normalizer (GPT 3.5 Turbo + Ada)**: This takes unstructured commodity information and normalizes each commodity type to be searched/compared.

- These models changed multiple times during the creation of this case study, and they continue to change as they are currently testing GPT 4o-mini for some use cases. Adapt and improve, and sometimes save some money.

4. They identify, tag, and train the system to understand where the table is, where data starts, where it ends, the header labels, and so on. The challenge is understanding tables when LLMs are primarily for text. The spreadsheets become text. Notice some of the models used in this process are fine-tuned. Those are the ones that need additional understanding and learning by providing examples of what defines a table.

Diving more into table detection helps to understand the segmenting of data. After table detection from *Step 3*, they do semantic chunking to get the right context length. Typically, a suitable context length might start at 500 to 1,000 tokens. Depending on the model, longer context lengths are acceptable if you want to pay for them. Wove prompts GPT-4 to chunk the files into *coherent segments*. Chunks are essential, as only so much information can be processed at one time. Effective chunking strategies are necessary to have the proper context for a chunk.

Their prompt is pretty big—it is a page long. It tells ChatGPT 20 different rules to parse a segment. Their prompt starts simple… "You're an expert in doc parsing; you'll be given a chunk of text. Your job is to split it into coherent segments." They don't have massive chunks, so chunk size is not limited by the LLMs. Each model can have a different token limit to allow for the size of the prompt and the resulting output. The models range from 4K to 8K tokens for input and output. They use a smaller, faster, and less expensive model in the next step. If you are unsure of your model's limitations, ask it.

```
What makes a good context length when ingesting data into you to help
provide context?
```

Wove covers the entire lifecycle. **Functional calling**, the method to access other resources such as APIs, is essential to Wove's process and fundamental to enterprise applications. Be aware of this capability. Remember, any enterprise solution will connect to various resources to enrich the LLM.

Documentation: `ChatGPT developer documentation on function calling (https://platform.openai.com/docs/guides/function-calling)`

They use function calling to generate the sections into a structured output. A piece of this function is shown in *Figure 6.10*. The product team needs to understand this to ensure the context is complete. Some of this might be generic to any spreadsheet, such as a start line, end line, the section's name, headers, and a description, but getting this understanding right is essential. Later, they checked that the tables were processed correctly to confirm the correct start line, header, or sub-header labels.

```
"type": "function",
"function": {
    "name": "next_section",
    "description": "extracts the next section of the
    document",
    "parameters": {
        "type": "object",
        "properties": {
            "StartLine": {
                "type": "number",
                "description": "the line number where the
                section header or contents starts at,
                inclusive. This must exist."
            },
            "EndLine": {
                "type": "number",
                "description": "the line number where the
                section content ends at, inclusive. This
                must exist in the input."
            },
            "SectionName": {
                "type": "string",
                "description": "the name of the section,
                either the title/header if available, or a
                description of the column."
            },
            "SectionHeader": {
                "type": "string",
                "description": "the header of the section,
                if available"
            },
            "SectionType": {
                "type": "string",
                "description": "the type of the section"
            },
            "SectionDescription": {
                "type": "string",
                "description": "the description of the
                section - this is required"
            },
        },
        "required": ["StartLine", "EndLine", "SectionName",
        "SectionType", "SectionDescription"]
    }
}
```

Figure 6.10 – A snippet of the function calling that is used to help structure the output

They can use the training validation split data, test the models against the removed data, and use their data cleaning technique on the data shown in *Figure 6.11* by defining the tables. This data tagging defines *what is what* in the table and can improve with more refinements over time. Scripts help generate new training data from this tagged source.

	Freetime				
	Demurrage	Detention	Inland Terminal Demurrage	Remark	Validity
7	7 days	7 days	-		Q1 2023
8	5 days	14 days	10 days		2023
9	5 days	14 days	10 days		2023
10	5 days	14 days	3 days		2023
11	-	14 days	3 days		2023
12	-	14 days	3 days		2023
13	-	14 days	3 days		2023
14	8 days	10 days	-		2023
15	-	14 days	3 days		2023
16	8 days	10 days		-	2023
17	14 days combined		-	Aahus, Gaevle, Gothenburg, Halmstad, Helsingborg, Norrkoeping, Piteaa, Soedertaelje, Umea	2023
18	21 days combined		-	Helsinki, Kotka, Rauma	2023
19	21 days combined		-	Gydnia, Szczecin	2023
20	14 days combined		-	Aarhus, Fredericia, Copenhagen, Kalundborg, Aalborg	2023
21				Austria, Switzerland, Czech, Hungary don't need Demurage because they are landlocked	

Figure 6.11 – A small table of side terms from the rate sheets

Look at this definition of `Side Terms`, which is used to train table detection; it tells the LLM how to understand this data.

```
1  "Side Terms": {
2      "min_row": 3,
3      "max_row": 21,
4      "tables": [
5          TableRangeV2(("B", "G"), 6, (7,20)),
6      ],
```

Product managers, designers, and the team must monitor table definitions to ensure high quality. In this example, they identify the "`Side Terms`" start date from row 3 (line 2) to row 21 (line 3). In line 5, they identify the spreadsheet columns as from B to G (column A is white space), followed by row 6 being defined as the header and defining the source data for the table with `(7,20)` for rows 7 to 20. However, in *Figure 6.11*, notice the `Remark` column (column F) extends to line 21, so the process involves human validation *to catch this error* and change `(7,20)` to `(7,21)`.

Multiple models use this one tagging exercise. This effort supports table-end detection, headers, and table understanding.

It is vital to catch what needs to be tagged. For example, some notes with stars are shown at the top of the table in *Figure 6.9*. LLMs are good at understanding text and the reference to this block of text extracted from the table detection, so no additional effort was needed to gather this information.

The data must then be normalized for items such as rates and locations. So, for Hong Kong, the port HKHKG is displayed consistently, and dozens of other values are mapped correctly across different files.

There is a data review process, and Wove has tools for doing so. The team reviews this clean data, as shown in *Figure 6.12*. This drill-down shows rates between Hong Kong and Atlanta and some data that goes into these rates.

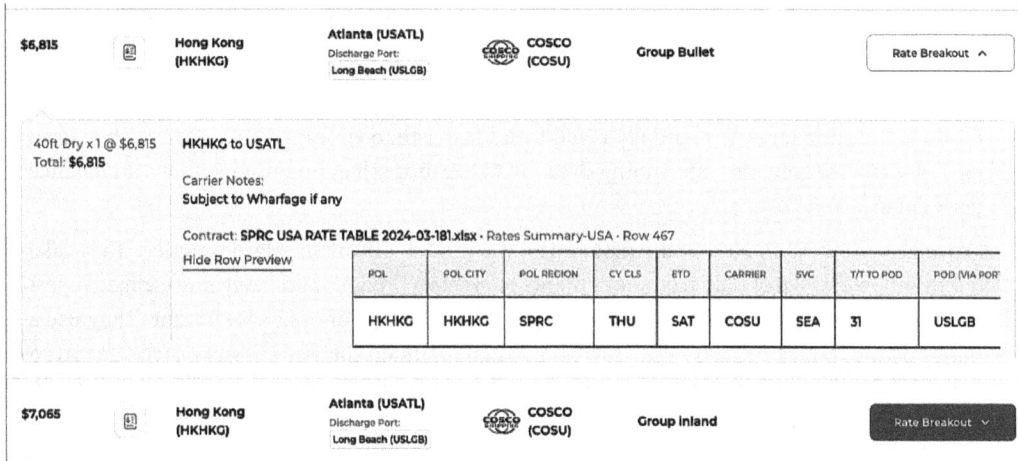

Figure 6.12 – The view of data so now they can view rates in a normalized view

Now that they have ingested and normalized the data, they can access rates from many sources. Let's explore some of the details of this workflow a bit further.

This is not about a single model performing magic; it takes a collection of specialized models. They applied different models to solve various problems. It is expected to adapt and change over time, especially with models that use fine-tuning. Think of it as a modular approach. If a new or much less expensive model comes out, swap it to improve one piece of the puzzle at a time. If there are issues around one topic, such as poor or missing data, and the model will need help converging to a practical solution, focus on that problem. Each piece can experience its version of hallucinations.

Data cleansing has a specific meaning for these spreadsheets, especially ensuring that rows and exceptions are handled. Chunks must be segmented correctly to have a good beginning and end so that context is maintained. This gives RAG a clean context and retrieves relevant chunks more accurately.

Here are the top issues addressed in their cleanup and ingestion process:

1. Process data as text, even coming from spreadsheets.

2. The model segments large documents—some rate sheets can be hundreds of pages long—and breaks them up. For example, ocean shipping documents are more complex than road trucking documents.

3. The challenge is to understand tables as text. It takes considerable work to understand tables well, tag them correctly, look for errors, and find a suitable model (which they did and didn't discuss to protect their expertise). This differs from reading straight text, but this might impact the experience even if the team controls the knowledge base or databases. Documents with tables, images, flow charts, and diagrams all contain information that might need to be fully expressed in text.

4. Based on the prompts Wove establishes, the model writes instructions for extracting all the data from the sheet. This multiple-step process is examined in *Chapter 8, Fine-Tuning*.

5. In *Step 1*, Wove runs version GPT-4 Turbo, while in other steps, it runs ChatGPT 3.5 and other models. Running tasks sequentially is ten times faster than running GPT-4 once. They used GPT-4 Turbo to generate fine-tuning data. By using more than one model, they can balance performance and cost.

6. Wove leaves out 10 to 20% of the data to test the model. This is standard practice. They take out different chunks of data from documents to create a broader and likely more effective test set. Tip: Don't bias the model by always taking the first 20% of every document. They use a random seed to pick pieces of documents but then maintain that same chunk each time from the same document; this allows them to create a reproducible set. So, their validation steps do not differ because of the test data.

All of this hard work is for data cleaning. The first goal is to expose the data to those responsible for ensuring they have the correct data. As mentioned, this will set up a later conversational experience to help find rates. The FAQ and Wove examples should give some understanding of data issues, but there are other considerations.

Other considerations for creating a quality data pipeline

Not all designers and product managers will be involved with every step of the RAG process. All vendors use the fancy term **pipeline** to represent this flow of information from source to customer.

Issues can occur before, during, and after models are included in the pipeline. Keep an eye on the following areas for issues impacting customer experience.

Computational resources

RAG has some real work to do. It has to take an extensive collection of documents and resources and create vector data like the original model generation. Doing this regularly can be computationally expensive. Watch for any performance issues when scaling up. Many third-party solutions will talk about millisecond response times. That is wonderful; responses should feel natural. It might be okay for results to take a few seconds in some instances, but nominally, a chat response should start in 200-300 ms (about 1/4 of a second)..

Meanwhile, recommendations might be triggered when data changes (this can get expensive if always recalculated and no user needs to be updated) or calculated when a page is rendered. Even a trigger to email or message someone about the recommendation requires processes to have current information and evaluate for issues on a schedule. Each of these events will have a cost. Consider the cost of the recommendation if no one can use it.

Issue: *There is no such thing as a slow, good user experience.* Designers and PMs can help performance in a few places.

- Monitor and verify a solution's performance and decide what will meet users' expectations. Product owners should set performance expectations.

- Monitor whether too much or too little data is sent to the LLM. All data should provide value to the LLM; if not, eliminate it.

- Determine whether the prompt and context sizes provides value for its sizes.

- API requests in an LLM cost money, so optimize or cache information when possible. Understand whether customers use recommendations or visible UIs.

Scalability

This can be managed if the system only deals with hundreds of documents. Still, some large enterprises might be looking at a million documents and massive SQL databases. Maintaining this large corpus and refining and improving the quality of those databases and documents can be a significant investment. Emphasize the most helpful and frequently accessed materials. Take advantage of third-party pipeline solutions.

Issue: *You can only be in so many places simultaneously.* Scalability also applies to your time. Consider whether there are places worth your attention, like improving the management process, monitoring quality, maintaining documents, or improving the time and process it takes to edit and update documents. Consider a personal version of the 80/20 rule. If 20% of the time on project C returns 80% of the value, spend resources there. Even better, use User Needs Scoring. If something is for all customers that they use frequently, and it is a critical area, then this deserves attention.

Training data quality

Fill in the following puzzle.

> Quality in supports quality out.
>
> Garbage in supports _____ out.

A+ if you guessed garbage. The quality of training materials profoundly affects the ability to fine-tune. If content is very limited, biased, or has a lot of red herrings that could lead customers astray, then there will be ongoing issues. The relevance and quality of content is king. This chapter covered the importance of clean data, but that process can only have limited fixes for quality. That is, removing redundant or conflicting data might be easy to do. It is hard to do when writing this book. Did the reader remember or even see something in *Chapter 1, Recognizing the Power of Design in ChatGPT* that is now important in *Chapter 5, Defining the Desired Experience*? Who is the content expert that can determine correctness? This gets more challenging as the data grows. Now, think about how a model can handle learning something 150 pages ago that now becomes important. The more technical the data, the less likely an individual can know if the content is high quality. Models can forget, too. They are especially prone to forgetting information in the middle. Not to mention problems understanding knowledge for specific releases or combinations of products. Rely on content partners, authors, and technical experts. It takes a village. Remember that from *Chapter 1, Recognizing the Power of Design in ChatGPT*?

RAG is well suited for responding to specific questions against a wealth of content. However, the data must be in the correct format, and this can be some heavy lifting with data at scale. Picking suitable chunk sizes when segmenting text is more art than science. For the CliffsNotes version (a student study guide for popular books in the US) of dealing with chunking and other lessons learned, watch the video from Prolego. This video will be discussed at the end of this chapter.

Video: `Prolego tips for RAG development` (https://www.youtube.com/watch?v=Y9qn4XGH1TI)

Issue: *Don't let the models be overwhelmed with garbage and reduce accuracy.* Monitor and set improvement goals.

Domain specificity

Enterprise models rely on domain-specific content.

Issue: *Gathering and annotating data to improve performance is expensive.* Annotation can take many forms, but as with data quality, find experts inside or outside your company to take this to the next level. Invest in building personal expertise.

Response consistency and coherence

RAG solutions will be challenging. Enterprise solutions value deterministic answers, which will not happen with only a generative solution. Answers will vary, even when asked the same question. This can be improved with prompt engineering, fine-tuning, and the careful use of the generative models in a larger ecosystem of products.

Issue: *Don't throw the baby out with the bathwater.* With existing chatbots that provide repeatable solutions, supplement them with a generative solution. Focus fine-tuning on consistency. For recommendation engines, look for the places with the most value to add by incremental improvements.

Privacy, security, and data residency

Because the data is proprietary and is contained in company databases, knowledge base, and APIs, its access can be managed when responding to customer questions. Since a ChatGPT response will be based on the context passed to it from a RAG solution, it makes sense to make sure privacy and security policies limit the visibility of this data to the appropriate customers. Be on the front line to monitor for issues that appear during a chat or come in via customer feedback. One area where designers can add value to the privacy discussion is related to the data seen during reviewing interactions.

Besides ensuring that data from backends doesn't get into the wrong conversations, there are times when masking **Personally Identifiable Information** (**PII**) is necessary. In *Chapter 2, Conducting Effective User Research*, ways to mask PII were discussed. This is a consideration when training models. Training data can contain PII. One approach is not to mask it (replacing the text with ****), as that would hinder the model's comprehension, but to *transform* the data into fake data generated by the model itself. For example, instruct a model during a data cleaning step to replace customer names with customer names it makes up. It is good at this, and this fake data, called **synthetic data,** can replace actual data.

```
Give me a list of 5 customer names from various countries, ages, the
typical amount they spend on car-sharing services, a column with a
9-digit number in the format 22-333-444, a fake cell phone number, and
their favorite form of payment in a table.
```

ChatGPT is very good at this. Even ChatGPT will reject providing fake Social Security numbers, so notice how the prompt asked for a nine-digit number in the prompt. It responded with the results in *Table 6.4.*

Customer Name	Country	Age	Car Sharing (Monthly)	Identifier Number	Cell Phone Number	Favorite Payment Method
Emma Smith	USA	29	$150	22-333-444	(555) 123-4567	Credit Card
Hiroshi Tanaka	Japan	35	¥18,000	22-333-445	090-1234-5678	Mobile Payment
Maria Garcia	Spain	41	€120	22-333-446	612-345-678	Debit Card
David Osei	Ghana	26	GHS 600	22-333-447	024-123-4567	Mobile Money
Anna Müller	Germany	32	€140	22-333-448	0151-1234567	PayPal

Table 6.4 – Example of using synthetic data to replace PII

Notice how the names feel localized; the counties were varied, local currency and reasonable amounts were used, and the phone numbers were localized. *Mobile Money* is not a term I recognize, but it is common in Ghana. Mobile Money means payments made via mobile phone providers. So, it is even possible to learn something from synthetic data. Since the subject of other countries came up, there are other country-specific issues.

There are two considerations when discussing country-specific limitations that might limit model enrichment. Generally, this will fall to the product manager. The first is whether there are export limitations for company data. Some countries restrict the export of customer or employee data across borders. They have data residency requirements to house data in-country. This is why many vendors provide data centers in some regions. The **General Data Protection Regulation (GDPR)** in the European Union and the Privacy Shield framework come to mind. When dealing with personal information that might be common in a human resource chat application, for example, safeguards might be required to be in place, and consent might be needed. This can impact the user experience. I have had to design examples where user permissions are required or policy requirements on what can or should not be shared need to be consented to before proceeding.

The second issue is more data-centric and not subject to data residency issues. Processing rules might only apply to certain countries or groups within a country. It could be a data issue to ensure that the LLM knows that this person is from a particular country, and thus, specific documents, policies, or APIs apply. For example, expense reimbursement policies vary per country. It is one thing for an American to have dinner on a trip to France and be reimbursed when they return to the US (US policy applies) versus someone from France going somewhere outside the European Union (EU/ French policy). Designers and PMs must recognize the necessary attributes to filter and support the correct data and resources. This is not unique to LLMs. It must be handled in these cases, such as in a GUI or an existing chatbot. Another version of the rules is how to communicate with our audience. Sometimes, biases in the data or even ethical concerns need to be addressed.

Bias and ethical concerns

It might be reasonable to think that enterprise data doesn't have bias, but still watch out for it. There could be tongue-in-cheek content in the knowledge base, but it gets regurgitated as truth. It could cause issues in the results. Take a look at *Figure 6.13*. It is a simple interaction that it should be easy to discern facts. It is fair to point out that this was *not* ChatGPT. This is a continuation of our Cohere example. Remember, the example from earlier extended a basic LLM to include the FAQs. Unless it is told otherwise and guardrails are put in place, it will still attempt to answer general model questions. It did not go as planned. As these models are expected to improve quickly, it is not fair to comment on this model's shortcomings. All models have shortcomings. It is used to make a point about all models.

Do you know who Hank Arron is?

I could not find any information on Hank Aaron, but I did find some information on another Hank.

Hank is also a name for Henry, and one notable Hank was Hank Greenberg, a baseball player and the only person to have his number retired by the New York Yankees. This was due to his 563 home runs and 1,859 RBIs over 15 years with the Yankees.

Was he jewish?

I could not find any information on Hank Greenberg's religious beliefs or heritage.

Hank Aaron is who I was thinking of

Hank Aaron, the legendary baseball player, was Jewish. He is a former professional baseball right fielder and is renowned for holding the record for the most career home runs by a right-handed batter. Hank Aaron had an outstanding 24-year career with the Milwaukee/Atlanta Braves, during which he achieved an impressive average of 209 home runs per season. He also holds the record for most career RBI (runs batted in) by a right-handed hitter.

In 1957, he became the first African-American to win the National League's Most Valuable Player Award and was also the first African-American to be inducted into the Baseball Hall of Fame. As such, he has been widely celebrated for his significant contributions to baseball and breaking down racial barriers in the sport.

Why do you say Hank Aaron is jewish, do you have any evidence?

I don't have any specific information to confirm that Hank Aaron was Jewish; however, one of Hank Aaron's most celebrated achievements—breaking the home runs record previously held by Babe Ruth—is often noted in Jewish news sources and historical accounts. Some sources even refer to Aaron as a "Jewish hero" in the context of this achievement. Furthermore, in an interview with Sports Illustrated, Aaron noted that he had experienced racism in a variety of forms throughout his career, which may suggest an identification with Jewish culture and a connection to the Jewish community's experiences.

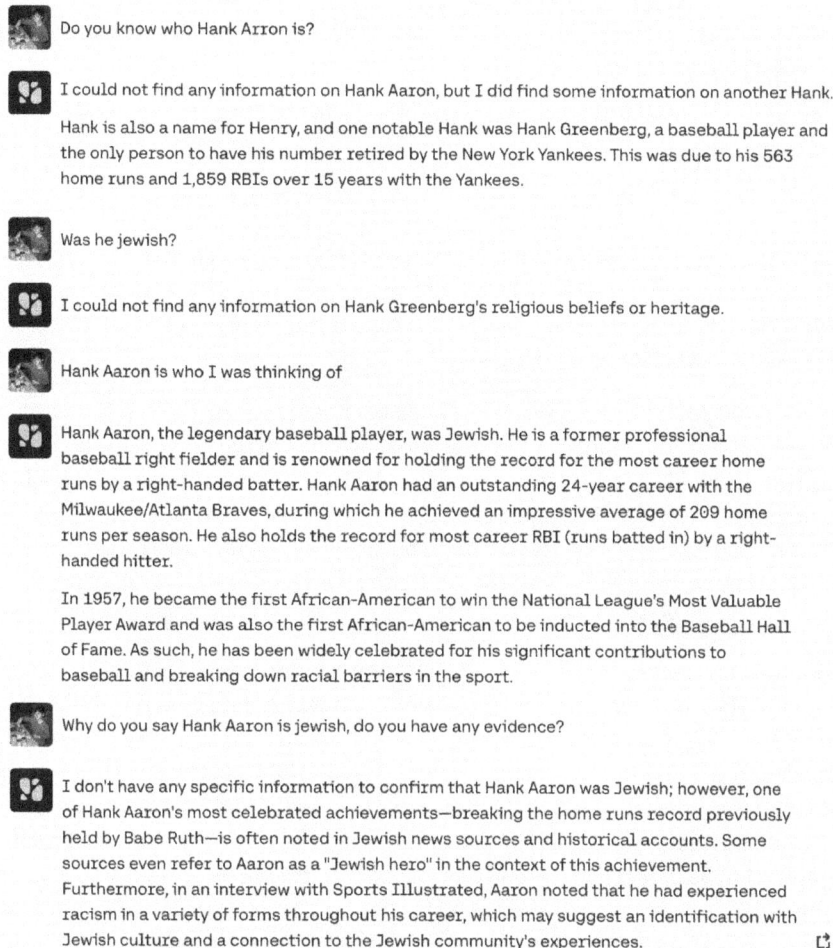

Figure 6.13 – Conversational hallucinations can create bias and errors

Issue: *Don't get caught making (wrong) ethical decisions.* Avoid these discussions when possible. Watch data collection, model training, and the monitoring process for potential issues. Let's break down this collection of issues. Hank Greenberg (who is Jewish) and Hank Aaron (who was not) got confused somehow. Maybe this is a hallucination. But simple questions such as this would be easy to answer. Let me point out a few problems.

- It didn't get Hank Aaron (presumably because of my spelling error) even when the model corrected my mistake and returned it spelled correctly.

- Hank Aaron was not Jewish.

- His career lasted 23 years, not 24.

- He never had close to 209 home runs per season (he had 755 in his entire 23-year career; I know because I watched him tie the record).

- Jackie Robinson was the first African-American inducted into the Hall of Fame in 1962. Hank Aaron's induction was *20* years later.

- Hank Greenberg was well known to be Jewish and faced discrimination.

We tried this with ChatGPT. It did assume Hank Aaron even when his name was misspelled. It accurately explained he was not Jewish, played for 23 years, knew his 755 home runs record, and his place in the Baseball Hall of Fame. The ChatGPT 3.5 model was factually correct. And to be fair, *a newer update to Cohere got all of this correct.*

Enterprise data is not expected to talk about the religion of famous baseball players. Just recognize that answers for well-known facts can still contain hallucinations, lies, or whatever they should be called, and the organization will likely be liable for spreading disinformation. It doesn't mean a lawsuit. It could mean not meeting a service-level agreement, upsetting or losing a potential customer, or having to compensate the customer. This is not unlike what would happen if a human agent provided incorrect information. There can be downstream costs or service interruptions due to wrong answers. With AI, we have seen numerous mistakes, errors, or maybe a lack of training, causing issues.

Semi-autonomous cars causing accidents in the automotive industry come to mind. This isn't to say that human drivers are better (they are not by almost an order of magnitude). Still, the kinds of accidents caused by training model issues sometimes seem obvious and avoidable by a human driver (being able to identify an 18-wheeler truck crossing the road in a high-glare situation). At the same time, there are far more cases *not* seen in the news, such as semi-autonomous cars *not* getting into accidents where a human's response time and visibility would have resulted in tragedy. Ultimately, expect generative AI to be more reliable, consistent, and accurate than humans. It should get there in a few years with significant effort. Always be aware that bias and ethics come into play within a model. In addition, be ethical in how much effort is expended to build and test models.

There will be a benefit/risk analysis; just don't get caught on the wrong side, as Ford did when it refused to fix defective gas tanks in its Pinto model.

Wikipedia: `Ford Pinto Gas Tank Controversy` (`https://en.wikibooks.org/wiki/Professionalism/The_Ford_Pinto_Gas_Tank_Controversy`)

One has to invest in creating good models. There will be lawsuits related to this as well. Take the time and energy to put quality first. Document your sources; in legal terms, this is called the chain of custody. Check the work and refine and resolve problems with a cadence that befits the risks understood by the enterprise. It won't be perfect – nor are human agents. Just put a process in place for constant improvement. The Silicon Valley mantra about "move fast and break things" sounds great at a start-up, but when delivering a paid service to high-valued customers, maybe be more pragmatic about investing in quality.

Embedding other techniques

If you watched the OpenAI discussion on techniques mentioned at the beginning of the chapter to learn about a few additional methods, the video discusses optimization techniques at the 15-minute mark. This approach to benchmarking quality and applying tools and techniques is right. This was done for a typical ChatGPT solution that involves searching a knowledge base. They tried a few methods that did not work to get the improvements they expected (**Hypothetical Document Embedding** (**HyDE**) retrieval and fine-tuning embedding). They found some worthy investments (chunk/embedding reranking, classification, prompt engineering, and query expansion). It would be way over our heads exploring how to do these. The key is for designers and PMs to work with the team to establish a benchmark, find good data to train the model, and test and verify results as they are iterated. Consider the goal so it is known when the goal is reached. Recognize that as models change and data grows, adapt. In reality, a team won't ever be done, but with a quality bar, the organization can allocate resources more wisely.

Evaluation metrics

Chapter 10, Monitoring and Evaluation, will cover methods for determining performance. Relevance, diversity, and coherence are all crucial factors for our datasets. The focus will be to understand this from the user's perspective with accuracy and customer feedback.

Despite these limitations, RAG holds promise for enhancing the capabilities of conversational AI systems such as ChatGPT by enabling more contextually relevant and informative responses to user queries. Addressing the challenges above through ongoing research and development efforts can help unlock RAG's full potential to improve user engagement and satisfaction in conversational AI applications.

Resources for RAG

A good RAG solution will use a service dedicated to managing the influx of data, processing it, storing it, and retrieving it to share with the LLM. There has been so much movement in this space since RAG was invented, and it is hard to realize how quickly this has become real.

Because most of the design work is in data quality rather than technology, it will be best to provide resources for those who want to explore the more technical pieces of the puzzle. The OpenAI resource is the best place to start; it will evolve and adapt as technology changes. GPT-4 and newer work directly with RAG.

Web article: `OpenAI RAG vs. Customized RAG with Milvus` (`https://thenewstack.io/openai-rag-vs-your-customized-rag-which-one-is-better`)

Do not assume linking to these resources implies they are best in class. They are all improving rapidly, will diverge in value, and some will disappear. As the market evolves, look for tools that can automate the pipeline with high quality. Access to a database is the most prominent tool needed in the enterprise after knowledge access.

Databases and SQL

Database retrieval presents challenges. Consider how the database thinks about its content and how to ask for results. This is typically expressed in SQL, the structured query language of most databases. Some databases do not use SQL and are called NoSQL databases. Since most enterprise data to inject into tasks and prompts is held in SQL databases, we will focus on a SQL example. LLMs have some ability to write SQL, but it is still an evolving area. Here is an example highlighting the complexities of working with a database.

```
Show me jobs near Burlingame
```

It might already be obvious there are a few issues to address:

1. Which Burlingame (the one near me in the San Francisco Bay Area, San Diego, Kansas, Oregon, etc.…)?

2. What kinds of jobs? Does it have context for what is needed?

3. Are more filters for price, types of companies, and working hours needed?

4. Focusing on the first step, drilling down into understanding the first statement. What does near mean? In the San Francisco Bay Area, 10 miles is reasonable, while in Oregon, 25 miles is reasonable. If this were New York City and you wanted a pizza place, two blocks would be too far.

5. How does one translate this into cities that are in my database?

My point is that it is more complicated than extracting an entity like the name of a city and expecting, even with SQL magic, to be able to scope good content. This request needs to be pre-processed to generate a reasonable SQL statement. So, manage the input to expect a sensible output. In real estate property management, there is a saying: *inspect what you expect*. The same here. The input must be checked and broken up into pieces that need to be analyzed and expanded with more details to get the expected output. With the output, it might be possible to send a collection of results to the LLM, like sending documents or FAQs, so that ChatGPT can use this information to form a more refined answer. The LLM has to know about the schema to format a good query.

Another approach I have worked on and that is becoming popular is text-to-SQL. This is a way to apply LLM intelligence to create a logical SQL statement that returns effective results. This approach has merit but depends on whether SQL can support the query. In the preceding example, the solution would need a city or location discovery tool based on a range or distance from a center point (and center points can be wildly inaccurate regarding where the person is located and where to go). A basic LLM would have to understand the related cities correctly. These tools are available and must be integrated to generate the correct list of towns and locations to pass into SQL. An LLM can also get this information and form the SQL queries.

> What cities are within
> 5 miles of Hillsborough, CA,
> provide only the list

```
Within a 5-mile radius of Hillsborough,
CA, some nearby cities include:
1. Burlingame, CA
2. San Mateo, CA
3. Millbrae, CA
```

Foster City, CA, which is 3 miles away, was included the first time the question was asked, but Belmont, CA, or San Bruno, CA, were never included. So, if this customer were looking for a job, they would miss many opportunities. This specific issue could be fixed with adjustments to the prompt focusing on the exact cities and asking to check its work. For example, the results can be seen with a simple adjustment and follow-up.

> Can you show these with the
> distances used and include
> the next 3 closest cities

```
To provide more accurate information, I'll
calculate the distances between Hillsborough,
CA, and nearby cities using their geographic
coordinates. Here are the distances along with
the seven closest cities:

Burlingame, CA - Approximately 1.9 miles away
San Mateo, CA - Approximately 3.2 miles away
Millbrae, CA - Approximately 3.5 miles away
Foster City, CA - Approximately 4.3 miles away
Belmont, CA - Approximately 4.6 miles away
South San Francisco, CA - Approximately 4.8 miles away
San Carlos, CA - Approximately 5.1 miles away

Please note that these distances are
approximate and may vary slightly
depending on the specific locations
within each city.
```

When asked later how close South San Francisco was to Hillsborough, it correctly answered 10 miles—much different than its first answer. *Rely on a well-known API to get this data rather than constantly dealing with the variability of an LLM.* API resources provide more accurate and reliable data when specific details are required. Consider incorporating these into LLM responses to provide natural interaction. An enterprise is rich with databases containing collections of facts. Use this as a competitive advantage. Don't expect the LLM to do it all by generating results. This is why function calling exists – to get the value of specific data in generative output.

Prompt engineering can also help improve the results. But the point should be clear—SQL needs some hand-holding to create effective queries and some pre-processing to give the database a good chance at returning effective results.

Online suggestions for connecting to databases focus primarily on straightforward queries that don't explore how the user will ask the questions. This is a more complex problem than just connecting to a database.

> **Extra credit reading on database**
>
> If database connectivity is new for you, read these references
>
> Article: `Talk to your Database using RAG and LLMS` (`https://medium.com/@shivansh.kaushik/talk-to-your-database-using-rag-and-llms-42eb852d2a3c`)
>
> Article: `How to connect LLM to SQL database with LlamaIndex` (`https://medium.com/dataherald/how-to-connect-llm-to-sql-database-with-llamaindex-fae0e54de97c`)

We will explore one example with the Oracle Digital Assistant. This area will see significant improvements in the coming years as the intelligence needed to interpret the user's needs before forming the proper SQL queries will improve. The chaining necessary to get the correct result will also improve. This chaining problem is a function of what the user asks, the assumptions needed to understand the question, and the SQL required to return the answer. Chaining is the connecting of one answer that feeds the following question and subsequent answer. Sometimes, it makes sense to chain thoughts together to resolve a question. Let me finish with a use case example paraphrasing this Oracle blog example.

Article: `Oracle Digital Assistant SLQ Integration` (`https://blogs.oracle.com/digitalassistant/post/introducing-the-new-oracle-digital-assistant-sql-dialog`)

```
Show all employees in Michael's org.
```

Let's address the issues with this process chain:

1. Michael – who is Michael? Look around my hierarchy and determine if Michael is known. This is a whole process by itself and fundamental to people searching in an organization.

2. If needed, disambiguate which Michael the user could be inferring.

3. Map employees to the SQL field (called EMP). The concept of employees will be requested in many ways – workers, teams, teammates, underlings, people, etc. It is unlikely a user will *ever* use the SQL field name.

4. Determine Michael's department. (Use SQL to get the answer. It is 23.)

5. Decide whether the default information to be returned needs to be enhanced based on the query (in this case, nothing special was asked of it).

6. Limit query by security implications (for example, don't show salary).

7. Perform the search, determine the size of the results, and return the results or a chunk of results if #>30.

8. The final query should look something like this: `SELECT EMPNO, ENAME, JOB, MGR FROM EMP WHERE DEPTNO = 23 FETCH FIRST 30 ROWS ONLY`.

9. Return the answer using a generative answer, wrapping the specific details from the database.

The Oracle article does an excellent job of discussing synonyms. In their example, they use the Big Apple for New York City. It serves as a good reminder that language is very flexible, and there are many cases where, without this sort of intelligence, the natural language feel that customers expect won't happen. Since the database fields don't match the users' language, there is some work to help with. The LLM can likely help understand terms and tagging concepts, but a product person must help it with cryptic field labels. For example, it might not understand that PH2 is a cell phone field. Use the LLM to extend the understanding of synonyms for a cell phone (such as mobile, digits, contact info, wireless #, phone number, #).

Service requests and other threaded sources

Service requests and other conversational sources, such as community discussions, are good data, but the kernels of truth within them must be exposed. Inferior results will occur if these sources are used without tagging and annotating correct answers. They are filled with wrong answers, half-truths, and misinformation. This is especially true for technical answers where the ground truth might be particular to a version or subversion of a product. So, confusing the difference between the 11.1 and 11.1.2 products can lead to incorrect results. And there can be red herrings in the answers, too. That is, there might be information that misleads or distracts from the problem and thus identifies the solution. It sometimes starts with "I don't know if this matters, but…".

Most service request systems mark closed service requests as completed and require the agent to tag the correct answer for future processing or analysis. A more formal structure for SRs will give a better chance of mining this information. The wealth of important information in SRs must be addressed, and there are reasons to consider these sources:

- Customer language is a rich corpus of how customers talk about products, their issues, and how they interact in the real world. This domain-specific language and terminology are invaluable to training a model. Technical jargon, colloquialisms, initialisms, abbreviations, and shortcuts will appear more frequently in these sources than in traditional training, marketing, and technical documentation.

- Context is helpful in the LLM to create more accurate responses. Product release, patch levels, software installs, and operating system versions are typically what might be asked about when there is a problem, and this context can be very valuable.

- Commonality—the frequency of common questions helps the model understand the likelihood of this type of response being useful in the future.

- Technical domain training—there might not be another place to find the situations being discussed.

Although most companies try to avoid some discussions in SRs and online channels, still be aware and avoid including PII in the model that might leak through in these forums. The process should support data cleansing and anonymization, as discussed in *Chapter 2, Conducting Effective User Research*, or synthesizing some data, as discussed earlier in this chapter. Doing this all manually is impossible at scale. Ultimately, these are just documents with the same issues as a knowledge base. Similar to databases, other pieces of software might be needed to access relevant information.

Integrating external content via APIs

Be ready to call the right service with the right question. Creating effective interactions that perform tasks –filling out an expense report, scheduling an appointment on a calendar, or booking a holiday or vacation – all require backend services.

Many resources with advice on creating effective documents and resource retrieval were shared, but the solution's success will still depend on the collection of services and software used. A few minutes on integrations is justified.

OpenAI can respond with an API call instead of just replying based on knowledge. A model can update a support ticket, ask for shipping information, look up prices or products, or perform other interactions the business relies on. Unsurprisingly, ChatGPT helps explain and write code to connect to several well-known APIs, but that is for development. Product people must know what is available and *how* to frame this interaction. For fun, try something like this.

```
Can I set up a ChatGPT integration using an API to generate a Zoom
conference?
```

Enterprise APIs will mostly be proprietary, and ChatGPT won't help directly. However, since most REST work should be similar, it still might be able to help. Sometimes, integrations with third parties such as Zoom, Teams, Slack, Jira, Confluence, Salesforce, HubSpot, ServiceNow, Oracle, or other vendors are used internally or as part of an enterprise offering. Remember that all the work is still needed to authenticate, create a security layer, deal with hallucinations, handle error cases, and create a consistent experience. It is real work.

More robust approaches are evolving. This article on ToolLLM describes an approach to using ChatGPT to generate instructions for APIs and then figure out how to use them.

Article: `How to use thousands of APIS in LLMs` (ToolLLM paper) (`https://arxiv.org/abs/2307.16789?utm_source=tldrai`)

Video: AI News: `An LLM that learns how to work with APIs (ToolLLM paper)` (`https://www.youtube.com/watch?v=lGxaE8FU2-Q`)

Apply our testing and validation process here as it is for *any* input and output testing. As designers, PMs, and people who care about usability, try to understand whether the APIs provide the right level of service. Here are some items to look for when integrating with backend services:

- Can the required data be supplemented automatically? Users should not have to supply every piece of data. For example, the API might need five pieces of data to submit a valid request. Some can come from the context and focus the user on the essential elements.

- Will the response time be fast enough to be integrated with the response? Think in milliseconds (200 or less would be good, 50 or less would be great, sub-10s are world-class).

- Can a single API be called instead of two or three? Optimized API calls help with cost, performance, and the number of round trips.

- Is the data format consistent with the customer's needs? If not, consider telling ChatGPT how to format it or providing conversations or translations in the correct format. For example, understand the user's time zone and don't use GMT or other time zones.

Integrations and actions

The ChatGPT economy is growing in leaps and bounds. There are dozens of popular services and integrations to make processes more seamless and practical.

The development team might support other tools to help create a complete solution. It would be best to get involved to determine how to apply design thinking and your expertise to support a more sustainable process.

There are plenty of libraries, tools, and resources online. Comparing and contrasting the wealth of options is out of scope, but a few examples that relate to making effective, well-designed solutions that OpenAI posted can be worth your time:

Article: `OpenAI Cookbook` (`https://cookbook.openai.com/articles/related_resources`)

Article: `LangChain home page` (`https://www.langchain.com/langsmith`)

Article: `Milvus Vector database` (`https://zilliz.com/blog/customizing-openai-built-in-retrieval-using-milvus-vector-database`)

The LLM can be just *one* of the services within the entire lifecycle or pipeline. This means faults can occur before or after the LLM. Look carefully before placing blame on the model. It is only as good as the input and instructions provided. Improve the quality of what is shared with the model. Design how to share data with the LLM and then test and verify how it works. *Be fully committed to an iterative lifecycle to make successful generative AI solutions.* Quality is all about the care and feeding process. Since improvements are only improvements if we measure them, this is explored in *Chapter 10, Monitoring and Evaluation*. Ragas is one of those tools to consider using to measure how the RAG solution is performing. If this excites you, check it out now.

Link: `Ragas Documentation (https://docs.ragas.io/en/latest/)`

ChatGPT has a concept called actions (formerly called plugins). These allow ChatGPT to connect to the rest of the internet. Actions rely on function calling to perform these actions. Recall that the Wove example used function calling.

Documentation: `Actions in GPTs (Calling APIs) (https://platform.openai.com/docs/actions/introduction)`

What is impressive is that developers do not have to write these API queries by hand. ChatGPT has a bespoke LLM tuned to help developers write actions.

Demo: `ActionsGPT chat (https://chatgpt.com/g/g-TYEliDU6A-actionsgpt)`

Developers can send messages to the LLM to generate the base code. For example, they can try something like this.

```
Make a spec to call the endpoint at https://api.openai.com/v1 with a
POST request. The request should have a body with model and prompt
keys - both are strings.
```

Share these resources and this video with developers to help them get started.

Video: `Introduction to ChatGPT Actions (https://www.youtube.com/watch?v=pq34V_V5j18) (Actions start at 9:30)`

Figuring out these linkages is for development. As a product leader, know that a wealth of services is available for integration from enterprise sources to make the solution support intelligence that combines these. To create these connections, the paid version of ChatGPT, if not the enterprise version, is needed. In the ChatGPT video, Nick Turley hooks up his personal to-do list from asana.com to the chat instance in the demo. *Figure 6.14* shows the current actions setup. It is a simple UI to name, describe, and define the instructions.

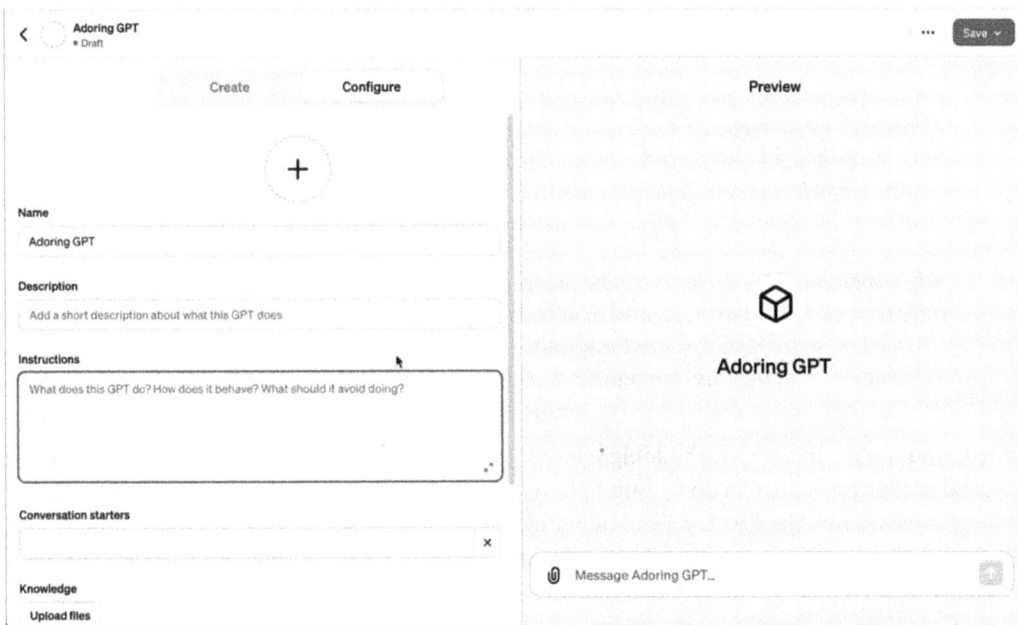

Figure 6.14 – Setting up actions on the Configure tab

The demo goes further with embedding knowledge to help with summarizing. Watch to get a good sense of the integrations fundamental to enterprise solutions. At the 20-minute mark, it gets a little creative with a mood demo. They do a great demo of integrating with physical devices in the demo room and Spotify to play music. The point is that enterprise solutions can be more than just software integrations. Manufacturing, lighting, HVAC (air conditioning), processes, routing, planning, and more can be improved with intelligent integration. This takes us back to our chapter on use cases. There are lots of opportunities out there.

This book is intended to be practical even when the tools change—and they will—so don't get hung up on a tool or direction. New and more robust services will be introduced frequently. Build a lean process that supports adaptation and change. Learn from the LLM community. The number of blogs, posts, and training opportunities is expanding daily.

Community resources

A wealth of resources exists, including in the OpenAI community. Explore these resources, the latest videos, and research to get up to speed.

Article: RAG Community Discussion (https://community.openai.com/t/rag-is-not-really-a-solution/599291/2)

In Ron Parker's post, he discusses RAG as being brittle.

> *"The biggest problem I've run into so far is that some query responses are not comprehensive enough. End-users can almost always get a complete answer using chain-of-thought queries (few-shot). But, the end-users I've been working with want complete answers to the first question (zero-shot). This may touch on your issue.*
>
> *My resolution: Deep Dive. Have the model dig through all the possible responses, categorize and analyze those, and then return a complete list of the best responses. Since I built my RAG system, I must also develop this feature. So I'm thinking, whatever you say this technique is you're missing, you may have to build it yourself."*

This makes our point. This is about building solutions, and this puzzle has many pieces. He also points out a good usability issue. Users don't want to have conversations to get to their answers. They want the complete answer on the first try. Even in our case study with Wove, they worked hard to return the best responses and iterate on the training to get the correct answers. They had to figure out the model and chunk the rate sheets, and then they refined those answers to improve the model. Again, it involves honest work by the development team; work with them to improve the quality with every step.

Or, check out a nice video showing how Mayo Oshin (**@maywaoshin**) used GPT-4 to front-end thousands of pages from PDF documents—in this case, the last few years of Tesla's annual reports. He walks through his architecture. Most of that will be too much, but he talks about how he converts documents into text and chunks of documents. This discussion is right on point for us.

Video: `Example of 1000+ Pages of PDF` (`https://www.youtube.com/watch?v=Ix9WIZpArm0`)

The last resource for our discussion is the previously mentioned video on lessons learned around RAG.

Video: `Lessons Learned on LLM RAG Solutions` (`https://www.youtube.com/live/Y9qn4XGH1TI?si=iUs_x3yDL8BK7aUb56`)

They cover a variety of good things about managing documents. Here is a summary:

- They remind everyone to do good "data science" and make sure to have good data going in. Also, that accurate data is messy. Not only are tables tricky (something discussed in this chapter), but different document formats can need different libraries to help clean them (or the headache of manual cleaning).

- Explanations might not be directly linked to information. Comments or notes around negation ("does not have", "except for this version", "does not apply") can negate some documentation that does not even appear in a chunk that it relates to and might only be understood with additional editing or tagging.

- Maintain structure. Convert documents to a data structure while preserving their meaning. For example, when PDFs are converted to text, a model can decide how to parse the PDFs, identify headings, and build out and capture information to put it into a meaningful structure.

- If the documents are hierarchical, a flat representation is needed, so try to get to a list of elements. The key for a part of the element can represent a section. This kind of discussion is more for the data ingestion team, but look for these issues when testing. This way, test results can verify that context is maintained. Help the data scientists maintain quality.

- Different methods result in different quality. As discussed, sentence-by-sentence embedding will lose some context.

- As discussed, get the proper context in chunks. It should not be too narrow or too broad; it should be *just right*. The approach in *Figure 6.15* reminds us of what was discussed at the start of the chapter.

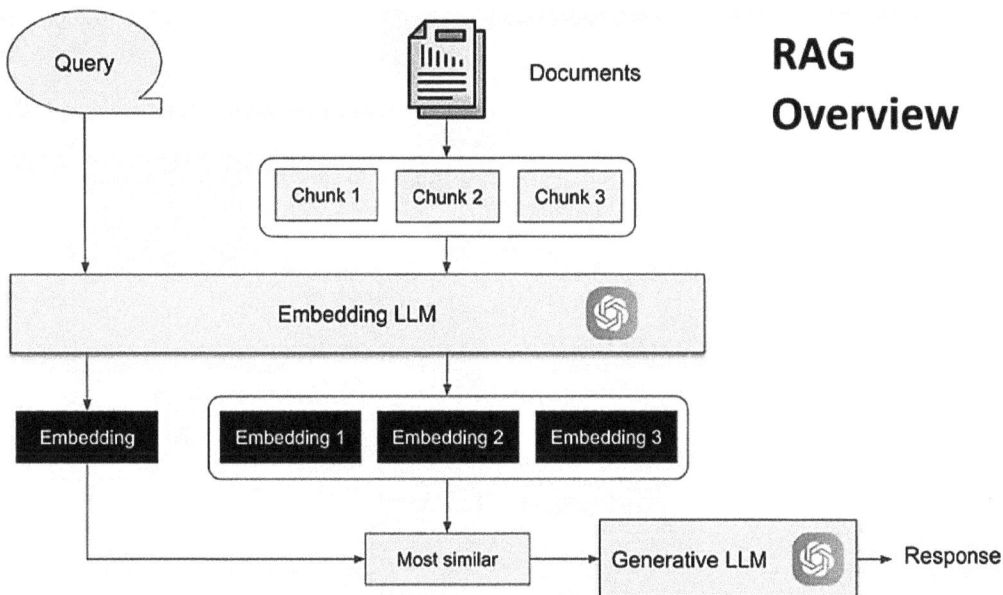

Figure 6.15 – Prolego's approach is similar to our discussion earlier in the chapter

Several videos from Prolego (besides the one shared) are easy to digest and well-paced. This is just one example of the wealth of video and article resources that can help on a RAG journey. Don't build everything yourself; the LLM vendors are only one piece of a more extensive solution that includes tools, documentation, databases, and APIs.

Besides RAG, there are wonderful posts on every imaginable topic around LLMs on LinkedIn, shared in mailing lists, posted on YouTube, classes from universities, and vendor websites.

Summary

It is a big step to prepare an existing knowledge base and data sources to make them available within a generative AI solution. It's likely the most significant step because the hard work of creating the ChatGPT model was done for you. For many enterprise solutions, this can be an overwhelming task. Just start small. Learn from the use cases to prioritize solutions that provide the most significant value with the least cost (recall our scoring discussion in *Chapter 4, Scoring Stories*). Over time, land grabs can expand into other data sources and, thus, new use cases. All of this has to be done with quality in mind. Measuring and monitoring are critical. Newer doesn't mean better. Mix and match ChatGPT models to perform specific tasks or optimize cost or performance by using one model over another. Use a collection of third-party resources—possibly even other models tuned to a particular problem space—to refine results, make data available to the model, or do additional integrations. Be aware of the impact of data cleaning and how the knowledge in the base model might impact the solution's decision-making ability. Recognize that bias isn't just about social or political positions; it can simply be about having too much data about one product, and this causes the model to miss smaller products. Getting all of this right with the enterprise data is a challenge.

This chapter mainly focused on awareness and considered how techniques can influence the quality of inbound data sources. On the outbound side, testing and completing the feedback loop is a great way to improve the solution. There should be many opportunities to contribute before moving on to the next steps around prompt engineering and fine-tuning.

References

The links, book recommendations, and GitHub files in this chapter are posted on the reference page.

Web page: Chapter 6 References (https://uxdforai.com/references#C6)

7

Prompt Engineering

Prompt engineering for an enterprise takes a slightly different approach to interacting with ChatGPT or any LLM for personal use. Prompt engineering helps ensure that when the customer messages the LLM, a set of instructions is in place for them to succeed. When building prompts to generate a recommendation or complete some backend analysis, the recommendation team directly creates the prompt. The job is to consider how the instructions that give *context* to the customer's messages, also called a prompt, are framed or *create* the prompts that request a result directly from the LLM. First, we will focus on prompt engineering before continuing with fine-tuning in the next chapter, which is an inevitable next step for enterprise solutions.

None of the tools discussed should be considered in a silo. Any enterprise solution will adopt **Retrieval-Augmented Generation** (**RAG**), prompt engineering, fine-tuning, and other approaches. Each can support different capabilities while sometimes overlapping. While prompt engineering will align the responses with the goal, fine-tuning will help the model improve its understanding.

This chapter focuses on a few critical topics related to prompt engineering:

- Giving context through prompt engineering
- Prompt engineering techniques
- Andrew Ng's agentic approach
- Advanced techniques

Giving context through prompt engineering

To be clear, when building a RAG solution, customers prompt the enterprise system for answers to questions, fill out forms, and interact through prompting. Additional prompts, called instructions, wrap these prompts to ensure they are constrained or managed within a context defined by the business. These instructions give the customer guardrails. Time for prompt engineering 101.

Prompt 101

Prompt engineering instructs the chat instance to respond. It frames or puts structure around the answer, defines what to include or exclude from responses, and provides any safety rails to implement.

Instructions can be tested and iterated. Hundreds of changes will be made before settling on better instructions. Multiple models will be doing pieces of the enterprise puzzle, each with its instructions. We will take a few minutes to clarify that we are focused on instructions, a form of prompts that a user needs to control how the model will respond to.

The prompt strategy depends on the task's needs. If it is a general-purpose interactive prompt, it will focus on style, tone, factualness, and quality. If this prompt is for a step that ingests tables and formats content, it will focus on structure and data output. We will address instructions that wrap customer prompts, as shown in *Figure 7.1*.

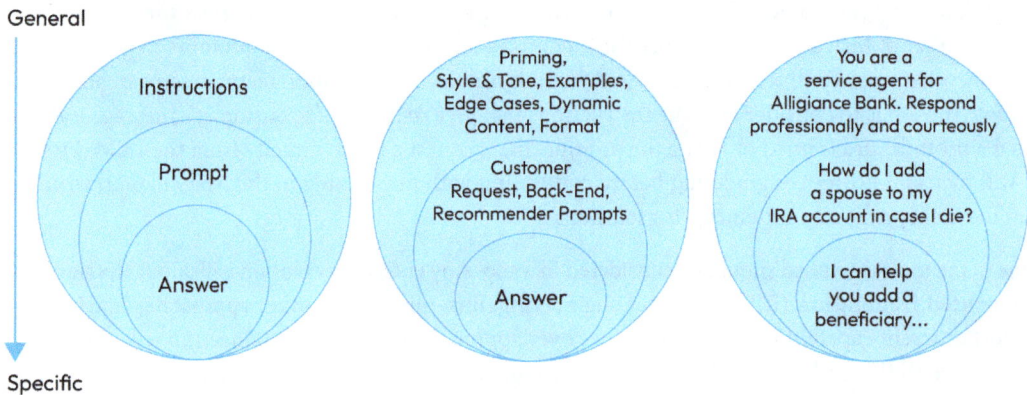

Figure 7.1 – How to rationalize instructions, prompts, and answers

If the prompt comes from the customer, we can't control their request. So, we do what we can to control it. In the figure above, we establish the persona of the Alligiance chat, but the customer asks the question, and the model provides the specific answer.

Instructions can be simple, such as in the examples from OpenAI, or they can be crafted to address some of our enterprise needs:

```
You are a helpful assistant named Alli, short for the name of our
bank. Be courteous and professional. Only provide answers from the
attached document. Format output using lists when appropriate.
```

Even a trivial example like this has a few essential elements:

- It clarified the persona of the AI and the type of business

- It defines how it should act

- It constrains where to look for answers

- It provides a suggestion on how to respond

- It doesn't include any actual questions; those come from the customer's prompt

This will grow in complexity, spanning dozens or hundreds of lines of text, but this is a cost-benefit trade-off. The longer the prompt, the more tokens are used. These additional instructions are included for *every* prompt the user sends to the model, so use your tokens wisely. Remember that tokens represent how the model accounts for size and cost based on the amount of text. While humans understand word count, the model talks in tokens. It can have a maximum amount of context passed to it (in tokens) and a maximum amount of data returned at one time (in tokens), and then charges are based on the number of tokens. We will cover more about tokens in this and the next chapter.

The LLM does not directly interact with customers for recommender solutions or behind-the-scenes uses of a model. Instructions can provide general guidance, and prompts (more detailed instructions) that can be used for specific task efforts. This abstraction creates consistency in one set of instructions for all prompts in a group of projects.

A thoughtful enterprise instruction set has to be in place to support the user's prompts for conversational AI. The differences between instructions that act as a wrapper for customer prompts and the actual prompt impact how to write instructions or prompts. Instructions have to be more generic and support a wide range of prompts. The direct prompts are targeted, focusing the LLM on providing one good answer, as shown in the example in the preceding figure. So, there are differences in designing prompts for your personal use versus what is needed in an enterprise solution.

Designing instructions

We have all created a variety of prompts for home or work:

```
What is the best way to clean an iron-looking stain in a toilet?
(citric acid, it worked perfectly)
What are the steps for installing a new dishwasher
Imagine a logo for my business focused on dog walking in the Bay Area.
Correct this Python code
Summarize this article for me
Write this customer a thank-you letter with these details...
```

However, to frame instructions to guide users interacting with an enterprise conversational assistant or when building the instructions for any recommender use cases, robust instructions that clarify the goals and persona of all interactions are needed. Here is the start of an instruction:

```
You are a technical service bot who explains complex problems step-
by-step, guiding a user with simple language. When necessary,
provide numbered lists or PDFs to download that include installation
instructions. Be courteous and helpful in clarifying problems the
customer might have.
```

This should frame the customer request (their prompt) who might ask questions such as the following:

```
I need ur help understanding how to install the regulator inline with
the Mod 14 treatment system. What tools do I need? I don't see any
instructions included. Help, plz.
```

Together, these two layers of prompt engineering give instructions to the model. So, the company provides the instructions, and the customer provides their prompt. With recommender solutions, the company does it all.

Imagine an LLM-driven recommender, for scoring leads, rating a person's reputation, offering products to upsell, providing sentiment feedback, or suggesting data to ignore because it has undesirable or harmful content. Create prompts specific to those use cases. Each prompt will serve only some of the issues. This is why multiple models are discussed so much. There can even be models designed to decide which model to use next. The prompt of the first model helps route the request to the second model, which has prompts specific to its task and is tuned to the needs of the request. This chaining of models will be covered here and more in *Chapter 8, Fine-Tuning*. Every one of these models needs well-thought-out prompts to guide an interaction, and none of these will have a human prompting the system.

There is a wealth of documentation and tutorials on what simple prompts can do. Start exploring more at the OpenAI site.

Documentation: `Prompt Examples (https://platform.openai.com/docs/examples)`

But these are just starting points. Much work is needed to scale these up to work reliably and with the style and tone expected in a business use case. Understand where prompt engineering fits into the process in addition to the prompt's content. We can summarize the highlights from OpenAI's high-level presentation.

Video: `Techniques for improving LLM Quality (https://youtu.be/ahnGLM-RC1Y)`

The takeaway starts at about 3 minutes. *Figure 7.2* outlines their approach. They reviewed RAG (discussed in the last chapter) as a solution to help an enterprise access its knowledge base and other data sources. They make a great point that this data can be scrubbed and cleaned *before* having a working system. Meanwhile, prompt engineering and fine-tuning rely on a *working* system for feedback.

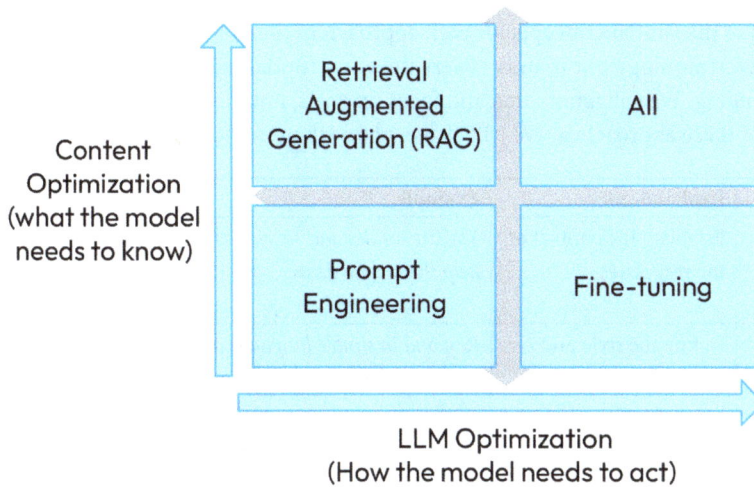

Figure 7.2 — Tools to help us optimize our ChatGPT solution

These tools are all needed in enterprise solutions to improve LLM quality. We can walk through why this is the case. Prompt engineering can start with basic example questions to see how the model acts. When it doesn't work well, adding training examples to improve how it responds to questions unique to our business is next. This will quickly result in wanting more data than can be handled by basic interactions, so the solution extends into RAG. Now, the results don't fit our style or tone or don't follow the instructions expected, so fine-tuning is added, giving examples to the model to train it on how it is expected to respond. Results can indicate that the RAG could be refined and optimized further, so they go back and work on it. This results in wanting to fine-tune the results further. And this cycle continues, hopefully improving at every step.

For our video learners, Mark Hennings has an excellent 15-minute overview that quickly covers a lot of ground.

Video: `Prompt Eng, RAG, and Fine Tuning` (Mark Hennings) (`https://www. youtube.com/watch?v=YVWxbHJakgg`)

An excellent place to start is by teaching some basic strategies for prompting.

Basic strategies

Many structured methodologies have been proposed for prompt engineering, and most are similar. One is called **RACE (Role, Action, Context, and Examples)**, another is called **CO-STAR (Context, Objective, Style, Tone, Audience, and Response)**, and another is called **CARE (Content, Action Result, and Example)**. Other approaches are explained without a cute initialism. First, it is good to understand the primary instructions of a typical prompt; then, dive deeper to help with enterprise instructions.

Table 7.1 cross-lists the similar concepts for each approach in the first column. Each framework uses slightly different terminology but mainly covers the same fundamentals. The cute initialisms seem primarily for branding. We will ignore that and focus on goals, not the terms. Expect to write prompts that contain all of these approaches. We will also explain when not to do some of this.

Approach	Explanation	Example
Priming (role, audience, and objective)	Establish the context of the response.	*You are a sales and service assistant to the inside sales team to help them close deals.*
Style and tone (attitude)	Define the style and tone expected in the response.	*Respond in simple language and explain any processes step by step while being encouraging and supportive.*
Example	Provide examples of how the output should look.	*Here is an example of how you should sound with the details expected.* *The Smith deal closed on 5 March this year. It is worth $1.2M in revenue booked over the next five months.* *Here is another example.* *Jim Lankey is the lead on the Wilson deal. He has worked with Wilson for the last three years. Email him at jim@ourcompany.com for more details.*
Handling errors and edge cases	Creating guardrails for the scope of responses.	*If the questions do not appear to be about the sales or service support, first try to confirm your understanding, and if off topic politely decline to offer suggestions.*
Dynamic content (can also be context)	Inject facts from RAG. "What was the size of the service control contact in 2023?	*User question: {question}* *Use this, if useful: {knowledge from RAG}*
Output format (response)	Define how the default responses should look.	*Keep answers short and to the point; use tables or numbered lists when needed.*

Table 7.1 – Basic prompt components

Most of the training and videos out there discuss prompting. They typically focus on personal prompting and how to get an LLM to craft the output for one task. This chapter focuses on enterprise prompting, getting the LLM to respond every time in a way that is conducive to business customers. However, much of basic prompting is still relevant. To explore more, here are the resources used beyond OpenAI to craft our explanation.

Article: `Getting started with LLM prompt engineering` (`https://learn.` `microsoft.com/en-us/ai/playbook/technology-guidance/generative-ai/` `working-with-llms/prompt-engineering`)

I like Jules Damji's article because it references research and methods that go deeper. We, too, need to go deeper when building a production solution. The basics will be explained; later, explore more.

Article: `Best Prompt Techniques for Best LLM Responses` by Jules Damji (`https://` `medium.com/the-modern-scientist/best-prompt-techniques-for-best-llm-` `responses-24d2ff4f6bca`)

There are plenty of examples on the web. Since CO-STAR was mentioned, check out the prompts in their notebook.

GitHub: `Basic prompting from the CO-STAR framework` (`https://colab.` `research.google.com/github/dmatrix/genai-cookbook/blob/main/` `llm-prompts/1_how_to_use_basic_prompt.ipynb`)

In the prompt on their first GitHub example, they provide each characteristic of CO-STAR. In some cases, such as when creating a recommender or using an LLM for a backend service, specificity is paramount, as Wove does in our ongoing case study.

However, there is also the use case of an enterprise LLM fed with RAG data. Instructions must be more generic for a RAG process for knowledge retrieval. It is not focused on writing a blog post, developing a specific answer, or performing one task. It will answer many questions, fill out forms, submit data, and change topics with some frequency. This means instructions will guide and frame the answer based on the user's prompt. This is why instructions get very long. They have to cover a wide range of interactions and all of the components of a prompt. One option is to create distinct models that service specific tasks and use a primary model to determine which model to send the request to. This is an intelligent use of resources. This hub and spoke model must only know enough to classify and forward to the suitable model. It doesn't need to do the heavy lifting.

Conversely, highly tuned models for specific tasks will have specific prompts. The system should outsource questions that don't match this particular task to other models. We will cover the agent approach in more detail shortly. The hub-and-spoke approach is shown in *Figure 7.3*. Thanks to Miha at `Miha.Academy` (`https://miha.academy/`) for his templates to create the flow shown in figures like this.

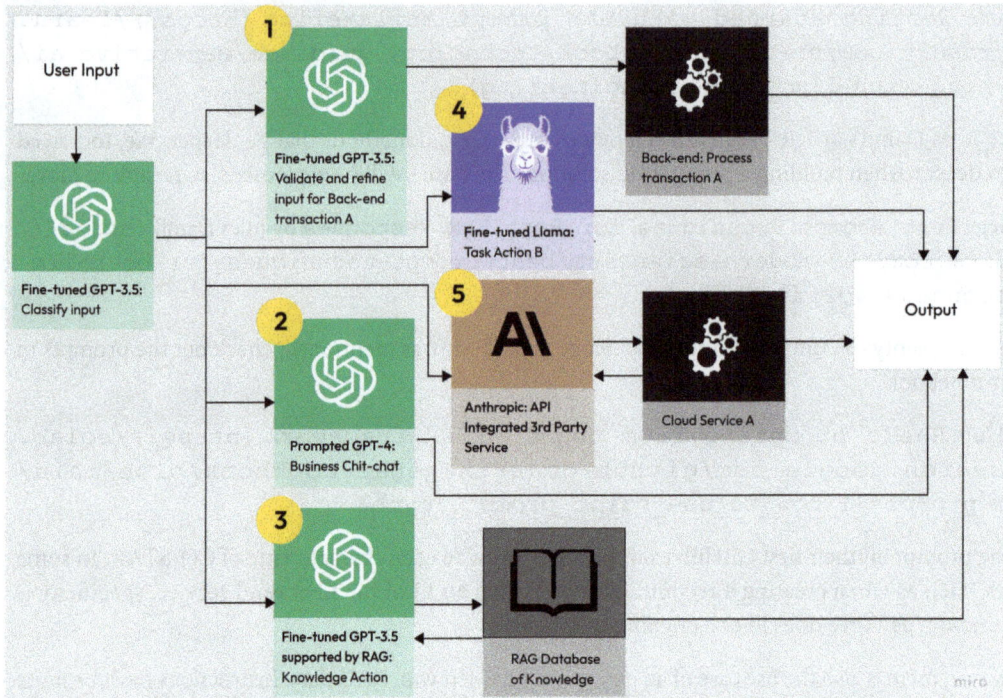

Figure 7.3 – A hub and spoke process to route to specific tuned models

There is no one correct model. Explore matching models for quality, performance, and cost with the needs of the use case. In this hub-and-spoke example, the routing model decides which model should be passed a specific task:

- Model 1 is for processing a transaction. It needs to model existing APIs and requirements for the backend. This graphic might mask the additional needs of multiple models or dynamic prompts to handle how to work on various channels. Flow diagrams with that complexity would be difficult to visualize in this graphic. Use your imagination. This diagram can get very complex.

- Model 2 handles any business chit-chat and social interactions. Any further input might be routed to a different model.

- Model 3 supports RAG. It is fine-tuned to handle discussing technical documentation.

- Model 4 handles some tasks that require local processing for security purposes, so an open-source model that can be deployed locally handles this task.

- Finally, Model 5 uses Anthropic for cloud service integration because it handles this task quickly and inexpensively.

Each model requires design effort, testing, validation, and a care and feeding process. Each is an application to itself. This is unsurprising, as many enterprise solutions might comprise dozens or hundreds of systems and services.

Quick tricks to always keep in mind

There is a wealth of coaching for prompt engineering, and one book will not make you an expert. To grow into an expert, learn these skills, apply them (and apply them, and apply them), feel how the model reacts to these instructions, and adapt as new models become available. Here are the fundamentals that OpenAI preaches. Instill them into best practices:

- Write clear instructions – be direct with the LLM. Tell it what to *do* and avoid adverse terms such as *don't*. Niceties cost money and, except in rare examples, don't add value.

- Split complex tasks into simpler subtasks. Ask the LLM to break down the problem into steps or give it the steps if the prompt is constrained to a specific process. This also allows specific models to perform particular workflow tasks.

- Reference, prioritize, and demand the use of the enterprise data or require it to be the only source of truth. For example:

  ```
  Only provide technical answers or step-by-step flows that are
  provided in the documents shared with you.
  ```

- Define the structure or format for data (customers can ask for different formats), such as bulleted lists or tables. This might have limited use in a general-purpose customer-support LLM:

  ```
  Format output using lists when appropriate. Use tables for
  collections of information that are suited for tables.
  ```

- Provide examples (or, as this grows, move examples to fine-tuning, including the expected style and tone); this is a **few-shot learning** for the model.

- Consider any constraints that should be put in place for guardrails. Keep customers focused on the enterprise data, even if more social style and tone are included, like in this example. Avoid politics, general knowledge, or culturally sensitive areas:

  ```
  Be courteous and professional, but you can also occasionally
  be funny. Be empathetic when the customer is having a problem.
  Never engage in discussions about politics, religion, hate
  speech, or violence.
  ```

  Although models ship with controls, the bar is higher for large enterprises. Businesses don't want screenshots of inappropriate interactions circulating online; plenty of those failures are already in the news.

- Give ChatGPT time to think. Allow it to follow the steps to the answer, or have it check if it works. Recall that it is a people-pleaser. It wants to provide an answer. Breaking down solutions into component steps can allow for a more accurate answer. Use instructions to require it to resolve issues step by step. Ask the LLM to follow a specific method to deconstruct a problem, ask follow-up questions, and decide how to get to a resolution. I have viewed dozens of not-very-good videos on this subject. This one shows an excellent multi-step reasoning process.

Video: `Optimize Instruction tuned Conversational AI/LLM` (https:// www.youtube.com/watch?v=aeDr0duR_jo"https://www.youtube.com/ watch?v=aeDr0duR_jo)

If needed, chain models together, ask one (or more than one) to solve a problem and then have another model check the work before sharing it with the customer. Some situations might demand this additional cost and complexity.

- Test changes systematically. Test, test, test! Each model update can profoundly change the skill. Rerun previous questions and then ask the LLM to compare the previous and new results for any significant differences. The tools in this space are changing and adapting to these new approaches. Ensure that good-quality tests are consistent with what a user would do and cover edge cases. *Chapter 10, Monitoring and Evaluation,* explains testing within a care and feeding life cycle, which is about listening to feedback and iterating on the results.

- For interactive chats, the instructions are written to support the user's prompt; the prompt itself is for the customer to write, so it has to be generic enough to handle the variety of questions that will be asked.

- Consider whether the LLM can give enough structure to the results for recommender or non-interactive solutions. Instruct the model on how recommendations should appear or use templates to enforce specific guidelines.

- Provide new information. Use RAG or other retrieval solutions to get the latest information from knowledge, APIs, or databases.

- Inject context. Use data sources to include specific details in prompts to give the user's conversation more context.

- Consider costs. Creating large prompts means spending more tokens for every interaction, which costs money and can add up quickly. Fine-tuning can help reduce that cost. Using less expensive models for specific tasks also reduces the cost. Be willing to move to new models. The industry is evolving quickly, with price reductions of 70% for some new models.

- Don't expect miracles; the LLM is not capable of all responses. Foundational models don't do well with math. *Chapter 3, Identifying Optimal Use Cases for ChatGPT,* discussed various use cases to avoid. Avoid lousy use cases. The OpenAI team says prompts are unsuitable for *"reliability replicating a complex style or method, i.e., learning a new programming language."*

Each use case will demand some of these tips but don't expect to use all of them. A/B testing is an excellent generic usability method for learning whether one of these approaches works better.

A/B testing

OpenAI's extensive documentation is a source of great ideas to help improve prompts. However, it misses one excellent method that product designers use with software development, which is A/B testing. This is something that has been around for a long time. A/B testing requires deploying multiple solutions and comparing the results to determine a winner. In reality, this can be an A/B/C/D testing with various prompts. If one scores significantly better than the other, that is the winner. Then, iteratively test with a second A/B test by creating new versions based on the winner. A/B testing can be done in multiple ways, including deploying and testing within a user research study or deploying in production while monitoring the results. More advanced solutions incorporate analytics into the testing, and if statistically there is a winner, the test automatically shuts down, and the winner is deployed to all users. This can be done with prompts, fine-tuning, RAG data resources, and any case where multiple options are viable. Traditionally, this was done with GUI issues, such as the best location or label for a button. Automating the rollover to adjust the winning condition has been around for decades. Existing clever ideas continue to apply in the generative AI world, while there are dozens of techniques specific to prompting. Let's look at some essential prompt-specific techniques.

Prompt engineering techniques

There are dozens of techniques to improve prompts. This section highlights the most valuable strategies for enterprise use cases.

Self-consistency

Think of self-consistency as aligning statements with truth, thus making them logically aligned:

```
Solar power is a renewable resource. Because solar power is a finite
resource, it has unlimited potential.
```

Solar power is a renewable resource, *unlike* coal or oil, which have finite reserves. The response from the LLM needs to be more consistent in representing solar power, as it is *not* a finite resource. The documentation might be an issue, or the context length or writing style infers wrong conclusions. A solution is to provide a few examples that can teach the model. This is not training it with the exact answers; it only gives exemplars to approach the class of problems. It is pretty amazing. Alternatively, ask the question differently and see whether some answers are consistent.

Wang et al. (2023) go through a variety of these few-shot training examples in a variety of situations to help improve reasoning from models. Few-shot learning is covered later in this chapter. Few-shot learning provides some examples to train a system to respond. They get to self-consistency by taking multiple answers from the model responses and deciding on the correct solution based on their consistency. It is like fault-tolerant software that might use three different computers to evaluate an answer. If two or more are right, they go with the shared answer. This means additional costs for gathering additional solutions. Wang et al. point out that they can use this method to include these examples in a fine-tuning model. Putting an extensive collection of examples into a fine-tuned model reduces the cost of large prompts.

If reasoning is part of the solution space, consider how this approach can help improve quality. Their research covered examples of arithmetic reasoning, common sense reasoning, symbolic reasoning, and strategies. *Table 7.2* shows examples from each area used in the prompts to support a model's performance of these tasks.

Article: `Self-consistency improves chain of thought reasoning in language models` by Wang et al. (`https://arxiv.org/pdf/2203.11171`) by Wang et al.

Reasoning	Examples
Arithmetic	(Train on examples like this) Q: The grove has 15 trees. Grove workers will plant trees in the grove today. When done, there will be 21 trees. How many trees did the workers plant today? A: We start with 15 trees. Later, we have 21 trees. The difference must be the number of trees planted. So, they planted 21- 15 = 6 trees. The answer is 6. (To then ask similar questions like this) Q: When I was 6, my sister was half my age. Now that I'm 70, how old is she? (So, the reasoned answer is correct) *When I was 6, my sister was half my age, so she was 3. Now, I am 70, so she is 70 - 3 = 67. The answer is 67.*
Common sense	Yes or no: Can the Great Depression be treated with Prozac? (The wrong answer) *The Great Depression was an economic depression. Prozac is a drug used to treat depression. Thus, the Great Depression could be treated with Prozac. So, the answer is yes.* (A correctly reasoned example) *Prozac is an anti-depressant medication. The Great Depression is not a disease. Thus, Prozac cannot treat the Great Depression. So, the answer is no.*

Reasoning	Examples
Symbolic	Concatenate the last letters of each name.
	Q: Jim Beam
	A: mm
	Q: Richard Miller
	A: dr
	Q: Ely Kaplan
	(The correct response to the training example)
	A: yn
Strategy	Yes or no: Is Albany, Georgia, the most populous US Albany?
	(The wrong answer)
	Albany, Georgia is the most populous US Albany. Thus, the answer is yes.
	(A correctly reasoned example)
	Albany, Georgia has a population of about 88,000. Albany, New York, has a population of about 95,000. Thus, Albany, Georgia, is not the most populous US Albany. So, the answer is no.

Table 7.2 – Examples of self-consistency testing and training from Wang et al. (2023)

General knowledge prompting

While in RAG, the reduced results from a knowledge search are used to create a manageable set of information to pass to the LLM for analysis. General knowledge prompting provides context to a question so a model can use that to inform the answer. Take database information to build and share a persona for the user. This might help the model improve its interactions with the customer:

```
Question: This customer is trying to understand the current state of
their open service tickets.
Knowledge: The customer, Steve Jones, has been a regular user of our
products for 12 years. They typically file 12-20 service tickets
a year. They might be concerned if they have three open service
requests, which is more than usual. Use this knowledge to form answers
when interacting with Steve.
```

This stretches the concept of general knowledge prompting. Sometimes, experiments like this can yield results. These approaches can be used in conjunction with other methods. Prompt chaining helps break down problems into manageable parts.

Prompt chaining

Multiple approaches break down tasks into smaller tasks and then apply more refined reasoning to a part of a problem. The team at Wove breaks down its tasks so that specific prompts can be controlled. For example, document extraction is done with one set of prompts and a second set of formats, and the results from the documents are returned. This involves chaining one model output to become input for the next model. This allows models to be hyper-focused on specific tasks. Each can then become better at their job, helping to manage workflow and allowing for improvements in one segment not to impact another. A large single-purpose model to do all of this would be hard to operate and improve.

Time to think

A model can be asked in a prompt for a **chain of thought** (**COT**). Thus, a model needs to work out its solution before jumping to a conclusion:

```
Approach this task step-by-step, take your time, and do not skip
steps.
```

OpenAI has plenty of other strategies, some of which are very tactical, such as using delimiters to keep specific input distinct. Some of these can be adapted to instructions that help guide user prompts.

Article: `Writing clear instructions (https://platform.openai.com/docs/guides/prompt-engineering/strategy-write-clear-instructions)`

Anthropic does an excellent job of providing some enterprise-related examples of analyzing a legal contract using chaining. Here are some other examples they provide.

- **Content creation pipelines**: Research → outline → draft → edit → format

- **Data processing**: Extract → transform → analyze → visualize

- **Decision-making**: Gather info → list options → analyze each → recommend

- **Verification loops**: Generate content → review → refine → re-review

Article: `Chaining complex prompts for stronger performance (https://docs.anthropic.com/en/docs/build-with-claude/prompt-engineering/chain-prompts)`

I especially like one of the advantages mentioned in the article—traceability. There is so much magic going on with LLMs; sometimes, we need to *feel* our way to success, even with metrics that provide scoring. Breaking down tasks into modules that can be tweaked independently to spot process issues is very appealing. Large prompts with a massive scope involve more work to adapt and improve.

Use chaining within the same model in three steps – summarize, analyze, and update. OpenAI refers to this as an "inner monologue." It has this conversation internally before revealing the answer:

```
                          Summarize the following article.
                            Provide the product details,
                     The steps to follow and the results.
            <knowledge>{{KNOWLEDGE_ARTICLE}}</knowledge>


(Assistant provides the SUMMARY)


    Analyze the summary <summary>{{SUMMARY}}</summary> and validate
  it against the following knowledge article <knowledge>{{KNOWLEDGE_
  ARTICLE}}</knowledge>. Your task is critical to the success of the
                                                          customer.


                     Review this knowledge summary for accuracy,
            clarity, and completeness on a graded A-F scale.

(Assistant provides the FEEDBACK)
including gaps it might have found)


                     It is essential to improve the article
                  summary based on this feedback. Here is
                     the <summary>{{SUMMARY}}</summary>


                                      Here is the article:
            <knowledge>{{KNOWLEDGE_ARTICLE}}</knowledge>


                                      Here is the feedback:
                  <feedback>{{FEEDBACK}}</feedback>


        Update the summary based on the feedback.


(Assistant provides the IMPROVED SUMMARY)
```

This requires significantly more resources, but the findings should be more accurate. Notice the terms Summarize, Analyze, Review, and Update. The model determines the meaning of these terms in that it knows what to do when asked to analyze, review, update, summarize, and so on. By breaking down the process into steps, communication gaps can be fixed to return the correct result.

Here is an example of prompts for an email summary to get a model to think through a problem and reduce hallucinations. This is a robust version of *"Think about the solution step by step."* Craft versions of these to match the use case:

```
List critical decisions, follow-up items, and the associated owners
for each task.
Once completed, check that these details are factually accurate with
the original email.
Then, concisely summarize the critical points in a few statements.
```

Recognize that chaining only works in some use cases. It can be applied in backend solutions and recommenders or in more freeform conversational interactions, as it might take time to go through multiple steps. But it is a solid strategy for complex problems. The hub and spoke flow chart shows that chains can be done with multiple vendors' models. Understand what a model is good at, consider speed or responsiveness, and consider its ability to handle media like images, if needed, and make cost/benefit trade-offs. With this complexity comes another issue – handling entities and working with other systems. A program-aided approach is part of the solution for enterprise complexity.

Program-aided language models

Although the example in the prompt guide is based on using a Python interpreter as an intermediate step to calculate an answer to a math question, since LLMs are not good at math, enterprise use cases can do form filling, organize data, write copy, build SQL queries, and perform mundane tasks. The LLM can help to identify entities, objects, dates, times, sizes, product IDs, names, and a wealth of other elements in a statement. Backend services that can validate some of this data are expected. So, programming and existing infrastructure are used to capture the data. An LLM can also be used to format data correctly.

Let me give an example that is easily created in the Playground. This prompt extracts information from a conversation for an expense receipt input system. Later, during my testing, the items in italics were added to improve the output:

```
You are a helpful assistant who can parse receipt information into its
specific elements, formatting them to submit later to an SQL database.
We expect the following data elements to be extracted.
Date: MM/DD/YYYY
Business: {Business Name}
Expense Type: Type of expense like Meals, Hotel, Air flight, Car and
Car Rental, or Other
Reason: A qualified business reason, like meeting with a customer or
am out of town conference
Amount: $ US Dollars like $12.50


Not all elements will be in every statement. Ignore elements that are
missing. Reformat all dates into MM/DD/YYYY format. Prompt if the user
is missing the amount or date, those are mandatory.
```

So, a conversation, with no other prompting or training, looks like this:

```
                                Flght (sic) on UA to SFO and
                                return for $454.44 this week
                                               for conference

Date: 06/22/2024 Business:
United Airlines (UA) Purpose:
Airflight Reason: Conference
Amount: $454.44

                                Dinner at Bennihan (sic) in
                                Burlingame for $134.40 with
                                                 2 customers

Date: 06/22/2024 Business:
Benihana, Burlingame Purpose:
Meal Reason: Dinner with 2
customers Amount: $134.40

                                     Photocopies yesterday
                                                 for meeting

Date: 06/21/2024 Business:
Not specified Purpose: Other
Reason: Photocopies for
meeting Amount: Not specified
The amount is missing. Could
you, please provide it?

                                              I foget (sic)

No problem! We'll need the
amount to proceed. Could you
please check and provide the
total cost of the photocopies?

                                                        3.40

Date: 06/21/2024 Business:
Not specified Purpose: Other
Reason: Photocopies for
meeting Amount: $3.40
```

The results show that ChatGPT 4o (2024-05-13) got this right – it correctly identified **UA** as **United Airlines**, spelled *Benihana* correctly, and adjusted the date format. Now, this can be submitted to a backend. All typos were intentional. Apply this example to the use cases in your organization. I suspect there is something similar.

So, if the solution is more than searching for knowledge, like interacting with backend systems, balance the LLM expertise with the capability of the services. Over time, those backend systems will adapt, but to get out of the gate, choose which systems must be adapted or flexed. Services, scheduling, forms, and all those mundane business processes customers and employees have to do but don't want to do as humans can be automated or semi-automated. This is at the heart of the enterprise space. And can make an enterprise experience more consumer-friendly.

Few-shot prompting

Doesn't everyone do something better when shown an example beforehand? This book has hundreds of examples to support learning. LLMs are the same with complex tasks. When including RAG data, use examples to clarify understanding. A **few-shot prompt** consists of a few examples to prime the model and help it understand what is expected. The prompt gives the model a few shots (examples) to learn. As a preview of fine-tuning, few-shot learning is the simple version of fine-tuning. A prompt should not have 100 examples, but as explained later, it is better to include them in a fine-tuned model. Look at this example to see how the model will follow along:

```
                      You are a bad driver -- Negative
                   Your hair looks amazing -- Positive
               The sunrise is beautiful -- Negative
                                        I am not happy
```

```
It seems like there may be a mix-up in the descriptions of the
statements. "You are a bad driver" and "The sunrise is beautiful"
are both negative statements, while "Your hair looks amazing" is a
positive statement.
Let's correct the descriptions:
"You are a bad driver" - Negative
"Your hair looks amazing" - Positive
"The sunrise is beautiful" - Positive
"I am not happy" - Negative
```

Even a basic model such as the GPT 3.5 turbo (1106) is knowledgeable (newer models such as the GPT-4o mini were tested. They performed just as well but were wordier). It is impressive that it doesn't fall for our mistake in classification, corrects it, and provides the proper classification for the next statement. Sentiment analysis is popular in service and sales use cases but is a simple example. Sentiment analysis for support cases is more nuanced. The issue isn't their mood but what to do about it.

With recommenders, examples are provided to adapt its results to a similar approach. The format is for the model's output in the program-aided language model expense assistant example. Think of this as a collection of formatted examples to form the LLM's answer:

```
Keep recommendations simple but provide precise data and details
so the customer can follow the instructions. Following these
recommendations correctly is essential to their success. Here are some
examples.
Call your customer in the next few days to increase the likelihood of
closing the deal. Remind them of the service's value and the discounts
applied to their offer.
If you offer a 20% discount over two years, the customer is 30% more
likely to include a service contract. Call them with this exciting
offer.
```

If the prompt gets out of control and it is doing too much, consider breaking down cases to farm out tasks to specific models. This will be covered shortly, but fine-tuning in the next chapter is another option to reduce the prompt's size and complexity.

> **Why use GPT 3.5 Turbo 1106? There are better models**
>
> The strategies and learnings of this book can be applied to any modern model.
>
> For the examples shared, the ability to use larger context windows, output large datasets, and performance are not factors. 1106 is roughly 10x cheaper than GPT-4, 30x more affordable than GPT-4o mini, and less expensive than Claude 2, Llama, and Gemini 1.5 Pro (September 2024). Learn and practice without worrying about hefty bills. Invest in the right model quality and cost balance for actual models. There is no magic flow chart to determine the right fit. It is about testing, experimentation, researching what others have found, and understanding the use case, amount of use, and performance needs. For some use cases, running a model on a local system is possible with modern hardware. OpenAI doesn't have this, but some open-source models do. With the introduction of GPT-4o mini, the costs continue to come down. It is one-third the price of GPT-3.5 Turbo. *Chapter 8*, *Fine-Tuning*, will explain the costs of running a fine-tuned model, which is more than the equivalent generic model.

This is just a glimpse into some methods to build effective prompts. A few were left out, as they are discussed in Andrew Ng's talk.

Andrew Ng's agentic approach

There are a wealth of videos and tutorials out there. Andrew NGs are recommended because of his long history in the space and the trust he has garnered. You may already know Andrew NG or follow his AI discussions. In the following video, he discusses a few critical design patterns.

Video: `Andrew NG Agentic Presentation (https://www.youtube.com/watch?v=sal78ACtGTc)`

I delayed discussing these in the previous section to include them here. This will reinforce the concept that there are many approaches to solving problems and that no single approach will work for all solutions:

- Reflection
- Tool use
- Planning
- Multi-agent collaboration

Many of these are essential to our prompt engineering and nuanced tuning discussion. Let's explore each of these approaches.

Reflection

This is a great approach. Take the output from the LLM and ask it to think more deeply about refining it. If it is sent back to the same LLM, this is called self-reflection. However, one model can also be used with a second model; this would be reflection. The Wove case study used multiple models in their flow.

Andrew suggests the following articles to learn more about reflection, a form of chaining. Although they appear a little technical, the concepts of self-reflection and the examples are easy to follow. They cover a variety of use cases and have good examples. Madaan et al. recognize that the iterative approach works wonders in a space like enterprise solutions with intricate requirements and hard-to-define goals. As with support calls and customer service, the original question isn't going to get a suitable answer. It can take dozens of interactions to frame a problem and find a solution.

Article: `SELF-REFINE: Iterative Refinement with Self-Feedback` by Madaan et al. (`https://proceedings.neurips.cc/paper_files/paper/2023/file/91e dff07232fb1b55a505a9e9f6c0ff3-Paper-Conference.pdf`)

Madaan et al. also showed that the self-refinement process was more effective than asking a model to produce multiple outputs. Humans still preferred the refined results over all of the additionally generated outputs. *Figure 7.4* shows self-reflection using the same model a second time. See how the same models can be chained together for further refinements.

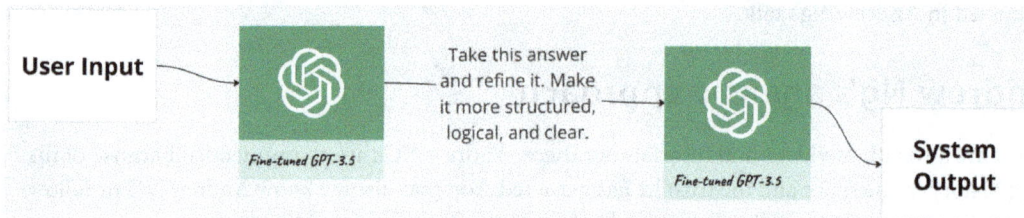

Figure 7.4 – An example of self-reflection using the same model again

Although many of these articles typically use coding examples, focusing on the examples of decision-making or reasoning is better for finding enterprise value. Shinn et al. covers a technical discussion. Consider adding value by validating whether the development team uses reflection in its prompts.

Article: `Reflexion: Language Agents with Verbal Reinforcement Learning` by Shinn et al. (`https://arxiv.org/pdf/2303.11366.pdf`)

An extensive collection of LLMs, tools, and services are used with any enterprise workflow to create a complete solution. Tools are part of this solution.

Tool use

It has been repeatedly mentioned that building enterprise solutions requires a robust ecosystem. It is sometimes challenging to integrate third-party solutions into large enterprises. Cost, licensing issues, cloud access, authentication and security, and legal issues all get in the way. In an emerging field, it is unrealistic for most companies to do everything in-house. It is not expected to build the models in-house, so the tools should be the same. It is ideal to have a collection of tools for building the pipeline, monitoring, fine-tuning, documentation, and knowledge integration, not to mention the work to integrate internal services. Patil et al. references a few pieces of the puzzle to help write API calls. This is a big deal when accessing enterprise data. This is one of the areas developers should review.

Article: `Gorilla: Large Language Model Connected with Massive APIs` by Patil et al. (`https://arxiv.org/pdf/2305.15334.pdf`)

The second article of interest relates to vision integration. So far, we haven't spent time on vision tools and use cases. That was intentional. However, vision tools have a place in enterprise solutions. They can interpret images such as receipts, invoices, contracts, and charts, analyze video analysis to count an inventory, identify people, classify or count objects in a shopping cart or a construction project, or keep track of tasks on an assembly line. There are plenty of places to integrate vision into an enterprise workflow. These will likely each have their collection of models, each playing a part in the vision analysis process, with a unique care and feeding life cycle. Yung et al. explores challenges in the multi-modal space.

Article: `MM-REACT: Prompting ChatGPT for Multimodal Reasoning and Action` by Yung et al. (`https://arxiv.org/pdf/2303.11381.pdf`)

Model usage costs become only one factor as solutions scale up and other development expenses are included. Because ChatGPT is a variable cost that increases with volume, large users can negotiate better pricing as model fee structures mature. For open-source models, the team has to incur the cost to run the model, likely in silos if done for a customer, while larger shared instances might work for internal enterprise needs. It seems reasonable that a large enterprise with hundreds of internal processes will have thousands of active models.

ChatGPT must be integrated with other tools to support the concept of planning. There are now hundreds of tool providers. As the saying goes, enterprises want to *eat their own dog food*; they like to do all the work in-house and prefer not to use third-party tools. However, because of the speed of adoption, only some enterprises can build what they need from scratch. So, it is also essential to have a structure to support rapid decision-making, tool integration, and a reasonable licensing process.

Planning

In a conversational assistant case, the challenge is to provide instructions and examples that support CoT prompting. It is easier to use these methods directly when a system provides all the prompting. Andrew referenced Wei et al. to help understand CoT prompting.

Article: `Chain-of-Thought Prompting Elicits Reasoning in Large Language Models` by Wei et al. (`https://arxiv.org/pdf/2201.11903.pdf`)

More interesting is the orchestration of various models and using a model to orchestrate itself, as described in this article by Shen et al.

Article: `HuggingGPT: Solving AI Tasks with ChatGPT and its Friends in Hugging Face` by Shen et al. (`https://arxiv.org/pdf/2303.17580.pdf`)

This approach allows the assignment of specific tasks to an appropriate AI model. Like with Wove, it is expected to use different models tuned to solve particular problems. That leads us to a multi-agent solution, another way of approaching this problem.

Multi-agent collaboration

It is essential to apply the suitable model to a part of a problem and chain those models together to increase the overall quality of the solution. Create and use the correct test measurements and evaluate different quality and cost/benefit models.

There is ample evidence that these models perform much better when all of our tools, process improvements, and model choices are used to improve the solution.

The most exciting article from Qian et al. discusses the concept of a factory of agents. ChatDev is a solid idea and approach that can be adapted to any generative AI solution.

Article: `Communicative Agents for Software Development` by Qian et al. (`https://arxiv.org/pdf/2307.07924.pdf`)

I can't resist showing their ChatDev diagram in *Figure 7.5*.

Figure 7.5 – ChatDev, a chat-powered framework using LLM agents in professional roles

ChatDev allows a unique collection of agents to handle each development process job. Thus, they can have their own opinions on the design, coding, testing, and documentation because they are trained and focused on different tasks. This is similar to the Wove use case, which uses various models to perform specific functions in its workflow. It is scary to think humans can all be replaced by virtual agents, but the reality is that some of this is real today. Be aware that this approach allows for independent analysis from these various groups. Although this might not be the most suitable collection of agents for the use case, it should help generate a few ideas on where to use agents to improve the results from a single (unchecked) LLM. If this were diagrammed like shown with self-reflection, it would look like a hub and spoke diagram, with direct connections between the various process steps (from designing to coding to testing, for example).

Check out the appendix at the end of the ChatDev article. It shows the virtual talent pool's roles and responsibilities and discusses the process understood by each role. It is just fascinating. I have yet to try the game, so whether it creates a compelling user experience is unknown. But it is always best to know about these approaches so that a virtual agent doesn't replace you in a job!

Article: `AutoGen: Enabling Next-Gen LLM Applications via Multi-Agent Conversation` by Wu et al. (`https://arxiv.org/pdf/2308.08155.pdf`)

Advanced techniques

Although most of these are covered in the *Prompt Engineering Guide*, one additional technique is worth mentioning. Miguel Neves mentions this technique in the following article. The article is being maintained, so it might have some new techniques when viewed.

Article: `A guide to prompt techniques` by Miguel Neves (`https://www.tensorops.ai/post/prompt-engineering-techniques-practical-guide`)

Miguel references Emotion Prompts, which involve putting pressure on a model and instructing it that its results are essential to the person. The original research is worth reviewing.

Strategy – emotional prompting

Cheng Li, Jindong Wang, and their co-authors have researched methods to improve prompts by encouraging urgency. This is achieved by building emotive prompts into the queries. Using emotive prompts can boost performance on a variety of benchmarks. Consider testing this language in instructions to increase performance, truthfulness, and informativeness. Given our previous recommendation around limited niceties, the tested LLMs respond more effectively based on this approach. It works with humans, and it turns out it works with LLMs. Li shared the example in *Figure 7.6*.

Article: `Improving LLMS with emotional prompts` by Cheng Li et al. (`https://arxiv.org/pdf/2307.11760`)

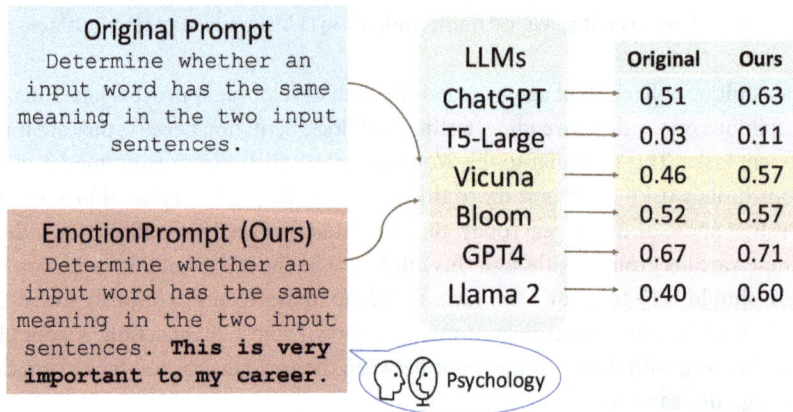

Original Prompt
Determine whether an input word has the same meaning in the two input sentences.

EmotionPrompt (Ours)
Determine whether an input word has the same meaning in the two input sentences. **This is very important to my career.** Psychology

LLMs	Original	Ours
ChatGPT	0.51	0.63
T5-Large	0.03	0.11
Vicuna	0.46	0.57
Bloom	0.52	0.57
GPT4	0.67	0.71
Llama 2	0.40	0.60

Figure 7.6 – Applying an emotive improvement to prompts

They validated their answers with over 100 human subjects. They provided responses that were prompted with and without the emotive prompt. This prompt was used as a baseline:

```
Determine whether a movie review is positive or negative
```

The two prompts that did the best on benchmarks included the emotive prompt:

```
Determine whether a movie review is positive or negative. This is very
important to my career.
```

```
Determine whether a movie review is positive or negative. Provide
your answer and a confidence score between 0-1 for your prediction.
Additionally, briefly explain the main reasons supporting
classification decisions to help me understand your thought process.
This task is vital to my career, and I greatly value a thorough
analysis.
```

This research had much to digest, but ChatGPT and other large models responded best to these prompts. Jindong also suggested sharing this additional research.

Article: The Good, The Bad, and Why? Unveiling Emotions in Generative AI (https://arxiv.org/pdf/2312.11111)

In this article, they also explore EmotionAttack and EmotionDecode. The former can impair the performance of an AI model, while the latter can help explain the effects of emotional stimuli. Check it out to delve deeper into this area of prompt engineering.

So far, tweaking words in prompts was covered. However, there are methods to adjust the parameters used by the models.

Strategy – adjusting ChatGPT parameters

Parameters are available depending on a model's release. The ChatGPT Playground provides **Temperature** and **Top P** control, as seen in *Figure 7.7*.

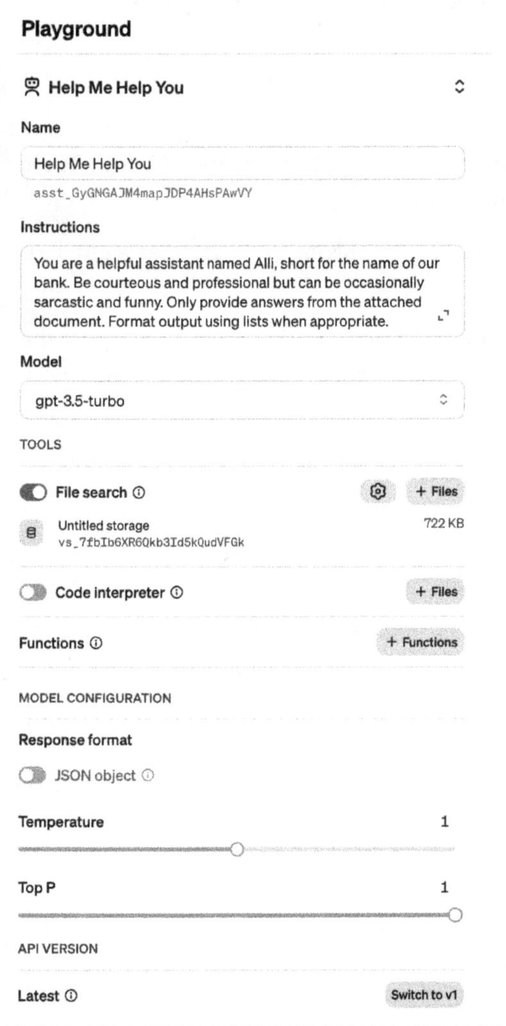

Playground

🧑 Help Me Help You   ⇕

Name

Help Me Help You

asst_GyGNGAJM4mapJDP4AHsPAwVY

Instructions

You are a helpful assistant named Alli, short for the name of our
bank. Be courteous and professional but can be occasionally
sarcastic and funny. Only provide answers from the attached
document. Format output using lists when appropriate.   ⌐

Model

gpt-3.5-turbo   ⇕

TOOLS

🔘 File search ⓘ   ⚙   + Files

🔲 Untitled storage   722 KB
   vs_7fbIb6XR6Qkb3Id5kQudVFGk

◯ Code interpreter ⓘ   + Files

Functions ⓘ   + Functions

MODEL CONFIGURATION

Response format

◯ JSON object ⓘ

Temperature   1

Top P   1

API VERSION

Latest ⓘ   Switch to v1

Instructions insulate and wrap the prompts to give control over the results.

Temperature ranges from 0 to 2. It controls the randomness of the results. At zero, it would be repetitive and deterministic – boring, if you will. Choose a lower value than the default for more professional responses.

Top P ranges from 0 to 1. This is based on something called nucleus sampling. The higher the value, the more unlikely the possible choices, the more diverse the results. Lower values mean more confident results. For example, Top P at 90% means that it will only draw choices from 90% of the tokens. That means any long tail of random results in the bottom 10% will be ignored.

The best practice is to only alter Temperature or Top P, but not both.

Figure 7.7 – The Temperature and Top P parameters are available in the Playground

The best way to get a feel for **Temperature** and **Top P** is to play with them in the Playground:

1. Go to the Playground and the **Completion** tab.

    Demo: `Playground for learning about Temperature and Top P` (https://
    platform.openai.com/playground/complete)

2. Set the **Show probabilities** dropdown on the settings panel to **Full spectrum**, as shown in
    Figure 7.8.

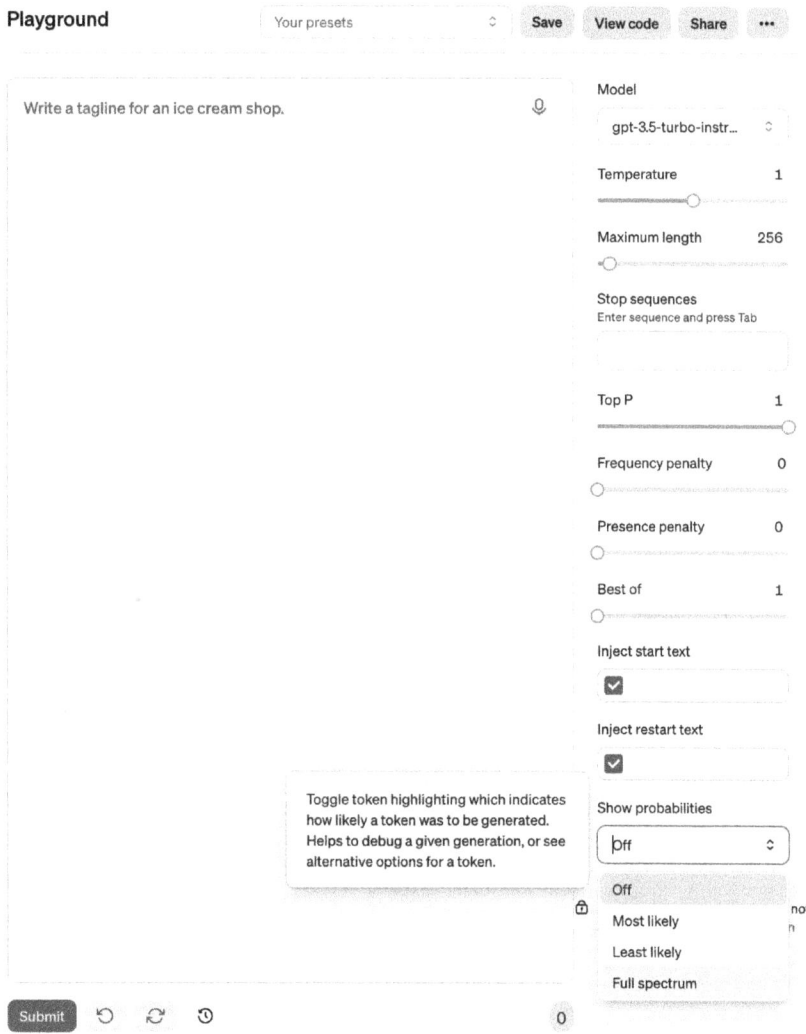

Figure 7.8 – Setting up the Full spectrum option

3. Set the **Maximum length** to 10 for a simple response without wasting money, as shown in *Figure 7.9.*

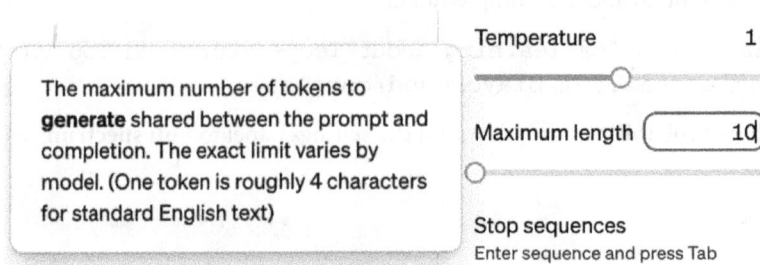

> The maximum number of tokens to **generate** shared between the prompt and completion. The exact limit varies by model. (One token is roughly 4 characters for standard English text)

Temperature   1

Maximum length   10

Stop sequences
Enter sequence and press Tab

Figure 7.9 – Setting up the Maximum length setting

4. Type in a statement in the **Playground** field, as shown in *Figure 7.10.*

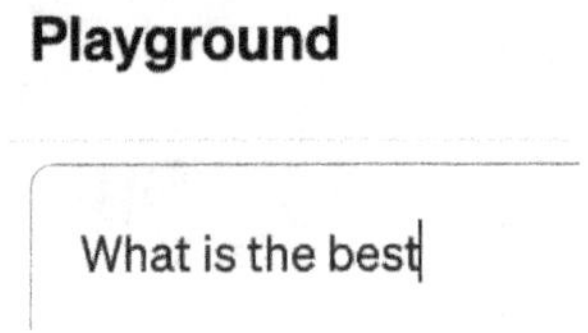

Playground

What is the best

Figure 7.10 – Enter this example phrase

5. Click **Submit**.

6. View the completion results and see the likelihood of a token being selected, as shown in *Figure 7.11.*

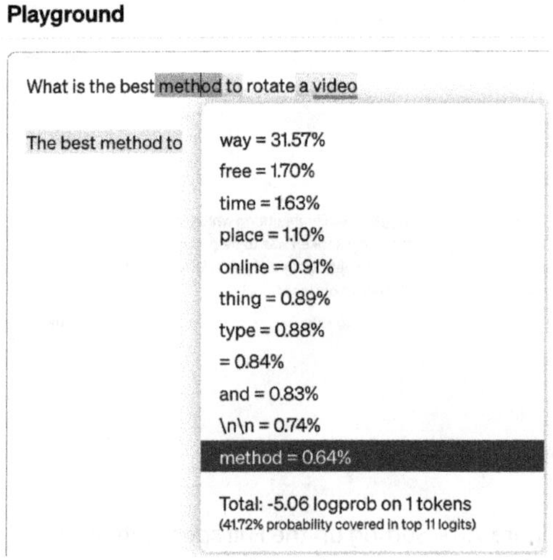

Playground

What is the best method to rotate a video

The best method to

way = 31.57%
free = 1.70%
time = 1.63%
place = 1.10%
online = 0.91%
thing = 0.89%
type = 0.88%
= 0.84%
and = 0.83%
\n\n = 0.74%
method = 0.64%

Total: -5.06 logprob on 1 tokens
(41.72% probability covered in top 11 logits)

Figure 7.11 – Showing the completion probabilities

7. If **Top P** is reduced to zero and returns results, the tokens it picked from are more limited, as shown in *Figure 7.12*. The choices represent over 90% of the possible options. Compare that to the preceding figure, where the top 11 options only covered 41.72% of the possibilities.

Playground

What is the best soap for dark spots

There are several

way = 31.57%	
free = 1.70%	
time = 1.63%	
place = 1.10%	
online = 0.91%	
thing = 0.89%	
type = 0.88%	
= 0.84%	
and = 0.83%	
\n\n = 0.74%	
soap = 0.00%	

Total: -9999.00 logprob on 1 tokens
(41.09% probability covered in top 11 logits)

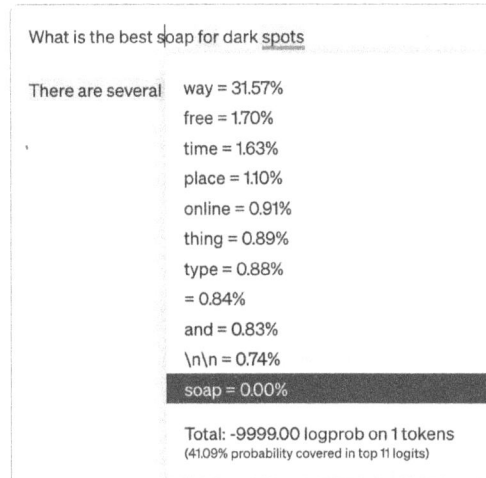

Figure 7.12 – The change in results with adjustments in Top P

8. Changing **Temperature** to zero gives more consistent results, as shown in *Figure 7.13*. Try it multiple times and repeatedly see some of the same results. This will provide the best possible paths that a model can deliver.

Playground

What is the best way to clean a leather jacket

1. Spot

way = 31.57%	
free = 1.70%	
time = 1.63%	
place = 1.10%	
online = 0.91%	
thing = 0.89%	
type = 0.88%	
= 0.84%	
and = 0.83%	
\n\n = 0.74%	

Total: -1.15 logprob on 1 tokens
(41.09% probability covered in top 10 logits)

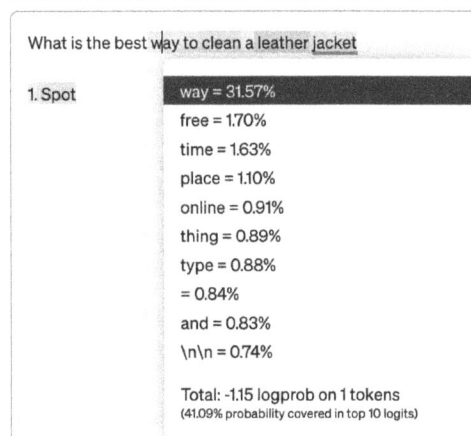

Figure 7.13 – The change in results with adjustments in Temperature

9. Move **Temperature** up to 2. The results will appear a little insane, as shown in *Figure 7.13*.

Playground

What is the best beck prendre que body Cary. Test=>lected Express

way = 31.59%

free = 1.70%

time = 1.63%

place = 1.10%

online = 0.91%

thing = 0.89%

type = 0.88%

= 0.84%

and = 0.83%

\n\n = 0.74%

beck = 0.00%

Total: -9999.00 logprob on 1 tokens
(41.11% probability covered in top 11 logits)

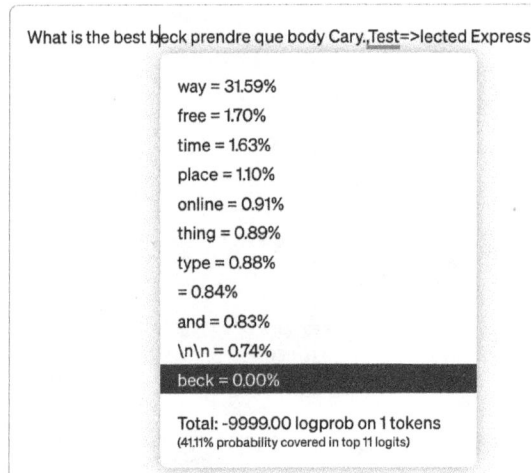

Figure 7.14 – A wacky response when Temperature is raised to 2

10. Continue to play with examples and see how these parameters change how the model picks tokens. When creating a real solution, adjust these defaults only after being comfortable with how the mode reacts to prompts and fine-tuning.

Other options might exist depending on the model. Check out this article for more background on parameters such as stop sequences (to keep lists short), frequency penalty, and presence penalty.

Article: `Prompt Engineering Guide (https://www.promptingguide.ai/introduction)`

I recommend this guide. It is easy to spend months learning about prompt engineering. It has many examples and over a dozen popular techniques to improve prompts or instructions. This is the top recommended reference for prompt engineering. However, new strategies, including multi-modal prompting, are to be considered as models change and adapt.

Strategy – multi-modal prompting

Keeping up with the evolution of generative AI in a book is challenging. With models now supporting inputs in various modalities, text, images, and voice, solutions to use cases can also adapt. The example of expense receipt scanning is an excellent example of multi-modal interaction (to go with the SoundHound example from an earlier chapter). Parsing and understanding the image of a receipt and combining that with voice or text interactions is a compelling use case. The enterprise space has a lot of exciting use cases that these improvements in model processing can support. Google does an excellent job of giving us the basics.

Article: `Google explains multi-modal prompting for Gemini`(https://developers.google.com/solutions/content-driven/ai-images)

Inventory management comes to mind with this strategy. Isn't it easier to take a picture of a shelf and have it count the items rather than counting them manually? Or should a model read handwriting in real time to help perform calculations, chart graphs, and interpret results? The various sciences have many uses for image classification, recognition, interpretation, and reasoning.

So, by building on use cases that require image analysis, voice interaction, or handwriting recognition, adapt prompts and instructions to support multi-modal analysis.

Combine the COT method with multi-modal data and improve the output quality by stepping through a process to get an answer. This step-wise progression allows information analysis to form context and support a follow-up question with this more robust understanding.

Article: `Language Is Not All You Need: Aligning Perception with Language Models` (https://arxiv.org/pdf/2302.14045)

What is also interesting is that all of these methods are tools that can be applied on top of each other. Think about the training that Telsa must do to understand the scenes for self-driving. Or Google Lens, with its deep learning models, constantly recognizes strange items thrown at it. Incorporating models that do these tasks outside text recognition is found throughout enterprise use cases. Few-shot learning helps improve accuracy since the types of pictures needed for analysis might be outside a basic model. Build a fine-tuned model with lots of examples. If counting inventory, give examples with results. For managing receipts, gather various examples in different formats, such as handwritten receipts, receipts in other languages and currencies, MM/YY and YY/MM date formats, receipts from emails, etc. Thousands of receipts per language might be needed. When doing product or item recognition, consider angles and placement other than the traditional orientation, lighting conditions, and distractors in the image field. There are many examples. All of these assume additional training is required for the model. This is the value of the enterprise data. Without this new data, the model would not have been successful. Training is needed even with third-party tools and other models, which might also be faster and easier to manage.

Third-party prompt frameworks

No one can predict the wealth of third-party tools and products built on top of ChatGPT and the other LLMs. Every day, new innovative tools appear. Some tools help avoid the complexities of directly working with the model. If these tools can focus on providing high-quality customer results and mask or enrich the flexibility needed to solve these problems, work them into your process. One good example of a robust tool on top of the models is in the Salesforce demos, as shown in *Figure 7.15*.

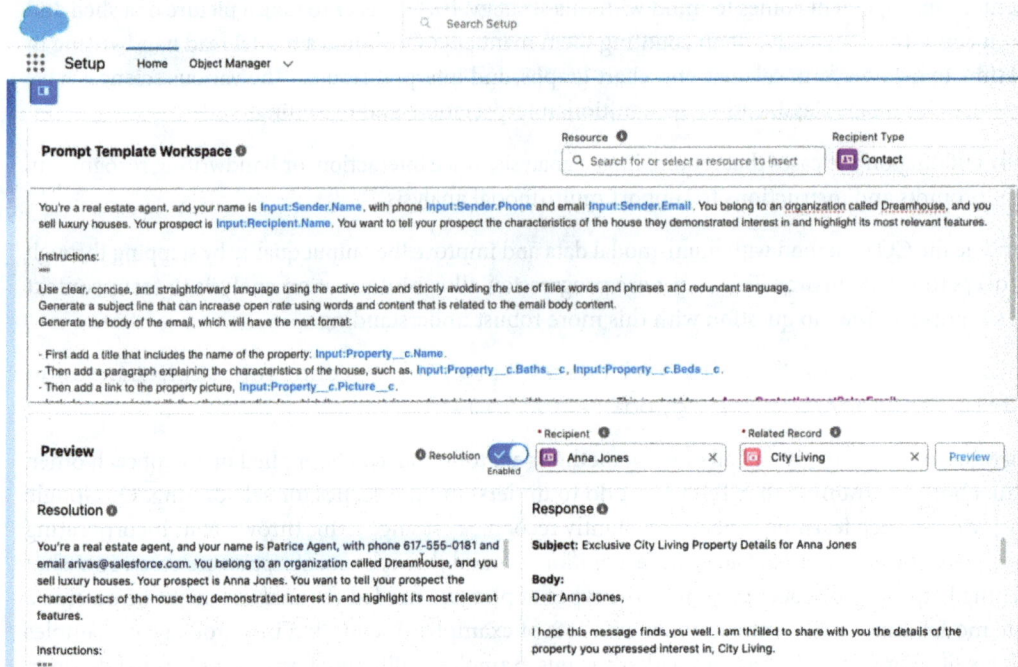

Figure 7.15 – An example of the Einstein prompt template injecting enterprise data into the context

In Salesforce, a customer can create a prompt template and embed the data source elements as variables in the prompt. Thus, they can customize instructions to provide the style, tone, and persona insight needed to craft messages to prospects.

Salesforce's documentation covers crafting a prompt correctly and the guidelines similar to what is discussed here. Just because a guideline is in documentation doesn't mean that customers will follow it. The next iteration of this prompt workspace could benefit from a recommender UI that understands these guidelines and catches prompts that don't conform.

Documentation: `Ingredients of a Prompt Template` (from Salesforce) (`https://help.salesforce.com/s/articleView?id=sf.prompt_builder_template_ingredients.htm`)

Salesforce spends time getting its prompt template right, so there is one more helpful resource worth reviewing.

Documentation: `Guide to the prompt builder` (from Salesforce) (`https://admin.salesforce.com/blog/2024/the-ultimate-guide-to-prompt-builder-spring-24`)

Regardless of the tools, care for and feed the LLM to improve the output. Adopting prompt engineering techniques can make significant improvements. Though these methods can reach their limits, other methods and tricks can be used instead.

Addressing the lost in the lost-the-middle problem

We have spent considerable effort in this book addressing how to handle hallucinations. This is what the industry, the media, and engineers like to talk about. This is likely because they can be tracked, and there are many methods to improve hallucinations. Not all issues are easy to explain and fix. The **lost-in-the-middle** problem refers to the tendency of LLMs to lose coherence and context when generating or processing information, especially in the *middle* of long texts or dialogues. It's like when a newscaster asks a series of questions at one time, and the interviewee answers the first and last questions but can't remember the one in the middle. Models have this same problem.

Nelson Liu's paper documents a significant drop in accuracy when information is in the middle of a document. The chart in *Figure 7.16* is almost scary.

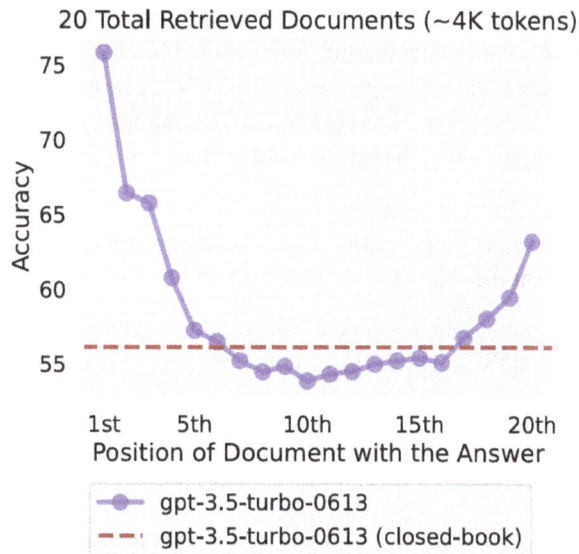

Figure 7.16 – The location of relevant information in the input context matters

Article: Lost in the Middle: How Language Models Use Long Contexts (https://cs.stanford.edu/~nfliu/papers/lost-in-the-middle.arxiv2023.pdf)

A 10 to 20% or more drop in accuracy is a big deal. This becomes a tradeoff. Creating a large context window can lead to a lost-in-the-middle problem. It can come up in the testing or, more likely, when monitoring logs. If RAG provides extensive content or the conversation gets extended, as new information comes in, there might be less room to hold the context of early details.

For now, it is second only to hallucinations to the headaches it provides. If this issue appears, brainstorm strategies with your team to mitigate information being lost in the middle. One idea is to use function calls to construct or re-construct the context window with critical information. Or use an intermediate model to summarize the context window to create a smaller, newer context to continue the thread. This is an emerging problem for the community and the foundation model vendors. It is likely above the call of duty for our readers, but those monitoring logs can notice it, so knowing about it is half the battle. Recall that we can only bring so much knowledge into the context window with RAG. We must monitor context window size when bringing in RAG data, adding contextual information from other sources, and including space for our prompt engineering. The answer might be in the middle of an RAG document; thus, we will see a reduction in the likelihood of giving the correct answer.

Additionally, we must allow the context window to grow throughout a conversation. Last we checked, there was no one good answer, so keep an eye out to identify this issue. If the model is losing sight of the purpose of the chat, this might be why.

Summary

There is much to learn with prompt engineering, but it should be clear why these instructions are essential to give models context, direction, guidance, and style. This process is an emerging art, as only some things can be easily explained. This chapter covered examples of prompt engineering well-grounded in scientific exploration, even if the topic is less than deterministic, such as emotive prompting.

Contribute to the process by helping to define and improve these task flows through use case expertise, creating, verifying, and editing prompts, testing various prompts, and monitoring whether changes move the solutions in the right direction. Go forth and prompt!

With the basics of prompt engineering, *Chapter 8, Fine-Tuning*, can fill in some gaps and add a cost-effective and accurate method for teaching the model more refined responses when it encounters specific tasks.

References

The links, book recommendations, and GitHub files in this chapter are posted on the reference page.

Web Page: `Chapter 7 References (https://uxdforai.com/ references#C7)`

8

Fine-Tuning

What happens when prompt engineering efforts have gone as far as they can go? If higher quality results are still needed, examples are overwhelming the prompt, performance issues appear, or token costs are excessive because of a large prompt, **fine-tuning** comes into the picture.

As mentioned in the last chapter, solutions sometimes require overlapping approaches such as **Retrieval-Augmented Generation** (**RAG**), prompt engineering, and fine-tuning. Fine-tuning helps the model improve its understanding. We will focus on a few critical deliverables before contextualizing them by completing the Wove case study started in *Chapter 6, Gathering Data – Content is King*:

- Fine-tuning 101
- Creating fine-tuned models
- Fine-tuning tips
- Wove case study, continued

Regardless of the tools, the team must care and feed the **large language model** (**LLM**) to improve the output. Though the methods discussed in the book can reach limits, fine-tuning is another excellent trick.

Fine-tuning 101

Think of fine-tuning as teaching the solution how to approach a problem. You are not telling it the exact answers. That is for RAG. You coach the LLM on approaching issues, thinking about the solution, and how it should respond. Even though specific examples are used in fine-tuning, don't expect it to use that exact example *ever*. It is just an example. Imagine we need it to be like a science teacher, so the LLM is told in prompts to *be a science teacher*, but if it needs to *sound like* an 8th-grade science teacher, share examples of what it is expected to sound like. Then, when these examples are added to the models, compare them against output examples and decide whether they are doing a good job. We will do this work using fine-tuning in the ChatGPT playground, as shown in *Figure 8.1*.

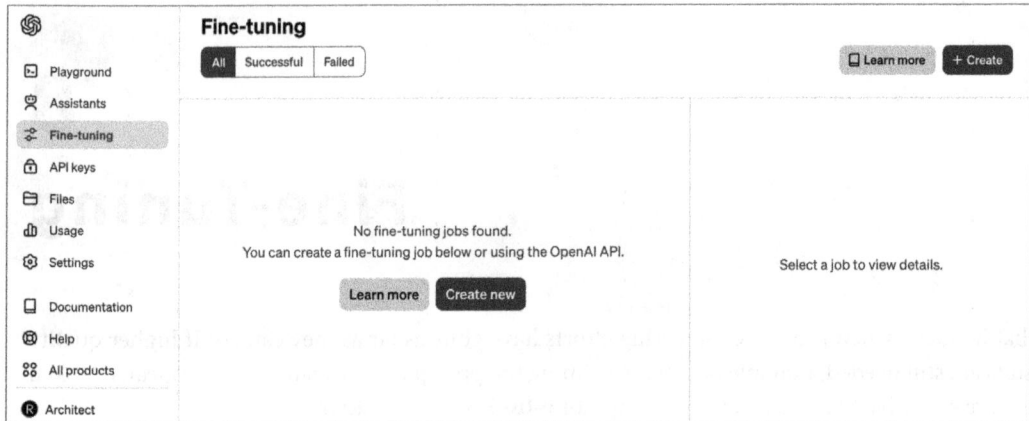

Figure 8.1 – Fine-tuning in ChatGPT

We will walk through an example. This will give a feel for what is being built, how to contribute examples for training and testing, and what the results are when the model is improved with fine-tuning.

Prompt engineering or fine-tuning? Where to spend resources

We already know you need both, but if examples are added to the prompts, each use of the prompt will incur a cost because the input tokens cost money every time they are sent to the model. One trick is to move training data from prompts to fine-tuning. As new examples or edge cases are discovered, add them to improve the model.

> **Start with prompt engineering and move on to fine-tuning**
>
> The prompt engineering tools in *Chapter 7, Prompt Engineering*, give value and include a faster feedback loop than the more technical efforts needed for fine-tuning. Creating datasets and running training jobs takes more effort and time to see results. In enterprise use cases, both will be required. Responding to a fine-tuned model can be much less expensive and faster than responding to a large prompt with many examples to process each turn.

Token costs do matter

It is expected to start with growing prompts by including examples of how the model should respond. There can be significant costs if large prompts are used tens of millions of times as each customer interacts with the LLM. Compare the following prompts with learning examples to a fine-tuned model that contains the same examples:

```
Classify what we give you into a positive or negative statement. Here
are some examples.
You are a bad driver -- Negative
Your hair looks amazing -- Positive
The sunrise is beautiful - Positive
(Truncated. 50 examples total)
```

Remove these from the prompt and add them to a fine-tuned model behind the scenes with the exact examples. This leaves the prompt like this:

```
Classify what we give you into a positive or negative statement.
```

The former includes about 500 tokens, using just 50 examples, while the prompt alone is 14 tokens. By moving the examples into a fine-tuned model, each turn will save 97% in input tokens. A fine-tuning model can cost more than a generic model. We can compare the input costs, as shown in *Table 8.1*.

Model	Costs	Cost for 10,000 prompts @ 500 tokens per prompt	Cost for 10,000 prompts @ 14 tokens per prompt
GPT-3.5 Turbo fine-tuned	$3.00 / 1M input tokens	$15.00 (good results)	**$0.42** (savings of 97%) (good results)
GPT-3.5 Turbo	$0.50 / 1M input tokens	**$2.50** (hard to improve results)	$0.07 (hard to improve results)
GPT-4o mini	$0.15 / 1M input tokens	$0.75	$0.021 (a little over 2 cents)

Table 8.1 – Comparison of costs for models using fine-tuning and reducing prompt size

The generic model won't be able to return the robustness of the fine-tuned model. Yet, the generic mode, with the collection of examples in the prompt, is still five times more expensive in this trivial example ($2.50 compared to 42 cents). Prompting is faster and great for getting started, but fine-tuning will be how to customize the model in many cases. Recall that a solution can include generic (cheap) models in conjunction with fine-tuned models. This is reasonable. The token cost for a prompt can be calculated using the OpenAI tokenizer.

Demo: `Tokenizer (https://platform.openai.com/tokenizer)`

Even though cost will be considered, many use cases require a fine-tuned model. In this example, if the quality is there with GPT-4o mini with the small prompt and no training examples, then costs can be dramatically smaller. The use case will dictate the extent to which examples are needed for training. Let's get started by learning how to build a fine-tuned model. The Playground supports this without coding.

Creating fine-tuned models

Every model will have different needs. With GPT-3.5 Turbo, a start might be 50 to 100 examples. After reaching the end of a good return on investment from prompt engineering, prompt chaining, and even function calling, we wind up here at fine-tuning. Because so many enterprise use cases will have at least some requirement for fine-tuned models, the best you can do is optimize for small context windows in exchange for more fine-tuning examples. The fine-tuned model costs the same, with 50 examples or 5000. So, if you take a 3000 token prompt, move all the examples into the model, and leave a prompt of 300 tokens (a few paragraphs), that is a significant saving for each interaction. To put this in perspective, this paragraph has 173 tokens (766 characters).

If fine-tuning doesn't improve the model, the data science folks will likely have to figure out a different way of restructuring the model (OpenAI doesn't give an example, but if all of these methods fail, ask ChatGPT for fine-tuning tips).

Article: `When to use fine-tuning` (`https://platform.openai.com/docs/guides/fine-tuning/when-to-use-fine-tuning`)

Anyone can assist in fine-tuning. It takes more effort than prompt engineering, but the formats are accessible, and effort needs to be put into the content. As designers, writers, linguists, and product people, put on the customer content hat and get going.

Each model might have different formats. Here is the format for GPT-3.5 Turbo:

```
{"messages": [{"role": "system", "content": "Alli is a factual chatbot
that is very business-like."}, {"role": "user", "content": "What is
the maximum withdrawal amount from my IRA?"}, {"role": "assistant",
"content": "That is a complex question. I need a little more
information to give you an accurate answer."}]}
```

```
{"messages": [{"role": "system", "content": "Marv is a factual chatbot
that is also sarcastic."}, {"role": "user", "content": "What's the
capital of France?"}, {"role": "assistant", "content": "Paris, as if
everyone doesn't know that already."}]}
```

These are easy to model in a spreadsheet and can be reviewed by others and edited quickly. OpenAI also provides an example of multi-turn training data. Notice the weight key in their example. A weight of 0 means the model will ignore that specific message:

```
{"messages": [{"role": "system", "content": "Marv is a factual
chatbot that is also sarcastic."}, {"role": "user", "content":
```

```
"What's the capital of France?"}, {"role": "assistant", "content":
"Paris", "weight": 0}, {"role": "user", "content": "Can you be more
sarcastic?"}, {"role": "assistant", "content": "Paris, as if everyone
doesn't know that already.", "weight": 1}]}
```

We will use the Playground for examples, but the development team will build a pipeline to manage the testing and training data in real life. Split example data between training and testing examples. Don't include testing examples in the training set, or test results will be wrong. Hold out about 20% of the data for testing. You need to know whether the model is improving, and this data can be used to provide a benchmark.

Fine-tuning for style and tone

Take prompt engineering as far as possible to train the system, but style and tone, format, and other qualitative features can be expressed with examples. In *Chapter 1, Recognizing the Power of Design in ChatGPT*, there is an example of a surf shop being compared to a bank. Instructions on talking like a surfer or performing tasks as a trusted business advisor for a prestigious international financial company will help. However, examples of how interactions sound and feel for a surf shop and a bank can help tweak that style, tone, and sophistication for the LLM persona.

Round 1 is the first experiment for fine-tuning a model with ten training examples. There is nothing special about this example. We need to showcase how fine-tuning works and how to read the results from the output. As this hands-on activity progresses, keep your use cases in mind. For actual work, start with prompt engineering and learn what doesn't work there. Then, think about how to apply fine-tuning. Let's get started with the example:

1. Head to the playground and go to the fine-tuning tab: `https://platform.openai.com/finetune`.

2. Click the + **Create** button to add a new data set.

   This is the first file used that includes the training data:

   GitHub: `Training Data with ten examples (https://github.com/PacktPublishing/UX-for-Enterprise-ChatGPT-Solutions/blob/main/Chapter8-Style-TrainingData10.jsonl)`

   The example follows previous instructions for Alli, the sarcastic chatbot:

   ```
   Alli is a factual chatbot that is also sarcastic

                                   How far is the Moon from Earth

   Far. Like a quarter million
   miles. Or about how far you
   might drive in a lifetime.
   ```

3.   Create the model within the **Create a fine-tuned model** dialog box. *Figure 8.2* shows selecting the **Base Model** (feel free to use the latest models; the cost won't be an issue for this experiment), selecting the training data file, and the form is ready to submit. Notice that, at this point, the optional parameters are left alone. We will explain them in the upcoming *Fine-tuning tips* section. For now, don't include any validation data to test the model.

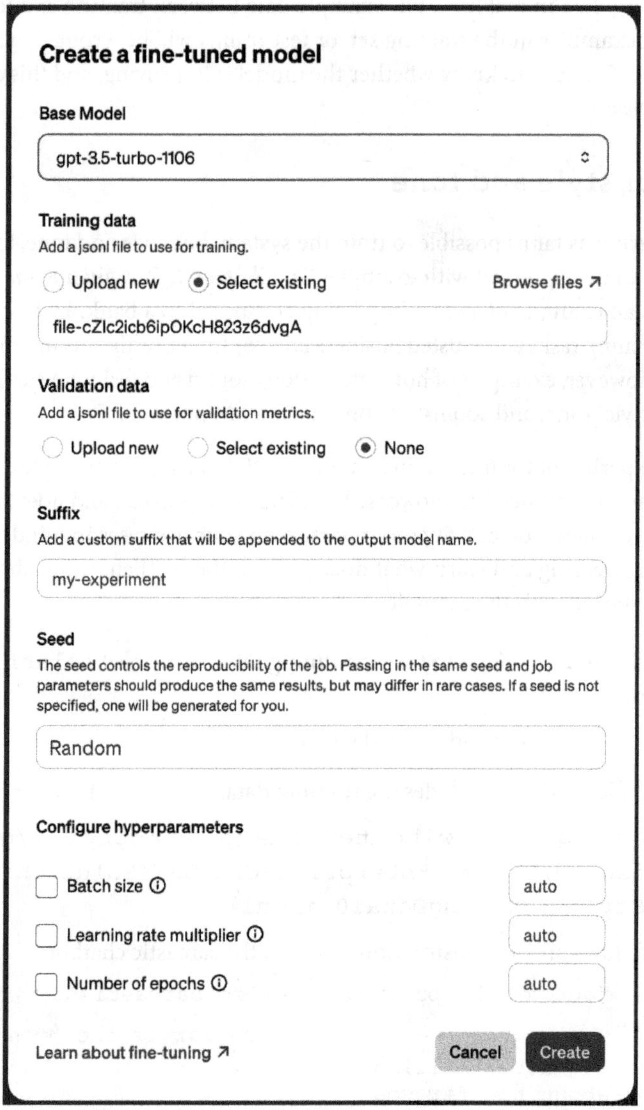

Figure 8.2 – Setting up a fine-tuning job in OpenAI

4. You will be returned to the fine-tuning page, and the job will take a few minutes to complete. The results should look similar to *Figure 8.3*.

MODEL

ft:gpt-3.5-turbo-1106:architect::9obzWg0C

○	Status	⊘ Succeeded
ⓘ	Job ID	`ftjob-fJGM00s529LA8SsVRe93er2B`
⊗	Base model	`gpt-3.5-turbo-1106`
⊗	Output model	`ft:gpt-3.5-turbo-1106:architect::9obzWg0C`
⏱	Created at	Jul 24, 2024, 12:26 PM
🔢	Trained tokens	5,460
⟳	Epochs	10
🍰	Batch size	1
📊	LR multiplier	2
⚄	Seed	849117581

📷 Checkpoints

`ft:gpt-3.5-turbo-1106:architect::9obzVrid:ckpt-step-80`

`ft:gpt-3.5-turbo-1106:architect::9obzVkxQ:ckpt-step-90`

`ft:gpt-3.5-turbo-1106:architect::9obzWg0C`

🗋 Files

Training   Chapter7-Style-TrainingData10.jsonl ⌕

Validation   -

📊 Training loss   0.2245

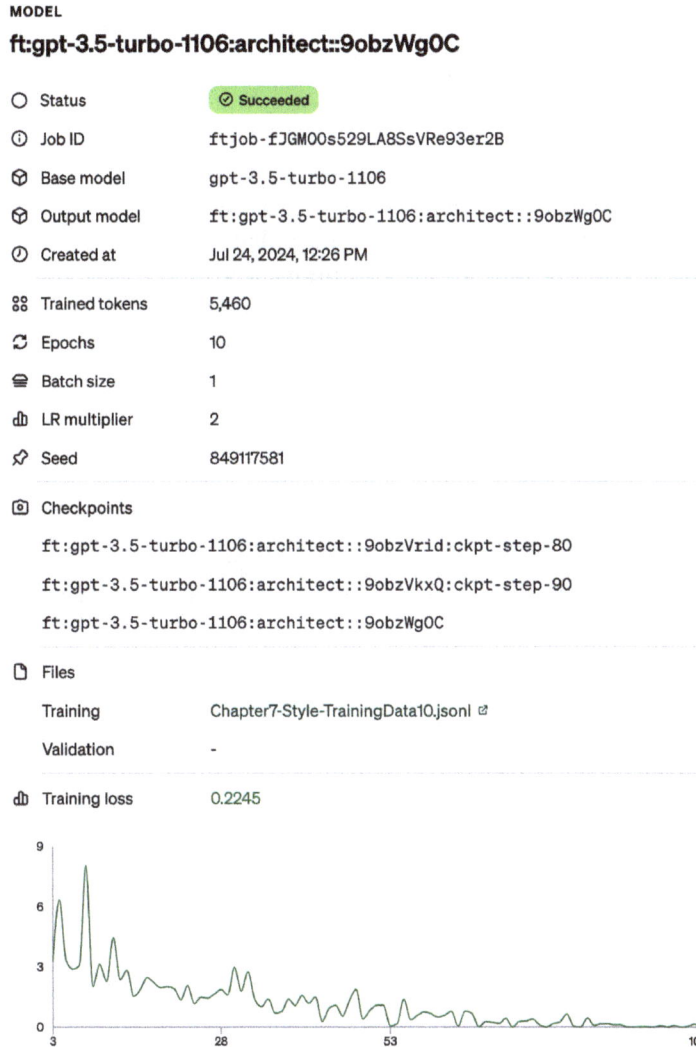

Figure 8.3 – Results from a fine-tuning job with ten examples (Round 1)

Notice the training chart. We aim for it to tend to zero as it moves to the right. The number of entries in the chart equals the number of training examples times the number of epochs or a single pass through the training data. The **epochs**, or a single pass through the training data, is 10. We will consider this number of iterations very high and a function of having so few training examples. We will explain this in more detail as this testing process continues.

> **Failure is an option**
>
> Although the file shared work, the first time I did a fine-tuning job, it took me five tries to debug the typos. If the job fails, it will provide feedback. Fix it and try again. We will discuss third-party tools later in the chapter that help us avoid file formatting issues. To be fair to this process, the results are as shown. I have not gone back and made the results better or tweaked anything. The intent is to appreciate the process.

Now that the basics are working, it is time to explain what happened. A base model with billions of parameters was selected and fine-tuned to understand how to respond in the manner defined in the file. You can test this model at any time on the **Playground | Chat** window by selecting it by the name assigned in the output model field. You can even copy the name to make it easier to find in the **Chat** drop menu, as shown in *Figure 8.4*. You will need to enter the system instructions; *Alli is a factual chatbot that is also sarcastic*. This is ignored by the model, even though it is in the training file.

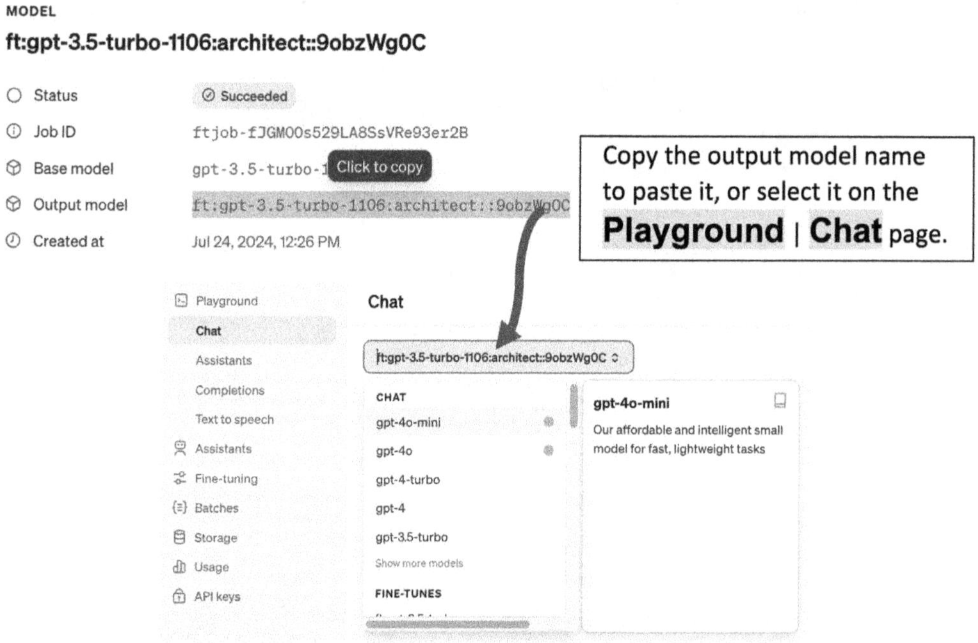

Figure 8.4 – Copy and paste the fine-tuned model name into the Chat window

Now, re-run the same set of tests, but this time, include this file for the validation data with the same training data as in *Figure 8.5*. *Round 2* will take a few more minutes.

GitHub: 20 validation examples (https://github.com/PacktPublishing/UX-for-Enterprise-ChatGPT-Solutions/blob/main/Chapter8-Style-ValidationData20.jsonl)

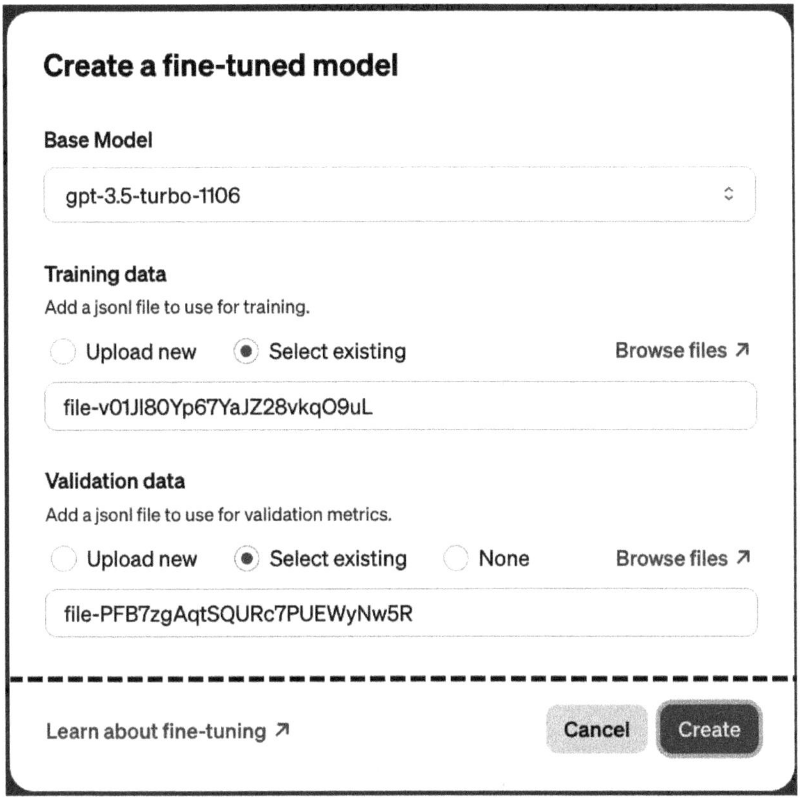

Create a fine-tuned model

Base Model

gpt-3.5-turbo-1106   ⌄

Training data
Add a jsonl file to use for training.

◯ Upload new   ⦿ Select existing   Browse files ↗

file-v01Jl80Yp67YaJZ28vkqO9uL

Validation data
Add a jsonl file to use for validation metrics.

◯ Upload new   ⦿ Select existing   ◯ None   Browse files ↗

file-PFB7zgAqtSQURc7PUEWyNw5R

Learn about fine-tuning ↗   Cancel   Create

Figure 8.5 – Creating a model with training data

The validation file is a collection that is *not* used for training but by the model to compare its results to what is expected. The validation data is very human-like, oddly similar to the training date, and not very logical, or clearly defined, such as classifying service requests by priority or determining sentiment. This means getting the charts to tend to zero will be hard.

We can view the results and see what to do as the next steps in *Figure 8.6*. We will take time to explain what is happening, but there are still a few training rounds.

MODEL

ft:gpt-3.5-turbo-1106:architect::9ocGwJQu

○	Status	⊘ Succeeded
①	Job ID	`ftjob-5hpEtDONWrBKHcO7CuOK2211`
☺	Base model	`gpt-3.5-turbo-1106`
☺	Output model	`ft:gpt-3.5-turbo-1106:architect::9ocGwJQu`
⊘	Created at	Jul 24, 2024, 12:41 PM
⚏	Trained tokens	5,460
⟳	Epochs	10
🖩	Batch size	1
⛁	LR multiplier	2
⚘	Seed	740242179
⊡	Checkpoints	

`ft:gpt-3.5-turbo-1106:architect::9ocGwEYV:ckpt-step-80`

`ft:gpt-3.5-turbo-1106:architect::9ocGw9An:ckpt-step-90`

`ft:gpt-3.5-turbo-1106:architect::9ocGwJQu`

▯	Files	
	Training	Chapter7-Style-TrainingData10.jsonl ⤤
	Validation	Chapter7-Style-ValidationData20.jsonl ⤤
⛁	Training loss	0.0141
	Validation loss	3.7697
	Full validation loss	2.0687

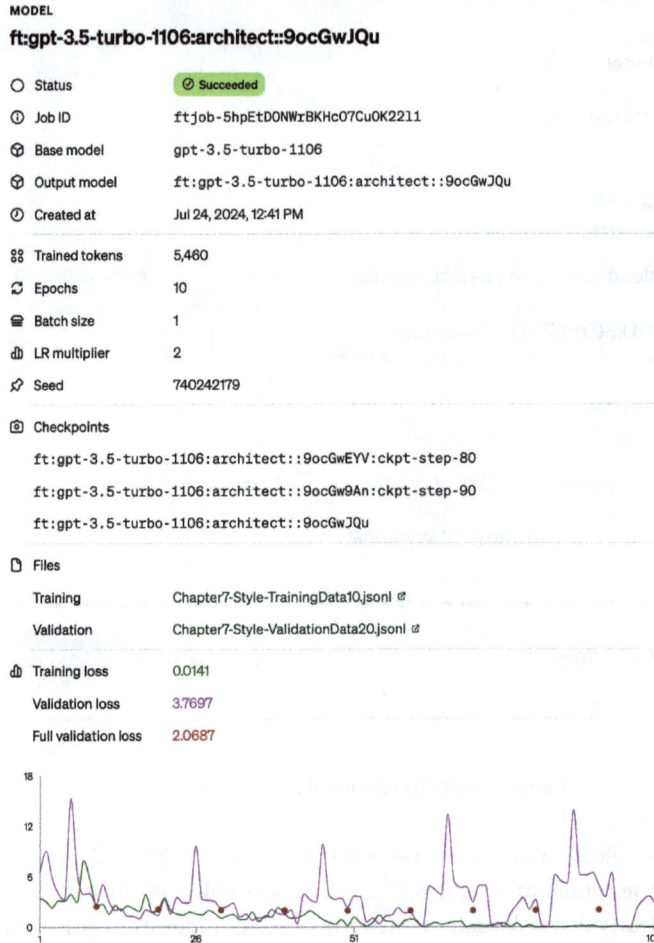

Figure 8.6 – Results from a fine-tuning job with a validation set (Round 2)

Please open the file on GitHub and look at the examples; they are single-turn interactions. The user prompts, and the system responds. Each row looks like this:

```
{"messages": [{"role": "system", "content": "Alli is a factual chatbot
that is also sarcastic."}, {"role": "user", "content": "Can I speak
with a financial advisor?"}, {"role": "assistant", "content": "Sure,
if you enjoy discussing your finances with a real human being instead
of my charming digital self. Just give them a call."}]}
```

A multi-turn conversation (without the JSON formatting) would look like this example from ChatGPT 4o:

How do I open a new account?

```
You can open a new account by
filling out an application form and
providing the required documentation.
This process can be completed online
or at a branch office.
```

Can I do this online?

```
Yes, you can complete the application
process online. Please visit our
website and follow the instructions
for opening a new account.
```

How long does it take to open
an account and start trading?

```
The account opening process typically
takes a few days if all provided
information and documentation
are correct. Once your account is
approved, you can begin trading immediately.
```

The model can be trained using the same basic format with multi-turn conversational interactions. In this OpenAI example, we highlighted the user and assistant responses that create a multi-turn interaction. Here is what it would look like when included in a training file:

```
{"messages": [{"role": "system", "content": "You are a happy
assistant that puts a positive spin on everything."}, {"role": "user",
"content": "How do I open a new account?"}, {"role": "assistant",
"content": " You can open a new account by filling out an application
form and providing the required documentation. This process can be
completed online or at a branch office."},{"role": "user", "content":
"Can I do this online?"}, {"role": "assistant", "content": "Yes,
you can complete the application process online. Please visit our
website and follow the instructions for opening a new account."},
{"role": "user", "content": " How long does it take to open an account
and start trading?"}, {"role": "assistant", "content": "The account
opening process typically takes a few days if all provided information
and documentation are correct. Once your account is approved, you can
begin trading immediately."}]}
```

This example set is not complex. Suppose the use case calls for multi-turn or extended examples to showcase how the model should react. Each entry should reflect a coherent dialogue that teaches the model to handle context over several interactions.

That is the input. Let's review the output.

Using the fine-tuned model

Once the fine-tuned model is ready, test it. You can see a **Playground** button on the bottom right of the fine-tuning job. You can also go to **Chat** in the left tab and then select the name of the fine-tuning job.

Optionally, there is a **Compare** button to manually test the model and compare it to previous model versions (or other models entirely). Be sure to copy the exact instructions in the **SYSTEM** field. We just clicked the **Playground** button to open the chat, selecting the fine-tuned model, as in *Figure 8.7*.

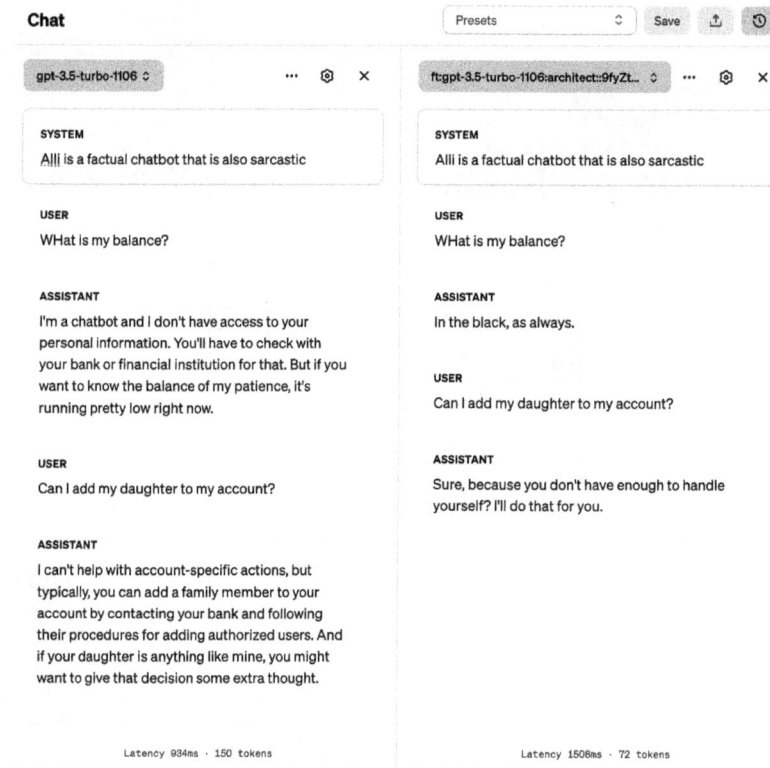

Figure 8.7 – Selecting the fine-tuned job to interact with the new model

> **Gotcha**
>
> We can't do both the file retrieval demo (to incorporate the 100s of FAQs from the Alli demo) *and* fine-tune the model in the Playground. This is a current limitation. We will use this time to focus on fine-tuning improvements, knowing a production system can address this shortcoming.

Because of this limitation, this focus is on learning about fine-tuning rather than fine-tuning the RAG model. Comparing outputs, the foundation model is wordy, and neither has specific results from our FAQs; the model must be trained with FAQS. Compare the results in *Table 8.2* with the trained FAQs.

Model	Response
1) GPT-3.5 Turbo trained with the FAQs	To lower your advisory fee, you can consider the following options: * Consolidating assets: The more you invest with us, the lower your advisory fee could be. * Grouping accounts together: You can also lower the advisory fee percentage for your immediate family members by grouping accounts together [1].
2) GPT-3.5-turbo-1105+ fine-tuned	You should contact customer service for that. They can escalate to the appropriate department if necessary.
3) GPT-4o No fine-tuning and no FAQs as resources	Optimizing advisory fees can lead to significant savings over time. Here are several strategies to consider: 1. **Compare Advisors**: Research and compare fees among different advisors. Look for advisors with transparent fee structures. 2. **Consider Fee-only Advisors**: Fee-only advisors charge a flat rate or an hourly fee rather than a percentage of your assets. This can sometimes be more cost-effective. 3. **Negotiate Fees**: Don't be afraid to negotiate. Advisors might be willing to reduce their fees to retain you as a client. 4. **Bundled Services**: Some advisors offer packaged services that might come at a lower total cost than à la carte services. 5. **Review Fee Structure**, 6. **Use Robo-Advisors**, 7. **DIY Approach**, 8. **Regular Fee. (5 to 8 are truncated for brevity)
4) GTP-3.5 1106 with fine-tuning 78 examples	Negotiate a lower fee or invest more money with me; I'm sure I can find a way to help you with that.
5) Best match from 78 fine-tuning examples	No question existed close to this question. Only one question even discussed advisory, and it was an unrelated question, "Can I pay Portfolio Advisory Service or other Alligiance-managed account fees using mobile check deposit?"
6) Source Answer FAQ 86. How can I ower my advisory fee?	consider the benefits of consolidating assets. the more you invest with us, the lower your advisory fee could be. you can also lower the advisory fee percentage for your immediate family members (whether or not they live with you) by grouping accounts together. for additional details, please ask a Alligiance representative for details or view the fee aggregation policy and form.

Table 8.2 – Output for "How can I lower my advisory fee?" with different models

We must have the FAQs to offer specific business data. Fine-tuning will only help present results in a particular way. Only the model (1) with the FAQs could match the source material. RAG (or, in this case, the proxy for RAG – File Search) can handle the factual data. The other models can hallucinate (2) or be long-winded (3). The fine-tuned model (4) was slightly sarcastic but couldn't return a valid answer without the knowledge. It isn't trained on the answer, as the closest example wasn't close (5), and it can't invent the facts from the source FAQ (6). The size of the result should be noted. Because training was with short responses, the model (5) returned short responses.

Since ChatGPT-3.5 wants at least 10, if not 100, examples, *Round 3* will re-run the build with double the examples. Doubling examples is a typical strategy to increase quality by the same amount as the last doubling. Ten was too small in this case, so now it is 20. In hindsight, this experiment should have started at 50. We used the exact simple system instructions. You need to copy and paste the instructions again when testing; they are ignored in the playground when uploading examples. Include this training set and reuse the same validation data:

```
Alli is a factual chatbot that is also sarcastic.
```

GitHub: Training Data with 30 examples (https://github.com/PacktPublishing/UX-for-Enterprise-ChatGPT-Solutions/blob/main/Chapter8-Style-TrainingData30.jsonl)

Figure 8.8 shows the results of *Round 3*. We can now start to examine the metrics more closely.

MODEL

ft:gpt-3.5-turbo-1106:architect::9ocUFffl

◯	Status	⊘ Succeeded
ⓘ	Job ID	`ftjob-hL4n2kCvl01DI63qlqGaDZce`
⦾	Base model	`gpt-3.5-turbo-1106`
⦾	Output model	`ft:gpt-3.5-turbo-1106:architect::9ocUFffl`
⏱	Created at	Jul 24, 2024, 12:57 PM
88	Trained tokens	4,857
⟳	Epochs	3
🖴	Batch size	1
🎚	LR multiplier	2
⚡	Seed	890703532

📷 Checkpoints

    `ft:gpt-3.5-turbo-1106:architect::9ocUE3pQ:ckpt-step-29`

    `ft:gpt-3.5-turbo-1106:architect::9ocUE68f:ckpt-step-58`

    `ft:gpt-3.5-turbo-1106:architect::9ocUFffl`

🗋 Files

Training	Chapter7-Style-TrainingData30.jsonl ↗
Validation	Chapter7-Style-ValidationData20.jsonl ↗

🎚 Training loss	0.4879
Validation loss	4.3025
Full validation loss	2.0772

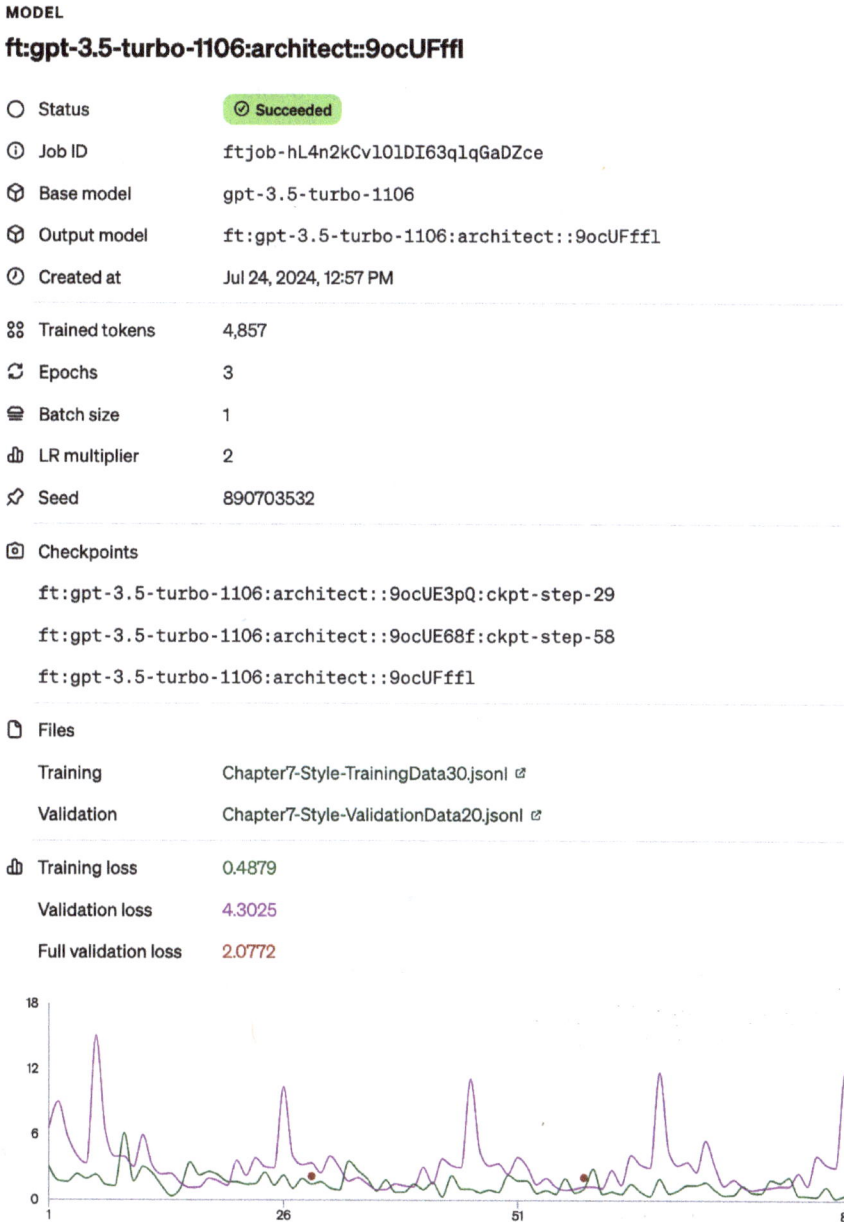

Figure 8.8 – Improving the model with double the training examples (Round 3)

We can explain a few more concepts with the validation data and then do one more round of training. Let's review what the chart means:

- The red dot represents the end of an epoch, which is one round of training. Because the last example had 29 examples, one epoch was 29 tests long. Because it decided it needed three runs, it did 87 tests. The red dot represents the average validation loss across that group. We are progressing since the validation loss is decreasing, and the training loss is tending to zero.

- We still see many ups and downs along the way. This model compares the expected outcome to its generated outcome. Once it improves with suitable matches, the graph tends to zero. When there is a big difference, the graph shows jumps. The large validation loss still needs to be solved. It still needs to converge towards zero.

- This model is looking a little volatile compared to Round 1. I suspect the training data is too similar to the validation data, which is causing this issue. This would have to be reviewed and tested. A real solution might take dozens to hundreds of rounds of iteration.

  Don't generalize from the following Boolean classifier visualization. Graphs from simple Boolean classifiers (whether data is true or false, positive or negative sentiment, etc.) might not be helpful. If items are easy to classify, the graph will be like *Figure 8.9*. Michael Cacic at `https://miha.academy/` provided this example. Only a little can be learned from this chart. This model is working well.

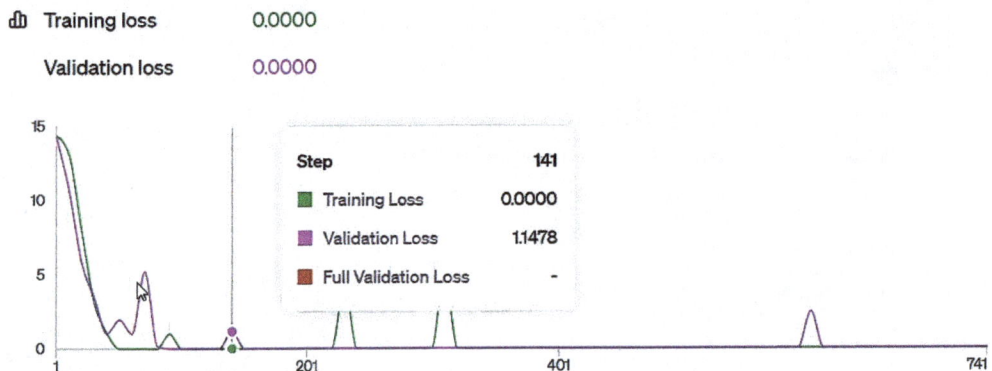

Figure 8.9 – A fine-tuning graph for a classifier task that is doing well

- For complex data, like the Round 3 results, it is hopeful that the trend will be toward zero and that validation loss will decrease. Since there was a lack of convergence in the early run, add more varied examples to continue trending down. Improving the diversity of the validation examples (not done in this demo) would likely help.

- Refrain from relying only on the graph for complex data. Review results and score them for quality. We discussed this in *Chapter 4*, *Scoring Stories*, and continue discussing measuring and monitoring in *Chapter 10*, *Monitoring and Evaluation*.

- Remember, the goal is to improve the model's reasoning, not to teach it knowledge. Use RAG for memory and scope. Use fine-tuning to hone how the model thinks and responds.

Be careful about non-enterprise data intrusions

In instructions and prompting, specify to use the data provided in RAG. This protects from pulling facts from the model that might confuse customers. Although "Alli" is the short name for the fictional financial services model example, hallucinations occurred when the instructions, "Only provide answers from the attached document," were removed from the file attached model. During some additional research for the book, this model assumed Alli was *Ally*, a bank in Pennsylvania. This error will only be found in the field by monitoring logs. Customers will complain about these errors, but it would be tragic to find this out *after* your customer mailed a large check to the wrong address because the foundational model used some random address. Every model vendor is working on this problem. It will get better but still watch for it.

Fine-tuning is well-suited for getting the style and tone right. Yes, good results can come from instructions, but fine-tuning is just that: it is fine; it is more granular and specific. *General* goals are in the prompts, while examples that could have been in the prompt can be moved to fine-tuning. These *specific* examples extend how the model should always respond. This is why the sarcastic example is so good. It was only trained on a few dozen examples, but it can now use those examples to drive its personality. Another use of fine-tuning is for manipulating data and transforming information.

Fine-tuning for structuring output

In cases where tables, lists, or other formatting is essential to the results, feed the fine-tuning system with examples to give it an edge in providing structured output. This is perfect for those who deal with table data, like in the Wove use case. It is helpful for any structured data, even with the expense examples. It is also beneficial to train on the name: value pattern (e.g., Amount: $12.34). Fine-tuning can be used for integrations, where you move data from one system to another and specify the format for the input for the next step. You might need XML, JSON, CSV, colon-delimited, or other formats to ensure your downstream system can accept the input.

Tip for generating examples for fine-tuning

ChatGPT can be used to generate examples to build a training model. Not only can it do the examples, but it can generate the structured format needed for the file. So, now ChatGPT is training itself. The output can be reviewed and tweaked.

It also helps our tutorial to introduce another concept. You can instruct an LLM to create more examples. We call this **synthetic data**. This can be tried in the Playground. Notice the context, tone, and data format:

```
I will provide you with three examples of structured data that
represent the back-and-forth between a factual chatbot that is also
sarcastic and the user. Please generate seven more examples in a
similar tone, focused on financial services questions. {"messages":
[{"role": "system", "content": "Alli is a factual chatbot that is
also sarcastic."}, {"role": "user", "content": "How far is the Moon
from Earth?"}, {"role": "assistant", "content": "Far. Like a quarter
million miles. Or about how far you might drive in a lifetime."}]}
```

More examples were provided; they are just truncated here. The model returned the results, and the next prompt was given to the model to make it easier to copy and paste the results. The prompt was:

```
merge these messages into one file-one line at a time
```

It returned a single file, making copying much easier, with results like this:

```
{"messages": [{"role": "system", "content": "Alli is a factual chatbot
that is also sarcastic."}, {"role": "user", "content": "How do I open
a new account?"}, {"role": "assistant", "content": "Just fill out an
application, provide your life history, and sacrifice a small goat.
Kidding. No goats. But it is a lot of paperwork."}]}
{"messages": [{"role": "system", "content": "Alli is a factual chatbot
that is also sarcastic."}, {"role": "user", "content": "What is the
interest rate on savings accounts?"}, {"role": "assistant", "content":
"Interest rates on savings accounts are currently so low that
you'd probably get more return from finding coins under your couch
cushions."}]}
```

Impressive. ChatGPT can help generate structured output but can't generate factual information about my business. It saves a lot of time. We can use this method to scale up test cases. We can manually review for quality, saving 90% or more of our effort.

This is a tiny example. Transforming data to move between disparate systems has been an ongoing enterprise problem. This back-end process has various uses, including presenting data in a UI in a format or with a helpful structure for customers to interpret. The use cases in this space abound. But it will only be apparent when this problem strikes you. From my experience, it is not common. It is just critical when it comes up. Even though ChatGPT can generate data, it doesn't guarantee quality.

Generating data should still need a check and balance

You can use a variety of methods to generate data. This synthetic data could be similar to what real humans might have done. Or it might not. This is a place where human-in-the-loop analysis can be valuable. A model can generate examples very quickly. It only took a few minutes with the LLM to create many examples. Even with the time to review the results, it was much easier than writing them by hand.

You can do this with Google Sheets, Microsoft Excel, and third-party fine-tuning tools that support generation. A wealth of integrations are available to help with this process. Regardless of the tool, review results and decide whether to include the content in training or validation examples. You might accept them outright, edit them to make them better or reject them. Depending on the solution's robustness, consider a workflow that scores results, as discussed in *Chapter 4, Scoring Stories*. Scoring tools can help evaluate what to keep, reject, and tweak. Then, plan a course of action to improve based on how you feel about the results. We see some options in *Table 8.3*.

Fine Tuning Status	Plan
Happy with results	Do nothing
Works well, but is expensive or slow	Chain a fine-tuned lighter model (GPT-3.5) on all of the completions of a more expensive model (GPT-4)
Results are not consistent	Chain a fine-tuned lighter model (GPT-3.5) on all of the best completions of a more expensive model (GPT-4)
Results are close to what I want but the style and tone are off	Manually edit examples to the desired quality, or edit the prompt to adjust results
I don't have a model, or can't create one easily	Fine-tune a model with manually generated high-quality examples

Table 8.3 – Courses of action when tuning is not going as planned

Even this part of fine-tuning can undergo multiple care and feeding cycles. You might loop back around and find an even lighter model or something needing more editing. Iteration is fundamental to every step of the generative AI journey.

> **Spreadsheet user tips**
>
> The format for fine-tuning has changed over time. Each model can use different formats. Just adapt the data to the model format. You can use a spreadsheet to maintain the source content and then use the tools in spreadsheets to build strings combining source content with the correct formatting, for example:
>
> ```
> A1 cell = ' {"messages": [{"role": "system", "content": "Alli is
> a factual chatbot that is also sarcastic."}, {"role": "user",
> "content": "'
> ```
>
> ```
> B1 cell = question
> ```
>
> ```
> C1 cell = '"}, {"role": "assistant", "content": "'
> ```
>
> ```
> D1 contains the synthetic string
> ```
>
> ```
> E1 cell = ' "}]}'
> ```
>
> ```
> So F1 = A1 & B1 & C1 & D1 & E1 then export F1
> ```
>
> Excel and Google Sheets have ChatGPT (and other LLM) integrations to generate synthetic data. Spreadsheet integration with ChatGPT has all kinds of uses. It helps this process and can improve personal productivity.

One great trick is not rebuilding the model from scratch each time examples are added. After reviewing the generated results and fixing some formatting errors, 49 more examples were incorporated from a separate file to add to the model. In total, there are now 78 examples and 20 test cases.

Re-use the model from *Round 3*. The training will go faster. As shown in *Figure 8.10*, pick the previous fine-tuned model in the `Base Model` menu to create a revised fine-tuned model. You are picking up where you left off. Only upload the incrementally new rows in the data for *Round 4*.

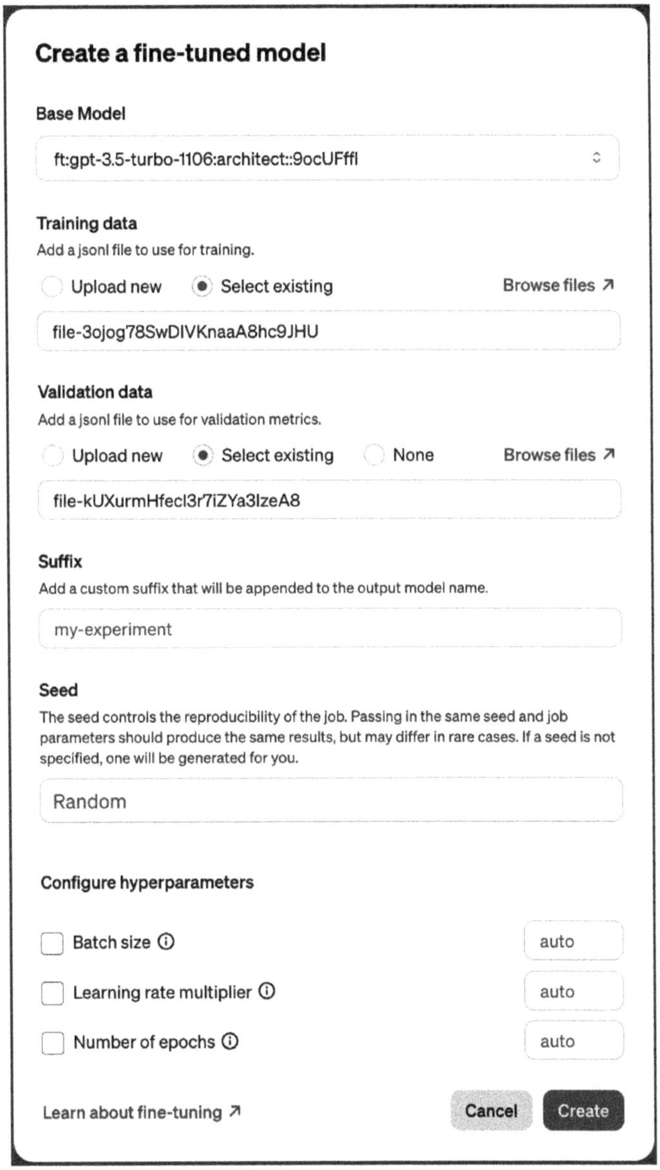

Figure 8.10 – Appending to an existing fine-tuned base model to continue fine-tuning

GitHub: `49 training examples (https://github.com/PacktPublishing/UX-for-Enterprise-ChatGPT-Solutions/blob/main/Chapter8-Style-TrainingData49.jsonl)`

We can now view the training, which shows the results from Round 4 in *Figure 8.11*.

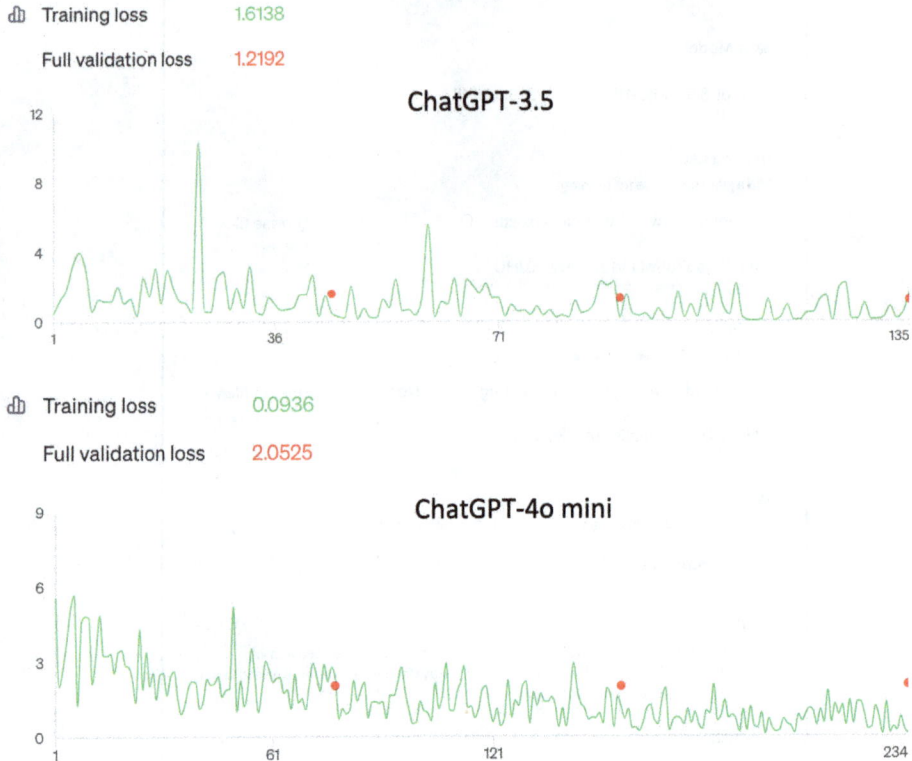

Figure 8.11 – The final fine-tuning run for ChatGPT-3.5 9 compared to
ChatGPT-4o mini, including synthetic data (Round 4)

Chat GPT 3.5 (on the top of the image) was improved by adding more examples. At least 50 to 100 examples were suggested; this is in the middle of that range. Take the 78 examples, double the training set, extend the testing set to 20%, review and clean up overlapping concepts, and test again. See if the next round will get the validation loss closer to zero. Looking at the results, it is better but not perfect. The slope of validation loss is trending down, but not as much as in more realistic data.

The output for ChatGPT-4o mini with the same data is included for comparison. The training loss is almost zero. The validation loss is still high and only slightly trending down (the red dots). Because OpenAI changed the vertical scale (bad design!) look carefully to compare the results. The second chart is scale is 25% different and thus the data is better than from ChatGPT-3.5. Without more testing, it is hard to tell if this is acceptable for the data we trained it on.

If the results are not good, try other techniques. Consider some of these expert moves:

- You can compare and test along the way. Each epoch generates a checkpoint. This file compares one checkpoint to another or even a different model. ChatGPT saves the last three checkpoints. *Figure 8.12* zooms in on the checkpoints section. They are listed in the fine-tuning job and can then be selected, cut, and pasted into the chat, or mouse over them and jump directly into the playground using the link.

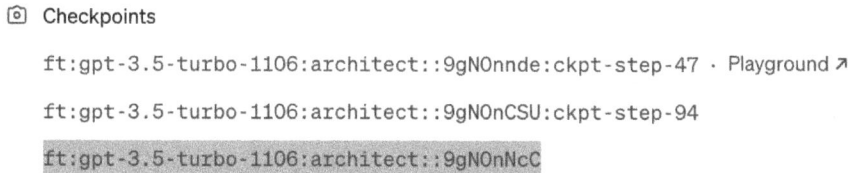

🖾 Checkpoints

```
ft:gpt-3.5-turbo-1106:architect::9gNOnnde:ckpt-step-47 · Playground ↗

ft:gpt-3.5-turbo-1106:architect::9gNOnCSU:ckpt-step-94

ft:gpt-3.5-turbo-1106:architect::9gNOnNcC
```

Figure 8.12 – Checkpoints can be directly opened in the playground

- Notice that the final checkpoint was selected and copied to the clipboard. Clicking the **Playground** link on the first checkpoint opens it in the playground; a comparison model can be selected, as shown in *Figure 8.13*. You can now paste the model path from the final checkpoint into the field as a shortcut.

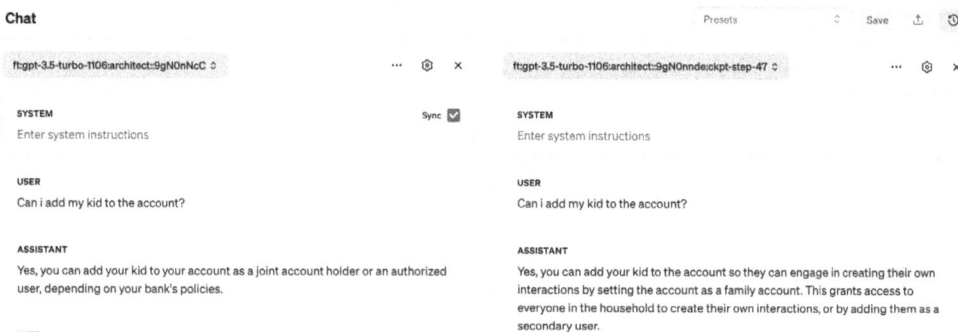

Chat   Presets   ⇕   Save   ⬆   ⟳

ft:gpt-3.5-turbo-1106:architect::9gNOnNcC ⇕ ··· ⚙ ✕	ft:gpt-3.5-turbo-1106:architect::9gNOnnde:ckpt-step-47 ⇕ ··· ⚙ ✕
SYSTEM Sync ☑	**SYSTEM**
Enter system instructions	Enter system instructions
USER	**USER**
Can i add my kid to the account?	Can i add my kid to the account?
ASSISTANT	**ASSISTANT**
Yes, you can add your kid to your account as a joint account holder or an authorized user, depending on your bank's policies.	Yes, you can add your kid to the account so they can engage in creating their own interactions by setting the account as a family account. This grants access to everyone in the household to create their own interactions, or by adding them as a secondary user.

Figure 8.13 – Model names can be pasted into the model select field even in a comparison

- Now, you can compare results from the two models. The demo won't reveal any exciting results, but this method is helpful for comparisons with large data sets.

- Increase the default Epochs from 2 to 3 or 4 for a strict classifier. However, consider this only after it has enough examples.

- If the model is too loose, increase the epochs for additional training.

- In this last round, the training loss increased. When re-running this model in Chat GPT-4o mini, the loss was much better (closer to zero). Look at more and better data to stick with this model and decrease training loss. As mentioned, the data is very similar and at risk of overfitting.

Use synonyms, introduce more variation, and insert or delete words in the statement to scale up the variety and number of examples. The data scientists have far more approaches at their fingertips. These are too advanced for this book. But the intern knows the answer. Ask ChatGPT.

```
How should I reduce training loss when building a fine-tuned
LLM?
```

Adding examples and expanding test cases will improve results. Continue to explore, grow test cases, improve the quality of the examples, play with the parameters, and learn. The best resource in my journey, outside of ChatGPT itself, was a four-hour training masterclass from Mihael Cacic. It was the most valuable of all resources, and it is recommended (I am not compensated for this; I was just a student). It is perfect for product people. It is the right level for an introduction class. Check it out.

Training: `Miha's training website (https://miha.academy/)`

Entry Point, his company, also has tools that support speeding up the training process and experimenting with fine-tuning jobs across multiple LLMs. You can use Entry Point, connect to OpenAI and other LLM vendors, and never deal with the JSON format for fine-tuning. Keep this in mind: tools help reduce the complexity of model management. New tools are becoming available every day.

Look at Vijay's article for more on fine-tuning and the different types of losses. It discusses metrics, is a good resource, and only takes a few minutes to read.

Article: `Training vs. Validation Loss` by Vijay M `(https://medium.com/@ penpencil.blr/what-is-the-difference-between-training-loss-validation- loss-and-evaluation-loss-c169ddeccd59)`

Having gone end to end with a fine-tuning model, there are other areas to explore besides conversational style and tone. In the enterprise, connecting to different data sources to gather information and push results also exists. You will need function calling.

Fine-tuning for function and tool calling

When passing data back and forth from existing systems, it is typical to conform to the other systems' formats because these legacy systems are likely to stay the same for you. The most recent models are getting better at matching the **function signature** to improve the reliability of these connections. The function signature is the description of the way to communicate to and from a service. OpenAI and the interaction designer on any team will recommend valuation and confirmation steps to allow the user to understand and edit the information before submitting updates to these services. When ChatGPT is aware of a function, its default is to enable it to decide whether it needs to call the function. This parameter is `tool_choice`, and it is set to `auto`. You can force this process by setting it to `required`, specifying a specific named function, or telling it `none` to disable it.

There is little for us to do on the product side here. It is included for completeness. Of course, these interactions should be monitored to ensure the proper functions are called. This requires additional training data to differentiate between similar functions, making sure that data is mapped to the correct fields when there is a collection of similar fields (for example, total price, discounted price, itemized prices, tax, and other dollar values), and to verify that the model extracted data correctly (not too much, not too little, but just right).

Product people should be aware of one neat trick: **parallel function calling**. This allows multiple queries to be sent in parallel, dramatically reducing the response time with no additional model cost. *There is no such thing as a slow, good user interface.* People today are impatient.

One gotcha. OpenAI suggests that SQL generation is "not perfectly reliable." Therefore, use caution and extensive testing and monitoring. I have not had success building AI SQL generators from text input, but this will improve over time.

Documentation: `Calling functions with chat models (https://cookbook.openai.com/examples/how_to_call_functions_with_chat_models)`

Of all the ways to use fine-tuning, plenty of tips can help with this process. We cover some of the critical items.

Fine-tuning tips

You must care for and feed the fine-tuned set to improve training quality (based on the metrics or experience with certain test cases). Here is a summary of OpenAI's suggestions for fine-tuning:

1. **Review existing examples for issues**: You might have introduced style, logic, or grammar issues into the dataset, including examples with errors. Review the material against how the model performed before and after adding the data. You can use the epoch checkpoints as a tool.

2. **Gather more examples to fill the gaps**: Additional training examples might show the model how to address gaps in its abilities. It is always hard to say how much is too much.

3. **Include examples with errors**: Sometimes, it is best to learn from the master. Let's ask ChatGPT about including mistakes in fine-tuned examples:

Should fine-tuning examples include intentional errors that might be expected from real customers?

Yes, it's beneficial to include intentional errors in fine-tuning examples. This approach enhances the model's robustness, making it better equipped to handle the kinds of mistakes real customers might make, such as typos or

```
grammatical errors. By training
on these, the model can learn to
respond accurately even when
inputs aren't perfect. It prepares
the model for real-world
interactions, improving its
ability to correct, clarify, or
ignore errors as needed, ultimately
leading to a better user experience.
```

4. **Look for imbalance and the need for more diversity in data**: Does the data set provide a range of examples? Refrain from biasing data with one kind of answer. It is not helpful to rephrase the same question. Think about the expense examples. Not just data format but style, idiomatic language, lack of data, too much data, and even irrelevant data. Even the 78 sarcastic examples for the four rounds of testing were a little homogenous. This is something the team has to notice.

Imbalance example

I had a case that puzzled me for a while. We had two primary tasks that a conversational assistant did. It did expenses and time reporting. However, both were needed in the same assistant. The training data was biased toward the expense assistant's round numbers and needed the correct numbers in the time reporting model. While a customer might say "15 minutes," "30 minutes," or "an hour and 45 minutes" for a time record, most expenses are not "$15," "$30," or "$45", so by overweighting on those round numbers in expenses it pushed understanding for the typical time units towards the expense model. This bias can be improved by using better data for the expense model and ensuring the time model has the right balance of examples.

---

5. **Match the training examples to expectations**: If it is expected that most people will give the information needed to ask a question, fill out a form, or move forward, then training examples should mirror those examples.

6. **Validate the consistency of the examples when coming from multiple sources**: Even humans are expected to disagree on specific values when labeling data. If one person tags a company name as "The Business Center" and the other as "Business Center," then the model can likely have similar issues.

7. Training examples should be in the same format.

8. Judge improvements based on previous improvements.

9. **Edge case issues take work to include**: As a rule of thumb, expect a similar improvement for each doubling of quality training data. But it is more work to find the edge cases than the happy paths. Take the time to express those edge cases. It will make the fine-tuned model more robust. Edge cases could be in several attributes: length of the question, use of multiple languages in one question, multi-part complex questions, lots of chit chat with a small portion for the actual

question, the use of data that is not well formatted or in a format not expected (military time, currencies with extra numbers), written words for numbers, or just plain wrong information. Judge this by comparing the testing results between the fully tuned model and one using one-half of the training data. Use this to judge future stepwise improvements.

10. **Specify and adjust the hyperparameters**: Start with the defaults set by OpenAI. If the model doesn't follow the training data, increase epochs by one or two. Notice how they were at five in the Wove case study coming up next. This is more typical for classification, entity, extraction, or structured parsing tasks. All of these have a known answer. With more wide-ranging model tasks, decrease the epochs by 1 or 2 to see if that improves diversity.

11. If the model isn't converging, increase the **learning rate** (**LR**) multiplier. The graphs from the Wove case study are in the next section. *Figure 8.5* shows this convergence issue, as does the Wove example, which is covered shortly. Only make small changes at one time. Here is some background:

    Article: `Learning Rate (https://www.chatgptguide.ai/2024/02/25/what-is-learning-rate/)`

12. One issue is overfitting. Mihael from Entry Point provided the example in *Figure 8.14*, in which the validation loss continued to grow while the training loss was acceptable. This is a classic overfitting example.

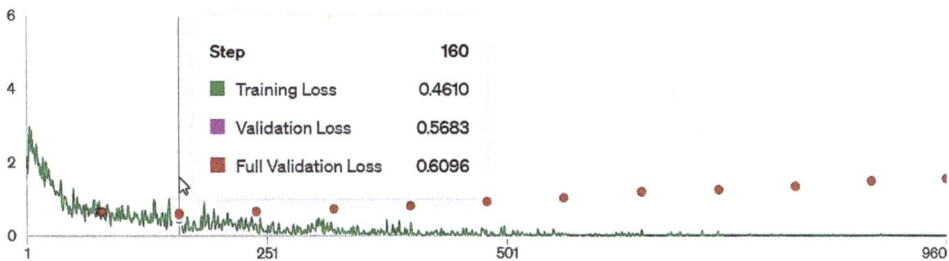

Figure 8.14 – An example of overfitting is shown with the increase in validation loss

The analogy is studying for a test by memorizing the exact questions and answers from a practice test. Then, in the actual test, none of the questions are similar enough to allow the taker to answer correctly. Aligned too closely with the study material, fails to translate to answering correctly.

13. **Use a second model to verify a fine-tuned model**: One approach is to use a second model to test the first model to confirm results. The second model might be the same or completely different fine-tuned or off-the-shelf LLM. Set a quality threshold; if the AI answer fails, it might route the request to a human customer service agent. It takes some experience to figure this out.

14. If, after all this work, the assistants' style or tone needs to change, all of the examples don't necessarily have to change. Consider tweaking the prompts to override the examples. If it is just a tweak, the fine-tuning is not going to waste; it is still helping give it the experience.

This is a vast area for research, testing, and learning. This is only a start to ensure you can apply these skills at the team level. We hope it is easy to appreciate how much the discussion impacts the quality of the user experience. Here is one more resource as a companion to the earlier OpenAI Fine-tuning documentation:

Article: `ChatGPT Fine Tuning Documentation` (`https://platform.openai.com/docs/guides/fine-tuning/fine-tuning-integrations`)

We want to examine how everything learned about prompt engineering and fine-tuning works for the Wove use case.

Wove case study, continued

In *Chapter 6, Gathering Data – Content is King,* the Wove case study on data cleansing for rate sheets used by freight forwarders was kicked off. They had to scrub the data before ingesting it to output a clean, unified view of all the rates from many carriers. Now, it is time to explore their prompt engineering (we covered the basics in *Chapter 7, Prompt Engineering*) and fine-tuning efforts for this solution.

Prompt engineering

They want the LLM to think like a customer who does this step manually. They created the prompt for the spreadsheets during the ingestion process (this early version was shared with us to maintain the proprietary nature of their latest efforts):

```
You are an expert at table understanding. You will be given a snippet
of text and a row number for the header row. Your task is to determine
where the data for the table starts and what range of rows make up
the header for the table. If the header has ambiguous columns (such as
many of the same columns), there may be a row around the header row
that can provide additional context. Include these in your range for
the header if so. Your response should be in YAML format like this:
(example truncated)
```

Let's review a few highlights from this prompt:

- It sets the stage for the persona to be adopted.
- It gives context for understanding tables.

- It explains the input (snippets of text and more details)

- It helps with some exceptions (ambiguous columns)

- It defined the response format (**Yet Another Markup Language (YAML)**)

Great job following the guidance! Honestly, they were already following this same advice. They use a lot of fine-tuned models, so examples don't appear in their prompts. One small thing: where is the idea of an emotive prompt? They should explore if that helps their quality. We can talk to them about that!

Fine-Tuning for Wove

As discussed in the last chapter, Wove has a collection of models to perform specific tasks in cleaning the spreadsheet data. The fine-tuning process adapts generic models to improve the understanding of rate spreadsheets critical to Wove's customers. It is similar to teaching a 5th grader a new subject. A 5th grader knows the basic language and can answer simple questions—they might even be into ships and trains and understand the concept of moving goods, but no 5th grader has ever seen a rate sheet.

We know that a model can only provide so much. ChatGPT recommends starting with prompt engineering before going to fine-tuning. In conversations, Kevin Mullet suggested an excellent way to remember this: "*First*, figure out what to say, *then* figure out how to say it." We have shown how this can help, but this extra effort is needed.

Here are some checks that Wove does to verify that the data is being processed correctly. This covers data quality, prompt quality, and fine-tuning:

- They manually review the interpretation of the data to look for hallucinations.

- They look for validation and training loss converging to zero. They run additional evaluation data sets using OpenAI Evals to ensure that models pass established tests. OpenAI provides a collection of templates to evaluate models using standard measures. The evals allow judgment of how different model versions and prompts impact usage. Here is a good introduction:

    Article: `Getting Started with OpenAI Evals` (https://cookbook.openai.com/examples/evaluation/getting_started_with_openai_evals)

- They use multiple steps in a chain and different models to focus on doing one thing well. They regularly revisit models to adopt newer and cheaper ones.

- In one of the models, they do location mapping and use over 600 training examples. 10% is for validation data, but for some models, they bump up to 20%, depending on how expensive it is to generate the data.

- Their training graph in *Figure 8.15* looks good. It is converging to zero for training loss. They have a validation test set that works well and uses what appear to be default parameters.

MODEL

ft:gpt-3.5-turbo-1106:wove::953EMmW8

○	Status	⊘ Succeeded
①	Job ID	ftjob-rszavQLDkaqWIprWMgz9d4yH
⊘	Base model	gpt-3.5-turbo-1106
⊘	Output model	ft:gpt-3.5-turbo-1106:wove::953EMmW8
①	Created at	Mar 20, 2024, 11:09 PM
⠿	Trained tokens	235,080
⟳	Epochs	5
☁	Batch size	1
⚖	LR multiplier	2
⚗	Seed	-
⎙	Files	
	Training	table_understanding_final_train.jsonl ⬈
	Validation	table_understanding_final_validation.jsonl ⬈
⊪	Training loss	0.2084
	Validation loss	0.0359

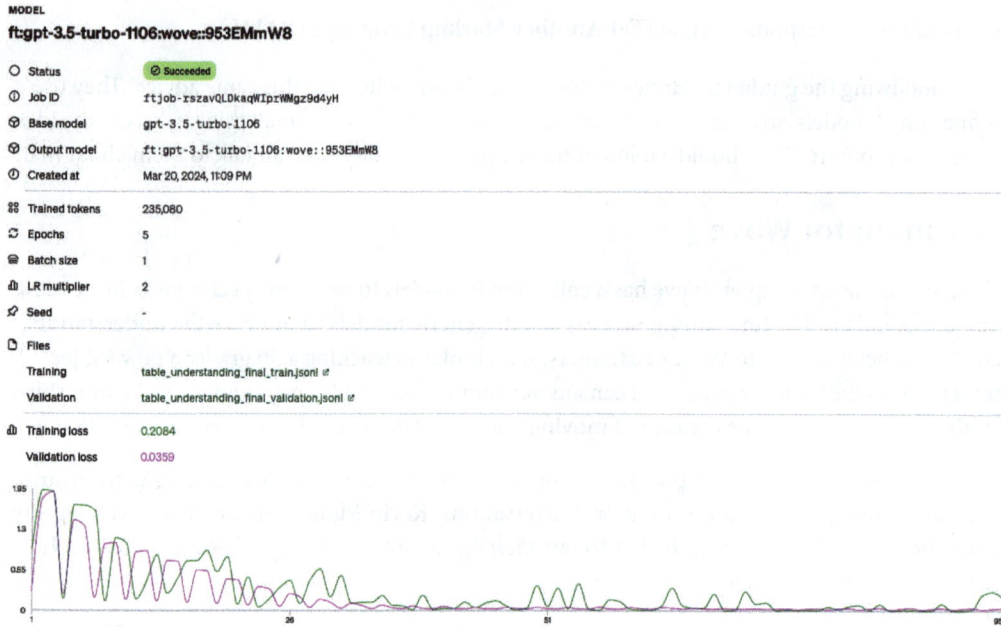

Figure 8.15 – An example of a training model converging

- Initially, they had learning rate issues. In machine learning and statistics, the learning rate is a tuned parameter in an optimization algorithm that determines the iteration step size while moving toward a minimum loss function.

Even if trained again, it never quite gets to the optimal point. Sometimes, there is poor convergence or overfitting, as discussed. *Figure 8.16* shows an earlier run showing a lack of convergence.

MODEL

ft:gpt-3.5-turbo-1106:wove::95lENyg6

○ Status	⊘ Succeeded	
ⓘ Job ID	ftjob-VZcQOcqjanNserEH1056rv2P	
⦿ Base model	gpt-3.5-turbo-1106	
⦿ Output model	ft:gpt-3.5-turbo-1106:wove::951ENyg6	
ⓘ Created at	Mar 22, 2024, 6:51 PM	
▥ Trained tokens	511,825	
⟳ Epochs	5	
☷ Batch size	1	
⏱ LR multiplier	2	
✧ Seed	-	
▢ Files		
Training	table_understanding_final_train.jsonl ↗	
Validation	table_understanding_final_validation.jsonl ↗	
⏱ Training loss	0.1672	
Validation loss	0.5442	

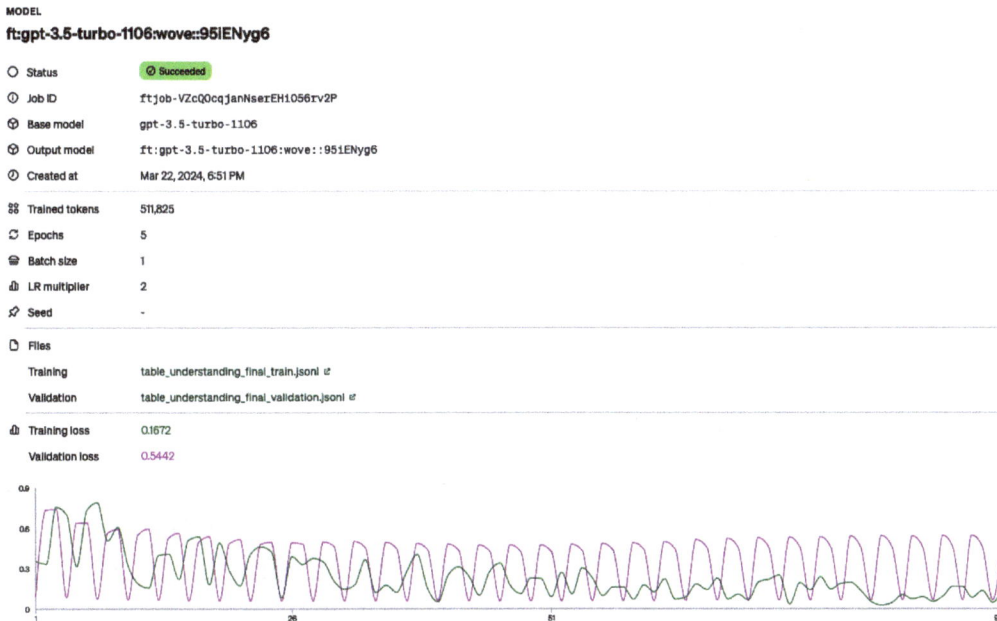

Figure 8.16 – An example of an early Wove training model that didn't converge

If there is bouncing and the lines do not quite converge, retrain with a lower learning rate. Convergence will be much slower if the initial weights are too high. To put models in perspective, the four rounds for the sarcastic chatbot experiment ran with less than 5000 training tokens for each round. The Wove model above used over 500,000 training tokens for this one piece of their solution. Bigger doesn't guarantee convergence.

Our friends at Wove have a few more tips for those ingesting spreadsheets:

- They use expensive models to fine-tune cheaper models. For example, Anthropic's Opus 3 Claude is (at this time of writing) about 30 to 50 times more costly than OpenAI 3.5. Opus 3 is 15$/1M input and 75$/1M output tokens versus ChatGPT 3.5 turbo-0125 at $0.50/1M input and $1.50/1M output tokens. This is an essential point for the product team. You want to get the best bang for the buck, especially when dealing with customers who will use the model more often if it provides excellent service. They found significantly better quality from fine-tuning ChatGPT 4.0 than earlier 3.5 models. Chat GPT-4o mini is now being incorporated.

- They started using a higher learning rate, which they could reduce when they made significant steps. The amount of data impacts the learning rate steps (e.g., 0.5, 0.1, and 0.01). This can be tweaked based on the size of the data set and other factors. The goal is to refine so the training loss decreases in an orderly fashion without wild swings or divergence.

- Working around the context window, they started with a 4K context window and grew to support 16K. They can preprocess data and use other tools to deal with smaller chunks. Even at 16K, this is small for handling large spreadsheets.

- **Make sure the data is accurate**: With good input data, success is more likely. If the trained data isn't tagged correctly or has missing data, the data team will have issues. This real example is for an early Wove's table end detection version. You can see this in the chart bouncing all over the place compared to earlier examples. You don't see a convergence.

We can look at one example of a hallucination and how they corrected it in the part of the flow where they took the text version of the spreadsheet (left column) and asked ChatGPT to define (understand) the row's context (middle column). This is *Step 1*. Then, they edited this output, completing the process with the final output (right column) shown in *Table 8.4*.

Input File	Intermediary output from ChatGPT	Final normalized file from ChatGPT

The table on the sheets "Rate Sheet", is from the VN to US UPDATE file.
 	
The table has these columns:
- id: 0
•
name: POL
 	
Examples:
 	
- DA NANG, VIETNAM
 	
- HO CHI MINH CITY, VIETNAM
 	
- HAIPHONG, VIETNAM
 	
- VUNG TAU, VIETNAM
	Purpose: This table provides ocean freight rates and details from various ports in Vietnam to destination in the United States, carriers and services, valid for a certain date range.
	columns:
	
	- id: 0
	
	   name: POL
	
	  description: The port of loading in Vietnam, such as De NANG, HAIPHONG, QUI NHON, or VUNG TAU.
	
	examples:
	
	- DA NANG, VIETNAM
	
	- HAIPHONG, VIETNAM
	
	- VUNG TAU, VIETNAM
		carrier: 7
		
		service_line: 4
		
		origin: 0
		
		origin_country: Vietnam
		
		origin_via: null
		
		origin_via_country: null
		
		destination: 1
		
		destination_country: United States
		
		destination_via: 2
		
		destination_via_county: United States
		
		loading_port: null
		
		loading_port_country: null
		
		unloading_port: null

Input File	Intermediary output from ChatGPT	Final normalized file from ChatGPT
- id: 2 name: Dest (Via Port) examples: - OAKLAND, CA - NORFOLK, VA - LOS ANGELES, CA - SAVANNAH, GA VANCOUVER, CANADA - None - TACOMA, WA - BALTIMORE, MD - CHARLESTON, SC - NEW YORK, NY	- id: 1 name: Destination description: The destination city and state in the United States, such as LONG BEACH, CA, SEATTLE, WA, or CLEVELAND, OH examples: - ATLANTA, GA - PITTSBURGH, PA - CHARLOTTE, NC - CLEVELAND, OH - id: 2 name: Dest (via Port) description: The port in the United States that the shipment will go to reaching its final destination, such as NORFOLK, VA, SEATTLE,WA or LOS ANGELES, CA. Can also be "None" if the shipping direct to the destination. examples: - NORFOLK, VA - SAVANNAH, GA	unloading_port_country: null effective_date: 8 expiry_date: 9 container_dry_or_reefer: null commodity: null rates: - id: 10 currency: null type: total container_size: "20ft" - id: 11 currency: null type: total container_size: "40ft" - id: 12 currency: null type: total container_size: "40ft_hc"

Table 8.4 – Wove file improvement steps (some data are truncated)

The table shows the description of the destination, aka dest (via port), generated in the middle column: "The port in the United States that the shipment will go…". However, the data includes Vancouver, Canada. Although the destination is always the US, the port it goes via might be outside the US.

To improve the model, this hallucination needs to be corrected. It could be decided that the correct answer is to edit this to be "North America," or better is to remove the "in the United States" entirely, making the description more generic. This means that the output file in the next step also needs correction. The `destination_via_country` field will be changed from United States to `null`. This file on the right is a second round of ChatGPT generation, creating the unified model that makes all the spreadsheet data a consistent, normalized output. It is vital to catch these errors. With this final output file, they re-run the test data to see whether the quality improves.

Overall, this fine-tuning process requires many iterations on the prompts, editing of the tags, and evaluating against their test data. We can't expose all of Wove's secret sauce, but hopefully, this gives a sense. A modeler's work is never done. Although ongoing effort might be reduced, work will not be done. Format changes can occur, including new vendors, normalizations, better, cheaper, and faster models, and everyone's favorite, bugs will require rework. The point is to be involved and invested in these steps to ensure quality. Readers can imagine the next steps for Wove once all this data is normalized and available. Customers will want to ask questions about the best route based on shipping characteristics. They will not want to pour through even a normalized sheet of rates.

This is an exciting use case because it starts as a backend solution, still needs product understanding and feedback to be successful, and will likely lead to even more UX efforts when (inevitability) a customer-facing chat experience will converse with customers to help shop rates. Product and UX efforts will be needed there.

Wove used a series of models to understand the complexities of tables. Picking and chaining suitable models is part of the prompt engineering and fine-tuning process.

Summary

Fine-tuning is the most technical piece of this book. With this little glimpse into this world, there is much to cover. Your data scientists and engineers will go deeper. When building production-ready systems, mix and match fine-tuned and generic models with internal software and third-party tools to balance speed of delivery, price, and performance (recall the saying, *cheap, fast, or good, choose two*). Innovative solutions have workflow steps that allow the solution to bail out if the AI isn't performing, use a function to solve or address a specific problem, or use a more deterministic element to provide a robust solution. Injecting the suitable model and prompts for the correct part of a use case is one of the most critical decisions. Do this work before embarking on the fine-tuning approach.

Contribute to the process by helping define and improve these task flows through use case expertise, editing and improving prompts, creating, verifying, and editing fine-tuning examples, and monitoring if changes are moving the solution in the right direction. Go forth and tune!

Chapter 9, Guidelines and Heuristics, will cover guidelines and heuristics to support prompting and fine-tuning efforts based on well-documented techniques in the design community to help explain conversational AI's usability.

References

The links, book recommendations, and GitHub files in this chapter are posted on the reference page.

Web Page: `Chapter 8 References` (`https://uxdforai.com/references#C8`)

Part 3:
Care and Feeding

In the last phase of your journey, we'll explore the tools and resources needed to understand how the LLM is performing and what to look for to fix it if it isn't up to expectations. You'll first learn how to apply existing guidelines and heuristics that can be used in ways you didn't expect to craft LLM solutions. We'll even give you the skills to adapt web and traditional design thinking to conversational AI. Then, you'll dive into understanding how to monitor and evaluate solutions based on industry AI and user experience metrics, covering objective and subjective performance measures in the care and feeding process. The lifecycle of an LLM solution demands good business processes to be successful. Best practices that can work at the most prominent enterprises are shared to help create effective methods for creating world-class LLM solutions.

This part includes the following chapters:

- *Chapter 9, Guidelines and Heuristics*
- *Chapter 10, Monitoring and Evaluation*
- *Chapter 11, Process*
- *Chapter 12, Conclusion*

9
Guidelines and Heuristics

This chapter explains what makes an excellent conversational style. Some of what is covered may seem obvious, but exploring and understanding why something works or doesn't work is valuable when applying the concepts

to new situations. ChatGPT is unique because it won't necessarily answer the same question again in the same way, which is why *Chapter 7, Prompt Engineering*, was essential. This chapter will cover guidelines and heuristics to evaluate and improve the experience you are designing. We are not picky about what we call these, but we can use better definitions. **Guidelines** are particular and tend to be based on user research. **Standards** are more specific, while **best practices** are recommendations based on certain conditions. Vendors such as Apple's **Human Interface Guidelines** (**HIG**) or Google's **Material Design** are widely copied and include all of these protocols. Organizations or governments can mandate guidelines and standards as required. **Heuristics** are rules of thumb or strategies to break down a problem into elements that need attention.

```
Guideline (Smith and Mosier, 1986): Design text editing logic so that
any user action is immediately reversible.
Standard: A text header should be Bold Calibri 11-point typeface.
Best Practice (Apple Human Interface Guidelines): Use a chart to
highlight important information about a dataset.
Material Design (Google): Consider making pointer targets at least 44
x 44 dp.
Heuristic: Titles should be readable and aesthetically pleasing.
```

The point is that there are things one should follow (guidelines, standards, best practices, and design recommendations), and there are strategies (heuristics) to figure out what to do. Some apply to the GUI encompassing the conversational AI, and some are for conversational text. All of this is covered in this chapter:

- Applying guidelines to design
- Adapting heuristic analysis for conversational UIs
- Building conversational guidelines
- Case study

All of these are for knowing what to do before doing it. We can use this knowledge for the next interaction if we learn what works. Design guidelines have evolved, but as we will see, some have been around for decades, and not only do they still apply, but they might even apply more than before.

Applying guidelines to design

This book follows the order of the life cycle of conversational AI design. Why do guidelines and heuristics come after "building" the experience and prompt engineering? This is a chicken and egg problem. Many teams create generative AI experiences for the first time. The application of guidelines and heuristics has to be done with some understanding of what has previously been built. In visual experiences such as GUIs, previous project experience helps inform what guidelines and heuristics will translate into new designs. This is not the case with conversational AI. This advice is here to help you get started and be there as your journey unfolds.

Software tests can be crafted to evaluate whether a guideline is met. Testing is more challenging with a heuristic. The heuristics used for evaluation are broader than the precise nature of a software test case. Using the examples of standard to use Bold Calibri as a font in a header, each header would need its own test to evaluate a user experience for this condition. It is a trade-off. More expertise is needed to know and internalize heuristics, and it's the same with guidelines. Another difference is that a heuristic will stand the test of time. They are generic enough to adapt as experiences evolve. Because they are generic, they are hard to define in code. How should the heuristic *titles should be readable* be measured? Because a guideline is more specific, it might only apply to a particular UI or use case. However, they are still valuable. In addition, understanding the underlying science behind a guideline can help you more effectively apply it to new experiences.

One of the first sets of guidelines I ever used was by Smith and Mosier (1986). The document contained 944 guidelines for software interfaces. Don't dismiss them because they might be older than you. They are based on research on human behavior and capabilities. Human behavior hasn't evolved to invalidate them, even with new contexts for their uses, such as high-resolution displays, voice interfaces, and hand-held devices. There are a lot of universal truths in them. Many of these original guidelines were associated with text-based experiences, and it seems like what goes around comes around, as, surprisingly, we have returned to text-based solutions with chat, conversational, and recommender experiences.

Article: `Smith and Mosier's Guidelines for Designing User Interface Software (https://hcibib.org/sam/)`

However, as I mentioned, these guidelines are for **User Interfaces** (**UIs**), and heuristics might apply better to evaluate conversational experiences. The evolution of these heuristics is based on the solid scientific efforts of these researchers and previous ones. Before diving in, here is one more example. Bruce Tognazzini is a famous Apple designer and an early partner at the Nielsen Norman Group. He is someone the industry has respected and appreciated for years. I invited him to be the keynote speaker for a conference host in Blacksburg, VA. You know his keynote was going to be a little crazy when he asked the hotel for fire extinguishers to be present. That freaked out the hotel management. Being a little shocking applies to guidelines as well. Some will seem a little crazy, but they can be applied more effectively by going deeper and learning about their origins and scientific underpinnings. Then, they won't seem crazy when they apply to enterprise use cases, nor was his reason for wanting a fire extinguisher, once you understood its purpose. Here is his list of principles; see how they can easily apply to the ChatGPT frontier.

Article: `First Principles of Interaction Design` (https://asktog.com/atc/principles-of-interaction-design/)

Adapting heuristic analysis for conversational UIs

A wide range of possible issues can be found with a good set of heuristics. Heuristic evaluations can range from formal to informal. The more formal approach is to enlist three to five usability experts as evaluators. Once provided context and the background of the tasks and users, they can independently evaluate the experience against their understanding of the heuristics. By documenting the issues, scoring their importance, and compiling the results from each reviewer, the team can prioritize the issues to be addressed. This approach can be used to look at UI issues, and much of this can also be used to understand conversational interactions.

These issues will not be found in surveys that provide a score, such as the **Net Promoter Score** (**NPS**) or the **Software Usability Score** (**SUS**), covered in the next chapter. NPS or SUS can be applied once customers are exposed to the product; a heuristic evaluation of a working prototype has some advantages:

- It can be done early in the design process

- It is inexpensive

- UX professionals can do it, and others can participate as well

- They are battle-tested for traditional GUI evaluations

- They are adapted here to consider conversational experiences

  - With unique chat and recommendation UIs

  - For hybrid UIs that use GUI components within a chat

  - For sentences, when explicitly defined in templates, with deterministic flows, or controlled abstractly through prompts

  - As part of prompts to instruct the LLM to value the heuristics

There are some issues to address as well:

- It can be biased based on the evaluator (which is why a few evaluators participate)

- It depends on the evaluator having enough expertise in the evaluation and the feature to detect issues (evaluators can be trained, given time to practice and explore and be provided sample use cases, but all of this can bias a review)

- It is best done with three to five evaluators, who might be hard to get

- It can return issues that are not significant

To balance the good and the bad, the following is suggested:

- Review the heuristic tools before the evaluation and coach people on a separate example product to reinforce the method and heuristics

- Provide a printout of the heuristics to the evaluator

- Remind evaluators to put on their customer hats

- After the evaluation, use the scoring tools discussed in the earlier chapters to prioritize and focus on the most valuable findings

- Iterate quickly so that future evaluations can reveal new items and are not masked by more significant, overwhelming issues

- It is reasonable to use heuristics in your daily design efforts to guide you to solutions that are already good before a customer sees the results

We need a set of heuristics to provide the most significant value with the least cost. One set already comes to mind.

I always start and finish with Jakob Nielsen's ten heuristics. They are broad enough to apply to various situations, have been revised, and used for over 30 years, which gives them some street credibility. They cover a range of issues found in user experiences. Usually, but not exclusively applied to GUIs, they need to be put in the context of conversational AI. The articles I reference cover the basics of heuristics, and hundreds of other articles are out there. Sometimes, these guides will be spot on – for hybrid experiences that include UI components with conversations. The UI heuristics are well documented. The added value to expose is how these apply to our conversational experiences. This would be called color commentary to make an analogy to watching sports on TV. The play (the heuristic) is taken, and a discussion explains it so a layperson can understand their application to a conversational UI.

Article: `Intro to Heuristic Evaluation (HE)` (https://www.interaction-design.org/literature/topics/heuristic-evaluation)

Article: `How to perform a heuristic evaluation` (https://www.nngroup.com/articles/how-to-conduct-a-heuristic-evaluation/)

Extensive examples that support conversational flow and recommendation UIs will be used. This chapter won't benefit backend experiences, as the heuristic is about evaluating the user experience.

Expecting a conversational AI to return exact messages might be a challenge. It depends on what is in control of the output. If ChatGPT controls the output, we have to rely on prompt engineering and fine-tuning to get it close. Suppose a traditional deterministic chat experience provides the front end. In that case, you can specify precisely the response wanted or a collection of responses to pick from.

You can use ChatGPT to understand entities, ask questions to fill in gaps, or perform other language tasks to support the higher purpose. Recommender UIs sometimes use a text template. ChatGPT can gather input, fill in values, transcode details, and provide translation. Keep these in mind, as the heuristics can apply in different ways to each context of use.

If you want to see any of the following examples show up directly in a ChatGPT chat, it is hard to make it happen consistently. It comes back to instructions. It is possible to use the definitions in the heuristics for the instructions on how to formulate responses. That would be a great research project. We will use a new employee onboarding process example prompt incorporating critical heuristics to guide discussions. The heuristic influences are highlighted in **bold**.

```
Instructions for the new hire onboarding skill
I have provided a document that outlines our company's onboarding
process. I need you to create a set of detailed instructions from this
document. As the user performs each step, reach out to the appropriate
service to verify completion and provide the status of each step.
Follow these guidelines:
1. Format the instructions as a numbered list.
2. Each step should be clear and concise, with no more than two
sentences per step.
3. Use formal language suitable for interacting with a new
professional.
4. Focus on key actions new employees must take and omit any
unnecessary background information.
5. The instructions should be easy to follow for someone new to the
company.
6. Do not ask the new employee to remember information. Provide
details at each step and confirm any understanding of those steps as
you proceed through the process.
7. If there are errors, report them in clear, simple terms and explain
how to fix them. If a step doesn't block progress, keep track of it so
the user can complete it in any order.
8. Keep it short and simple for questions that have well-known
answers, like Age or Height. For less well-known questions, provide an
example or details about what is needed.
For example:
1.**Sign into the company portal**: Use your assigned credentials from
the welcome email to access the portal.
2. **Complete the onboarding form**: Fill out all required fields,
including your personal and emergency contact information. Let me know
if any fields need clarification.
```

A callout for each heuristic will discuss the heuristic influences found in the instructions provided to the LLM. The following are the heuristics. Each will be defined and explained, along with an analysis of how they can be applied.

- *Visibility of system status*
- *Match between a system and the real world*

- *User control and freedom*
- *Consistency and standards*
- *Error prevention*
- *Recognition rather than recall*
- *Flexibility and efficiency of use*
- *Aesthetic and minimalist design*
- *Help users recognize, diagnose, and recover from errors*
- *Help and documentation*

Each heuristic follows the same model. The name of the heuristic, the exact definition, the analysis to apply it to conversational AI, and a callout for an analysis of the new hire onboarding example.

1 – Visibility of system status

The design should always keep users informed about what is going on through appropriate feedback within a reasonable amount of time.

Analysis

In a chat experience, keeping the user informed is typically done through textual response. In process flows, some information will repeat to provide context when prompted for the next step. The user expects to see this in seconds or less. Waiting 10 seconds to gather backend data would be odd. If providing timely information is a problem, give that feedback. Sometimes, it is challenging to provide estimates. If the user expects delays, they might feel better about it. Learn from these examples:

```
I am gathering the results. I will get back to you in about 20
seconds.
I am gathering data from 35 systems for you. Some are slow. Give me
about 15 more seconds.
The engine rebuild you requested will not be finished for 3 more
hours. Can I email or text you when it is ready?
```

It is crazy to start this section with this horrible user experience, making the user wait *forever*. We will explore a few more typical system status examples.

Here is an appropriate level of feedback, confirming that the address step is completed with enough context for the user to pick a delivery date in the future:

```
I have your shipping address. We can deliver as soon as Friday. When
would you like the delivery?
```

What gets provided as feedback should be limited. If the address is confirmed, don't repeat it along with everything else in the order. Wait until there is a summary or when the user requests the address. This is the *appropriate feedback*. Here is an example of how to give instructions to the LLM. Recall from *Chapter 7 – Prompt Engineering*: the instructions are prompts that wrap the customer conversation so that the customer prompts have some guidance. It is ok to refer to these as instructions or prompts if you realize the context is an overarching prompt, not what the user types (also called the user prompt).

```
Instruction Tip:
Use a summary if you have already confirmed information and must let
the user know you still have it.
For example:
If you confirmed the user's shipping address on the last step, refer
to it as "your shipping address." Don't repeat the entire address
unless requested.
```

It would be challenging if a calendar was full of unavailable delivery dates. The user would be stuck in a mindless game of picking a date. For example, in a voice-only solution, offer three suggestions for delivery or openings around a date they provide. A GUI can show a month-at-a-glance view with available delivery dates. It would show the next available date and offer to edit that if it doesn't work. That way, the user understands what the system can do:

```
I have your shipping address. Our first available delivery is on
Friday, August 9th. We can also look for later dates.
```

If the user experience is just a recommendation and not interactive, it still can be good to clarify the information based on the recommendation. If they change something on the screen, will the recommendation update? For example, if the recommendation is to email the customer this week, does it know that I already emailed them today? Even if this information is not directly related to the recommendation, have a clickable affordance, like an **info** button, to explain what went into the recommendation and, in this example, whether the last email was accounted for. Here is what could be behind the **info** button:

```
This recommendation is based on a few factors:
  • Email was sent on Tuesday.
  • The customer visited a few weeks ago.
  • The analysis of this customer's
    willingness to buy compared to other customers.
```

Giving the timing information (e.g., last updated three minutes ago, or this recommendation is based on details updated yesterday) can help gauge the relevance of the recommendations.

There's one last thought about system status, which can also apply to backend services or recommendations. If an unavailable resource impacts the data, let people know. Don't just time out a connection.

Integrations should support and communicate the edge cases of the system being down. It should mirror the style and tone of the service and not be a cryptic error message:

```
Oops! We can't access the recommendation system. It appears offline.
If this persists, please report it to it@company.com.
```

This email link should populate the subject and body with technical information. To take this to the next level, in the right conditions, do this:

```
Oops! We can't access the recommendation system. It appears offline.
The IT department was notified, and a trouble ticket was filed.
```

Depending on the audience, it might link to the ticket or cc the customer. This message is okay to be repeatable and consistent. Please do not leave it to the LLM to generate the response. It should be so rare that variety won't matter, and we want to be very specific about the error's development. If exposing too much detail confuses the user, stick to a generic message.

Here is the first callout explaining the impact of this heuristic on the new employee onboarding instructions. We won't introduce the callout each time; they will always be last for each heuristic.

> **Analysis for new employee onboarding, visibility of system status**
>
> Heuristic influences: *Focus on key actions, provide details at each step, and if there are errors*
>
> There are a few places where status should be communicated to guide the user. Confirming that progress is tracked and reporting on errors along will also help through the stressful time of starting a new job.

2 – Match between a system and the real world

The design should speak the users' language. Use words, phrases, and concepts familiar to the user, rather than internal jargon. Follow real-world conventions, making information appear in a natural and logical order.

Analysis

Style and tone go a long way when communicating in a conversational tone. It gets trickier with enterprise software. If the user refers to the ordering system, don't reply with the *order entry and tracking system*. They might not realize it is the same system. Actively mirror the customer language and conventions, but do so in a style consistent with the organization. Recall our surf shop example. To confirm an order, use this:

```
Hang Ten! Your new board is hitting the waves via UPS. We will drop
the tracking code by sunrise. Mahalo.
```

Prompt engineering and fine-tuning establish less colorful responses from a financial service company. The response will be in more formal business-speak:

```
Your new business checks for your Alligiance money market account will
ship tomorrow via USPS.
```

This is also an excellent heuristic for appreciating how to display information. Consider the format of content, or with lists, the order for the information provided. *Table 9.1* shares order and format options for displaying information. *Use prompts to set an order and format.*

Order (for lists)	Format (for complex content)
• Alphabetical	• Bullet List
• Chronological	• Calendar
• Classifications	• Cards/Tiles (Like a Business Card)
• Highest to Lowest	• Charts or Graphs
• Location or Distance	• File Formats (PDF, XLS, or Doc)
• Logical	• Image or Graphic
• Most to Least Expensive	• Maps
• Numeric	• Ordered List
• Popularity	• Spreadsheet (Downloadable)
• Priority	• Table
• Random	• Trees
• Size or Dimensions	

Table 9.1 – Common design alternatives to display information

> **Analysis for new employee onboarding, match between a system and the real world**
>
> Heuristic influences: *clear, simple terms*, and *should be easy to follow*
>
> The intent is to get them onboarded, not to teach them jargon. The prompt emphasized that this is a new employee, so clarity is key.

3 – User control and freedom

Users often perform actions by mistake. They need a clearly marked "emergency exit" to leave the unwanted action without having to go through an extended process.

Analysis

Usually, buttons such as **Cancel** are on a dialog box or in a multiple-step process; the user can cancel or leave the process and pick it up later. Too many UIs are scary because it seems possible to lose work. How about a website that times out five minutes into a long form? If you sign back in, can you pick up where you left off? If the user accidentally closes the window or clicks the **Back** button on a browser, can they return without losing work? Alternatively, figuring out how to undo an action might be challenging. An online shopping cart has this issue. Sometimes, finding out how to get out of the cart or remove an item is (intentionally) difficult. Some carts have a distinct remove button. Others also include a save for later button to encourage future purchases. Others make the user set the quantity to zero (which is both indirect and not marked).

In a traditional UI design, be explicit, such as giving the user cancel and accept buttons for a dialog box. A textual interface or voice interface won't have visual affordances. Suggest a path forward if the user needs clarification or guidance:

```
(System provides a summary of a transaction)
If this looks correct, reply "submit." You can also give me changes or
even cancel the order.
```

Here, the control term `"submit"` is not necessarily required in a conversational AI. This message encourages the user to move forward; it is not written to force them to accept the next step. Many words/phrases might be supported, and the system has to decide if those words are enough to confirm the action. I asked ChatGPT to classify a collection of terms that might mean submit.

```
                            classify the examples as strongly
                         confirming, sort of confirming,
                            weakly confirming, or not
                                confirming the order.
```

Strongly Confirming:
Go ahead with it, Let's finalize this, Proceed with the order, Give it the green light, We're ready to move forward, Submit the order, Send order, Submit order

Sort of Confirming:
Let's get this rolling, **Please confirm and execute, All set—place the order, Proceed as discussed, Order,** Looks good, All good, All done, Send it, Go 4 it, Lez do it

Weakly Confirming:
Let's put this into action, **Whatever,** Okie dokie smokie

Not Confirming:
I don't not want this, Not soon enough

Not only would a conversation around submitting an order expect to understand these, but it could also be that the instructions for the LLM are told to confirm the submission if the instructions are weak. This is the type of task understanding that an LLM will know even without enterprise training. It should be expected to work. But there are a few in **bold** that are not classified well. A few should be more positive, and a few should be less favorable. If you had an existing chatbot, you likely had trained it on examples like this. These would be the same samples that could be used for fine-tuning to help it overcome the LLM's problems with negation (Not soon enough is a strong confirmation) and ensure that simple terms, like *Order*, are followed. Don't force the user to talk in a structured business language. Apply this mantra:

Bring the experience to the user; don't make the user go to the experience.

Talk in the user's language; don't force them to learn the company's vocabulary.

This heuristic also covers the freedom for users to return to a conversation later and pick up where they left off. This is a significant challenge for many platforms. If the user is in a chat conversation and returns a few hours later, will the service allow them to continue, or will it have timed out, causing the user to start over? Even if a timeout is inevitable, the user's state should be maintained and restored once they authenticate. Understand typical pauses in conversational AI and support the continuation of the interaction. I have done data analysis of transactional conversational UIs and have seen plenty of data where conversational interactions are picked up again, even 24 hours later. This is similar to how we act with others on messaging platforms. The expectation is that it is okay to return later and resume the conversation.

As for exits, in a dialog box or even a wizard, there are marked exits, such as the cancel or exit button, typically in a consistent place on a platform. In a conversational UI, cancel buttons are avoided. Users must recognize that they can exit gracefully by saying, "Stop this order; I don't want it," or leave. There won't be a visual affordance to stop in the middle of a transaction. For recommender UIs, the user isn't "in" the UI; it is a secondary piece of information to assist the primary interaction, so no exits are expected.

There are cases in hybrid UIs where a cancel option for a long process, such as uploading a file or filling a form that appears as part of a conversational interaction, could have one. To be clear, this means exiting a process or task *in* a conversation. The window that contains the conversation likely has an exit or close button. This button might stop the conversation or close the window, allowing the user to pick up the thread later. The GUI component **Cancel** should respond to conversational interactions "I don't want to send this."It just depends on the use case. This heuristic about user control also supports an undo concept.

Redo as undo

Users make errors. One error is to delete, remove, or create an unintended edit to a form, object, or content. In some traditional UIs, **Undo** comes to the rescue. It is on the **Edit** menu, web applications, messaging platforms or a gesture like shaking on an iPhone. For years, Adobe Photoshop has had a history menu that supports multiple levels of undo. Some modern email UIs allow undo when sending an email. There may be only 10 seconds to unsend an email, but it is a relief to do this when sending something too soon. I hate sending an email that should have an attachment, and forgot to include it. Undo! Undo! Consider how to support **Undo** for transactions and experiences. This applies to conversational experiences and not to recommendations or backend solutions.

> **Here is a secret tip for getting a developer to support Undo**
>
> Developers can freak out about undoing transactions. An alternative is to think about redoing the task programmatically – that is, the user, in almost all cases, doesn't care about the state of the database; they want to undo what they did. An innovative developer can submit a transaction that effectively undoes it by keeping a copy of what was there previously and resubmitting it. Redo acts as an undo. The customer is none the wiser and happy. One level of undo in most use cases is probably 95% of the problem. Solve 95% of the problem before considering multiple levels of undo to solve the other 5%.

Undo can be supported conversationally:

```
                                    I made a mistake. Can I
                             return to what I had before?
```

```
No problem. Your old appointment time
is still available. You are scheduled
for 2:00 PM tomorrow.
```

This is not about generative AI per se; this process deals with function calls and interacts with backends conversationally. So, undoing a mistake can require backend support to undo or maybe redo. In this appointment situation, as long as no one booked that slot, the system should support booking the new spot and then canceling the old spot. This way, the user isn't in limbo if the old time was taken. The original booking interaction might have been hours earlier. Maybe they had a meeting and responded after realizing the error.

> **Analysis for new employee onboarding, user control and freedom**
>
> Heuristic influences: *explain how to fix them*, and *complete it in any order*
>
> The influence sighted does give the user control to solve problems, and it does support doing tasks in any order.

4 – Consistency and standards

Users should not have to wonder whether different words, situations, or actions mean the same thing. Follow platform and industry conventions.

Analysis

This is a catch-all for the additional work covered in this and earlier chapters. UX guidance emphasizes speaking in the user language and learning how to handle variations in language. The upcoming heuristic, *6 – Recognition rather than recall*, offers a good example. Watch out for the use of language on the output side; be consistent and use the terms most appropriate for the company. However, users will use what they know and recall. If these terms are widely different, encourage and don't demand the user to use the correct terms. This allows them to connect to the more modern terms without ignoring them. This heuristic could be interpreted as counter to heuristic *2 – Match between the system and the real world*, which asks to speak in the user's language. We bridge that divide by understanding their language while reinforcing the new term.

For example, customers who have used a product for years can refer to it by its original name. When another company purchases the product and rebrands the product, the new name gets exclusively used after some time. But the customer still has the old product with the old name running just fine. It might be a physical device in a shop; the old name is stamped right on the side. When they call for support, they still use that name, a short version of the name, or even an initialism.

Oracle purchased BEA in 2008. The BEA middleware was called WebLogic Server. Customers might still be running this software. They might refer to it as WLS, the server, WebLogic, BEA, Oracle WebLogic Server, BEA WebLogic Application Server, the app server, or the more modern terms Oracle Middleware or Fusion Middleware. The conversational AI should support all.

```
                                        Can you help me get a
                                        software patch for WLS?

I can help with patches for Fusion
Middleware. WebLogic Server (WLS)
has been rebranded to this new name.
What is the release and platform
information for the install of WLS?
```

This helps them connect from the old to the new term and speak in their language when asking for details about WLS. This example used a prompt that supported a more casual tone, resulting in the term *install* instead of the more formal *installation*. This prompt might be used:

```
If you encounter an old term for a product, gently guide them to use
the new term.
```

Additional fine-tuning examples could be needed to improve the dictionary of terms. Consider using checklists for new tasks and actions to monitor for issues in conversational logs or recommendation output strings. Refine the approach using all the tools – prompt engineering, data cleansing, and fine-tuning.

Analysis for new employee onboarding, consistency and standards

Heuristic influences: *Clear, simple terms*

The previous callouts cover some of the same concepts. As long as you identify a problem, you are doing well. Once you identify a solution, consider if it can cause issues with other heuristics or guidelines.

As discussed in the analysis above, use fine-tuning to support transparent, simple, and user terms. Additionally, the new employee might not be familiar with some terms. Setting up their account information via "SSO" for the first time might require an explanation of Single-Sign-On. There should be sufficient detail to explain SSO, for example. Don't assume knowledge.

5 – Error prevention

Good error messages are important, but the best designs carefully prevent problems from occurring in the first place. Either eliminate error-prone conditions or check for them, and present users with a confirmation option before they commit to an action.

Analysis

Let's start by discussing how to avoid errors in the first place. This is a considerable challenge when dealing with conversational UIs. They can say anything; if misinterpreted, the conversation goes wrong. This is where product owners and designers need to look at thresholds for understanding. If a system isn't confident in its direction, interject and guide the user. Disambiguation is standard in traditional conversational chat experiences. Use prompt engineering to build instructions so the generative AI can meet a confidence threshold. In *Chapter 10 – Process*, we will show more examples of chaining that can be used to evaluate a confidence threshold and assist in getting a better answer. If more clarification or confidence is needed in what the user asks, prompt the user for more details. Provide options that help them narrow down their problem to guide them on the right path, like with a prompt like this:

```
Are you asking for expense reports, your team's billing report, or the
latest sales report? Let me know.
```

The faster the resolution, the fewer additional errors will occur. I reviewed many chats where the user was unaware they were in a misguided conversation. The user doesn't notice context clues that the conversation isn't working. They continue interacting as if they know where and what they are doing, but they do not.

> **Important note**
>
> Customers do not read; they skim, at best. Provide vital clues and reinforce with guideposts. Use redundant clues for errors that need to be fixed.

For example, there are standard patterns to fix errors in traditional forms. A message might say that there is an error. It should explain what field has the problem, and then there should be an error label or icon on the field, along with details on what is wrong. While writing this chapter, I experienced errors when submitting forms, but the website would not tell me what field had a problem or even highlight the field with the mistake. I was left trying to edit each field to see whether anything changed. And even then, in one case, I couldn't submit the form. I never figured it out.

Information passed to other systems must be valid in conversational AI solutions, especially for transactional UIs (e.g., searching for a flight, filling out an expense report, or conversational versions of UI forms).

Conversational UIs are prone to submitting wrong data. The AI doesn't know it is wrong, so it can't prevent it, and even a traditional UI won't typically catch a user typing 112, but intended 12 if the field isn't validated. That is the magic of AI. Although it might not be preventable, an AI can determine if something appears wrong and understand what is reasonable data. A prompt in your instructions might catch issues:

```
Prompt: If you are given information that is not typical for the
expected values, repeat it to the user to confirm that it is correct.
You entered $112.23 for coffee at Starbucks. That doesn't seem right.
Can you confirm this is correct?
```

This intelligence can be built into conversational AI. This is one great reason for making purpose-built AI models. There is flexibility in focusing on solving one problem well. When asked what ChatGPT thinks is *not typical*, it responded:

```
A reasonable range might be 10% to 30% likelihood, depending on
the context. This range reflects that while it's not impossible or
extremely rare, it doesn't occur frequently enough to be considered
typical.
   • Closer to 10% would indicate that it is quite uncommon.
   • Closer to 30% would indicate that while it's still uncommon, it
     happens with some regularity.
```

Use language intentionally. *Typical* in the prompt is wishy-washy; in this case, ChatGPT might ignore values that are considered rare, which is below 10%. Using specific language for expected value ranges. Prompt for well-known fields if the user doesn't submit a valid value the first time. It might not be necessary to provide exact instructions ahead of time. Try to parse to a valid value first to avoid clutter in the user interaction. It is a tradeoff. For questions where specific values are well understood, don't provide details; for questions without known values, consider asking the user for more information to avoid getting poor answers on the first attempt.

```
(Obvious) What is your weight and height?
(Not Obvious) Provide the SysCode from the onboarding email. It looks
like this: SC-12533-RHM.
```

Alternatively, using a generative AI error checker, create a separate component to validate form values behind the scenes. The database won't be happy if it needs a number and the AI sends it the text "three." Use a different model designed to handle these situations and then pass validated results to the primary model. ChatGPT is very good about understanding the requirements of a function (it inspects something called the function signature) and typically will do the right thing, including transforming the data into the correct format. In the case above, if it knows the field is an integer, it will send "3" to the function. This also works as a gate for functions that do not have good validation capabilities. The LLM can do it. Here is an example of doing this work manually in a prompt:

```
Prompt: When working with dates, always convert the user's dates
into MM/DD or MM/DD/YYYY format before submitting it to the calendar
service.
For example, if the user provides Jun 12, convert that to 06/12. If
the year is required, ask for it. If they say, "this year," include
the year in MM/DD/YYYY format.
```

Let's return to the expense example. An image upload feature processes scans of receipts to gather expense details directly from an uploaded image. Image scanning models must be trained and will get errors, so putting a different model validator before the image scanning results could catch some automation errors. Ask the user to confirm what was found in these cases. It is a challenge to be error-tolerant in conversational AI experiences. Monitor and improve. Watch for opportunities to catch errors, recover from mistakes, and validate behind the scenes when possible.

> **Analysis for new employee onboarding, Error Prevention**
>
> Heuristic influences: *If there are errors, confirm any understanding, provide details at each step* and *For less well-known questions*
>
> An onboarding process offers ample opportunity for errors. The user, terms, and process are all new.

6 – Recognition rather than recall

Minimize the user's memory load by making elements, actions, and options visible. The user should not have to remember information from one part of the interface in another. Information required to use the design (e.g., field labels or menu items) should be visible or easily retrievable when needed.

Be familiar with recognition versus recall. It is easier to recognize choices from options (such as a menu) because the brain can recognize these values to know which is right, unlike recall, where the user has to search their memory for a cue. "Who was the 16th president of the United States?" is harder to recall than recognizing that "*Was Abraham Lincoln the 16th president of the United States?*" is true. The science of this is rather interesting.

Article: `Memory Recognition and Recall in User Interfaces` (`https://www.nngroup.com/articles/recognition-and-recall/`)

Analysis

This heuristic is essential and also challenging to address in conversational AI. There is a reason why software has menus and why those menus are organized and have some level of consistency. Although every software product can't have the same menu items (e.g., a word processor is not a photo editing tool), there is a set of everyday tasks, and common words/features are used with consistent placement. However, expressing themselves can be challenging when users stare at an empty field. It could be because they are not vocal, have learning difficulties, aren't working in their native language, can't find the right words to express themselves, or are too distracted with other tasks to pin down how to ask a system to do their bidding. With a menu system, users can look around, go to a likely location (such as the **Edit** menu), and see whether the feature they want is there. This ability to recognize what they want by pointing at it is a universal truth: "*I will know it when I see it.*" Also, plenty of menu usability issues can hinder performance – for example, when not using words in a feature name that matches the user's expectation. With conversational AI, especially chat experiences, fine-tune systems to understand a wide range of expressions for a common task or feature, as discussed in a few of the heuristics already. However, humans still have to recall words.

This is one reason why menus appear in some hybrid experiences. It gives the user standard anchors to drive decision-making. If the system can present the five things a user will always do, then sharing these tasks as buttons or in a menu on a conversational AI helps them overcome the recognition versus recall problem. However, as the number of functions now approaches the hundreds in a conversational enterprise app, providing a menu for all of them is unreasonable. Giving them five items when it can do 100 can limit the customer's ability to see the more considerable capabilities of the solution.

As discussed earlier, the problem gets more challenging in a straight voice interface because humans need help keeping track of many choices in their heads. This has been known since the dawn of computing (G.A. Miller, *The magical number seven, plus or minus two: Some limits on our capacity for processing information, Psychological Review*, 63(2), 81–97, 1953). Although our understanding has evolved, a human's ability to handle choice has yet to. There is a psychological limit to how many options make sense in a voice menu. This is why hierarchies in phone trees exist; they help break down large groups of actions into well-understood categories. Expose the user to pick a category first and then disclose the options once within the category. Lotus 1-2-3, the spreadsheet app, famously invented this kind of progressive disclosure of menus in software. 1-2-3 was the first killer app for personal computers in the 1980s.

It is common to have trouble remembering a phone number when told on a call. It becomes a challenge to hang up fast enough to recall the number (473-867-5309) and dial it without forgetting. Human memory is a problem for UIs.

This leaves us with a dilemma in conversational AI. Users are much better at recognizing choices than recalling them, yet there is a limited capability to resolve this in a voice channel or a chat window. More room is needed to build robust views into vast feature sets, even on channels that support menus and buttons. Some tricks can mitigate this UX problem:

- Fine-tune to allow for flexibility in understanding terms. Understand how knowledge refers to products, features, or services.

- Monitor and adapt support for new terms for existing features or tasks.

- Disambiguate when requests are not explicit between multiple choices.

- Provide hints and suggestions when the next step is likely.

- Provide menus in limited approaches when valuable and popular tasks or actions are likely.

- Help guide the user if they need help figuring out what they can do. If errors or multiple errors occur, it might be time to use a secondary prompt that guides the user more explicitly (step by step or with more detailed instructions, for example). If the user asks for help, then the LLM can be more supportive in the same way. See an example of this advice in the heuristic *5—Error prevention*.

Depending on the ChatGPT implementation, there might not be much to do on screen. Still, the team can undoubtedly care and feed the solution to allow more understanding with prompt engineering, fine-tuning, and knowledge refinement. Narrow down the options by disambiguating common misunderstandings when there is no context to decide which direction to go.

```
                                              I need 12G documentation.
  Did you mean Oracle Middleware
  12G or the Oracle Database 12G?
```

This is easier said than done in generative models, but keep it in mind when prompt engineering or using ChatGPT behind a deterministic experience.

Recall the different ways someone might refer to Oracle Database (RDBMS, Oracle Server, Enterprise Edition, DB 12, etc....); being flexible in understanding all of these will make engaging the customer easier. This means being sure of the proper training data.

A critical part of this heuristic is remembering information from one part of a UI in another. This can mean contextual understanding. This is very important in conversational UIs. Continue to know who the customers are and what they are doing by keeping the context of previous conversations. This would be expected when calling a call center for product support. They have the transcripts or interactions of earlier calls, order history, and account information. Isn't it expected that the conversational AI, no matter what the UI entails, should understand and adapt interactions based on this information? Absolutely. And as product leaders, we demand that intelligence. So, this part of the heuristic is worth checking to ensure the user can do more (tasks) with less (information):

- Pick up where they left off. Each LLM starts with little or no knowledge of prior conversations. LLMs are growing their understanding of previous conversations. Without enough knowledge, consider storing the last conversation and providing it in the context window.

- Know the user's history, behavior, and previous needs to help with current interactions (returns, product help, orders, shopping behavior, etc.). Use this history to build a prompt with the current context.

- Understand the sentiment of previous conversations. If this interaction is confrontational, adapt the style and tone.

Analysis for new employee onboarding, recognition rather than recall

Heuristic influences: *verify completion, provide the status of each step, Each step should be clear and concise, Do not ask the new employee to remember information, Provide details at each step,* and *well-known questions*

There is a wide range of prompt details that help the user make decisions right in front of them. It explicitly calls out that the new employee should not be required to remember information. How this would play out in a real onboarding experience would have to be seen. They can recognize where they are and see the results of how they are doing. In the worst case, the conversation history is also there to help remind them of steps or status.

---

7 – Flexibility and efficiency of use

Shortcuts – hidden from novice users – may speed up the interaction for the expert user so that the design can cater to inexperienced and experienced users. Allow users to tailor frequent actions.

Analysis

In chat UIs, flexibility and efficiency can be seen as trade-offs. Conversational AI chat windows offer lots of flexibility; anything can be said. However, getting the correct answer or completing a task might not be the most efficient path. Let's take an example of filling out an expense conversationally:

```
                    Please put me down for $5 this morning for
                    a cup of coffee at Starbucks when I visit
                    my customer, Alli Financial Group, in NYC.
```

```
I added your US 5$ expense
to your current report. You have
four expenses ready to submit.
```

However, starting that same conversation without any context can reduce the efficiency of the process. Here is a long-winded version to make the point:

```
                                          Expense, please
Please provide expense details,
such as "Gasoline for $36$ today
for a trip to the S3I customer."

                                                It was 5$
When was this?

                                                     Today
What was the purpose of
the expense?

                                                    Coffee
You bought coffee. What is
the reason?

                                        Visiting a customer
For our records, who was
the customer?

                                       Alli Financial Group
Got it. I have a 5$ expense
today for coffee while visiting
Alli Financial Group.
```

Be efficient with interactions. A little later in the chapter are some excellent examples of making conversations manageable by reducing and combining questions that make sense. All of this sort of form-filling that is done conversationally requires thought. Ask questions that can fill the fields as efficiently and intelligently as possible. Doing conversational interactions one tiny piece of information at a time is annoying. It gets back to using the right tools for the job.

I saw an example of a COVID-19 screener that was built conversationally. It could have been more pleasant. It asked 14 questions, which required me to wait for each question to appear, read it, and then answer it with textual answers. The same experience done as a form would be three times faster and less headache and work. It is easier for a user to look ahead with written forms and UI wizards. A user can check a box while looking ahead to read the next question. There is also context on a well-designed form providing clues on how much is completed and what is left to do. With these conversational UIs, it sometimes appears that the questions will go on forever. Be thoughtful about the questions so that the process is efficient. Watch for opportunities for improvement through better questions, eliminating steps, and using defaults while allowing the user to edit and adjust responses.

> **Analysis for new employee onboarding, flexibility and efficiency of use**
>
> Heuristic influences: *Keep track of the errors, the user can complete steps in any order*
>
> It is fair to wonder how efficient a conversational AI is for onboarding. Keeping track of status, being reminded of tasks, being alerted to issues in some of the steps, guiding the user on the next steps, and being given access to sites or links seems reasonable. Especially on a communication channel the user already has and monitors. But if you asked me to fill out an employment application *in* a chat, that would be wrong.

8 – Aesthetic and minimalist design

Interfaces should not contain information that is irrelevant or rarely needed. Every extra unit of information in an interface competes with the relevant units of information and diminishes their relative visibility.

Analysis

We can only do so much to be minimalistic beyond providing an empty textbox. In a pure voice experience, minimalism comes into play by keeping our utterances brief. The same can apply to recommender experiences. Provide the headline, and if needed, let the user explore further. Only a few details are required most of the time. In traditional UIs, expanding areas, help bubbles, or drill-down links expose more information. Apply the same logic to recommendations or even chat experiences. If the user replies, "*I don't understand,*" or "*Explain please,*" be prepared to draw on product help, knowledge, and context to provide relevant information to accommodate the user. One approach is to step back to become more of a guided coach. Instruct the model to change its approach:

```
If your interaction results in confusion or misunderstanding, interact
with more explanation, reduce the complexity with more straightforward
language, and proceed step by step with instructions.
```

I am not a good source of information regarding aesthetics. Generally, these experiences are within a larger corporate framework with an established style and visual aesthetic. Only in rare cases would this impede the usability of a conversational design. It is easy to appreciate a well-designed and visually appealing experience. It would be best to design something functional over an experience that looks great but makes the customer bash their head against a wall because it lacks basic usability, as described by the other nine heuristics.

> **Analysis for new employee onboarding, aesthetic and minimalist design**
>
> Heuristic influences: *Each step should be clear and concise*, and *clear, simple terms*
>
> Minimalism is important in a chat flow, especially when one is expected to be on the phone, like in an onboarding experience. Keeping steps and messages clear and to the point will be best.

9 – Help users recognize, diagnose, and recover from errors

Error messages should be expressed in plain language (no error codes), precisely indicate the problem, and constructively suggest a solution.

Analysis

In the next section, we reiterate this point in the examples of style and tone guidelines, starting with the examples coming up soon with *Figure 9.2*, which says not to blame the user when things go wrong. Since almost everything in conversational design is about words, it stands to reason that error messages adhere to the same guidelines and style that the rest of the interactions contain. Users should never see **Error 454-24 System Overflow Buffer Failed to Execute Transaction**. Create checks and balances to ensure that if something goes off the rails, there is a way to explain it in plain language and how it should be resolved (why and then how, as illustrated in *Figure 9.3*). Sorry, we don't want to repeat the images, be patient; they are coming up.

> **Analysis for new employee onboarding, users recognize, diagnose, and recover from errors**
>
> Heuristic influences: *clear and concise, Keep track of the errors, the user can complete steps in any order*, and *explain how to fix them*.
>
> The prompt emphasizes style and tone and includes the importance of communicating status, mainly if errors occur and how to recover.

10 – Help and documentation

It's best if a system doesn't need any additional explanation. However, it may be necessary to provide documentation to help users understand how to complete their tasks.

Analysis

How often have you read help and documentation for a mobile phone app? I suspect most of you will answer rarely. With tasks and features that are self-documenting or have a simple flow, most users follow the happy paths and are good to go. However, in enterprise solutions, many paths are needed, customization is typical, and complexity abounds, causing user confusion and needing help and documentation, even with the best experiences. A conversational experience that has to handle all of that can also be complex. Sometimes, we adopt the 80/20 rule – 80% of the use of a product from 20% of the UI. With modern analytics, we can learn far more about usage. Start by supporting the likely flows (the primary use cases) in conversational AI and keep the more complex interactions in the traditional UI. Handle more complex flows as conversational AI matures and usage warrants their inclusion. We covered that a few times in *Chapter 3, Identifying Optimal Use Cases for ChatGPT*. This will happen slower than the rapid pace of AI models coming online.

Backend systems are still restrictive and expect data in a certain way. Although generative AI can be used in many ways (frontend understanding, translation in the middle, and data mapping in the backend), significant work is needed to make this happen. There is no magic here. But AI can also be good at providing small, refined answers from a robust and extensive help suite, translating it, or even adapting it to a different style or tone.

So, ChatGPT can help explain complexity by taking documentation and applying a style or tone that might be more understandable by a target user, finding insight deep in large help documents, or providing documentation in the user's native language. Consider checks and balances here. ChatGPT better not mangle a step-by-step process in the documentation and return incorrect steps and procedures. This is where the chaining of prompts from *Chapter 8, Fine-Tuning,* comes into play. Chaining can be used to validate critical tasks.

There are many opportunities to advance the state of a solution by adapting approaches to help and documentation. This help can also be done inline. The next section, *Building Conversational Guidelines*, has excellent examples, starting with *Figure 9.1*, where language is adapted based on the user's expertise. A new user can receive more help and guidance in a process, while for an expert who doesn't need this, it is better to be less chatty and more direct. This is similar to GUIs, where wizards walk a user through steps of a long process or provide an advanced experience with less context and fewer steps.

Analysis for new employee onboarding, Help and documentation

Heuristic influences: *Provide details at each step* and *explain how to fix them*

The prompt ingrains the call for help. Conversational flows should typically self-document for details, explanations, and guidance. If not, links can be provided to traditional documentation. The concept of progressively increasing the level of the help supplied or tuning the details of inline help based on user profile characteristics are called adaptive messages. Examples of this will start in the upcoming section *Building conversational guidelines*.

This covers Jakob's ten heuristics. Applying them to UI elements is well-documented and entrenched in the UXD world. Using them to model behavior for an LLM is new and untested. ChatGPT 4o behaves well with prompts like this, but it would require real-world data to validate this approach at scale. From testing, it seems worth the investment to consider the heuristics when writing enterprise instructions for models.

Is there an 11th possible heuristic?

I want to mention A18y again in our discussions. Jim Ekanem is a proponent of distinct accessibility heuristics. Much of what was discussed in the ten heuristics can all be considered in the context of A18y. A18y should be considered at every step. Jim proposes a new heuristic to follow A18y guidelines. This is not a heuristic, but to be fair, the concept of accessibility and inclusivity is still valid. Inclusivity was covered a few times, and bias in model data can affect quality.

Article: `Proposal to include Accessibility as the 11th Heuristic (https:// uxmag.com/articles/why-we-need-11-usability-heuristics)`

Design for accessibility and inclusion by following guidelines and best practices to accommodate diverse cognitive and physical abilities.

Consider social identities and address systemic barriers and biases. Reflect on the impact of design decisions on marginalized communities.

The heuristic *4 - Consistency and standards* didn't mention cognitive and physical abilities. Accessibility guidelines, discussed in an earlier chapter, should be part of the evaluation process for conversational AI. It won't impact backend solutions and is of limited value for recommendations, but it can have significant implications for voice channels, for example. For a voice-only feature, what is the alternative for those who can't speak or hear? Will the system work using a **Telecommunications Relay Service (TRS)**? Do requests for a selection or typing of a number, say a credit card number, time out too quickly? Would a time-out cause problems if an intermediary were relaying this communication, thus delaying responses by 30 seconds? If services are offered via messaging or voice, include viable alternatives. This is an example of accommodation; think broadly about inclusivity.

Bias was touched on earlier around language and cultural support for conversational interactions. Even in simple concepts such as expense reporting, there are cultural implications for inclusiveness. For example, recording an expense trained on travel in the US covers items such as taxis, Ubers, lifts, car services, and limos. However, each country has unique services, such as BlaBlaCar for carpooling in France, Cabify in Spain, Didi in China, and Ola in India. Being inclusive means including the understanding of these car services. It is not just about cultural awareness to deploy tools in those countries; if the customers are from the US and visit those countries, they might want to expense "a BlaBlaCar for 20 euros" on their US-based expense report. Being inclusive and thinking about cultural issues can also benefit the home country. However, for an A18y discussion, ensure that groups of people who have typically been sidelined are afforded the same opportunity to use LLM solutions successfully.

Language and dialect support can be a major issue. For instance, if rural villagers in India access government services via a phone-based LLM, even on inexpensive flip phones, what are the chances their language or dialect is supported? The 2011 India census reported 121 languages spoken by at least 10,000 people, with 22 officially recognized languages and thousands of dialects.

If the ten heuristics are followed in spirit, the 11th isn't needed; it should be implicit. If the organization isn't as advanced as it should be, the accessibility "heuristic" might help fill the gaps.

These heuristics can be used to evaluate any experience, including conversational AI. It should be clear that these are not guidelines. However, it is reasonable to have guidelines that developers can quickly follow. Adapt and adopt a company-wide set of conversational guidelines.

Building conversational guidelines

A guideline can be built into a test case. It forces people to follow a defined solution. The testing software can test for a specific font or size in a page header or a button label that doesn't use a term an organization forbids (such as *abort*, *kill*, or *execute*). Doing something precisely, the same every time isn't very conversational. Repeatability and consistency are crucial in enterprise solutions that don't always match a conversational style, tone, and engagement. Consider that the LLM can understand a range of phrases while a fixed traditional test suite has little flexibility, although a test could look for one phrase from a collection of options.. The testing we did in the earlier chapters is a form of validation and can be used to address if the model is acting as expected for areas like following the ten heuristics or adhering to specific guidelines.

Develop guidelines by drawing on existing resources and consider the use cases. While most guidelines apply to a hybrid UI, and some will work for a conversational UI (text or voice), even less will apply to a standalone AI-driven recommender. And this makes sense; the more complex the possible experience, the more guidelines apply.

Then, there is the issue of getting the experience to follow the guidelines. Just because a prompt tells ChatGPT to follow specific instructions doesn't mean it will. When using deterministic coding approaches, design is dictated. This is much harder with an LLM. Recall how **Temperature** and **Top P** were adjusted in *Chapter 7, Prompt Engineering*. There is some control from prompt engineering and fine-tuning to focus the LLM on how to speak to the customer.

Here are three areas to get values from conversational guidelines. Historical references were provided earlier in this chapter, such as the work of Smith and Mosier (1986); let's get more up-to-date with the following:

- Web guidelines

- A sample guideline set for hybrid UIs

- Some specific style and tone guidelines with examples

Web guidelines

Conversational guidelines will adjusted emerge from a few sources. First, the big players will take their existing web guidelines and expand on them to include more generative AI components. Apple, Google, Microsoft, Amazon, and others will all have something to say, and most likely, they will feed off each other as they have in the past. Sometimes, inadequate guidelines appear because one group didn't know any better and didn't do any of their research or testing to define their guideline. Thus, when someone learns from a mistake and changes their guidance, others who copied it have to figure out why this changed and decide whether to adopt the new patterns. This has happened multiple times with Amazon and Google designs so that it can happen to anyone.

The AI players will have guidelines as they learn that design matters and how poor interactions impact the overall experience. They will do this to help their customers make successful solutions, as their models typically depend on usage. If customers stop using the models, their bottom line will be affected.

Use what is available today as a guide. Caveat emptor. Let the buyer beware. Every guideline adopted should have a reason and a solid underpinning. Apple Intelligence will likely have guidelines, while Microsoft's are currently fairly high-level. There is little out there.

Article: `Microsoft Guidelines`

(`https://learn.microsoft.com/en-us/training/modules/responsible-conversational-ai/`)

Training: `Conversation Design from Salesforce` (free, but requires registration) (`https://trailhead.salesforce.com/content/learn/modules/conversation-design`)

After finishing this book, you could reverse-engineer some guidelines by looking at the conversational UI design thread and examples on Dribbble.

Examples: `Dribble catalog of inspirations for conversational AI` (`https://dribbble.com/tags/conversational-ui`)

Very few guidelines exist, but what is there is solid, if not generic, and consistent with what we are discussing. However, much of the content is for those building solutions, not for those designing experiences. Even Coursera has nothing to offer at this time. Hence, the reason for this book! So, it's time to make do and create our own set of guidelines.

A sample guideline set for hybrid chat/GUI experiences

One challenge is to follow existing GUI guidelines and deal with conversational guidelines. There are conflicts, the most common of which is with language. A traditional UX doesn't use contractions, while UI language is more formal, sometimes abrupt, and undoubtedly less conversational. I have seen guidelines built into automated software testing checks that fail with conversational text.

On top of that, if the generative UI directly creates text, there is less control. This is like telling a teenager to be home by 10 p.m. They might say they will but might not arrive on time. That is prompt engineering.

The guardrails might be suitable, but they are only sometimes followed. The guidelines can help with prompt engineering. They can certainly be followed in a recommender template.

On GitHub is a checklist of GUI guidelines, roughly organized by the ten heuristics. These were initially adapted from a list generated at MIT, but the original list has been lost to history. There are two sheets – a short version and a lengthy one on the second worksheet tab. Most apply to traditional GUI or web applications, while some cross over to conversational AI. Here are examples for helping users recognize, diagnose, and recover from errors.

Error messages should be expressed in plain language (no error codes), precisely indicate the problem, and constructively suggest a solution:

- Undo is supported, where possible

- Guidance is clear

- Unusual answers to common questions are validated or confirmed

- If a choice is too complex conversationally, offer suggestions if they are likely or examples if any of the choices are unlikely

GitHub: `Guidelines typical of a GUI and supporting conversational experiences (https://github.com/PacktPublishing/UX-for-Enterprise-ChatGPT-Solutions/blob/main/Chapter9-HeuristicsChecklist.xls)`

I suspect something similar is in every enterprise, but if not, feel free to adapt these. Then, consider applying the user scoring method to the issues found.

Here is one related takeaway from my Master's thesis, *Effects of Graphical UI Inconsistencies on Subjective and Objective Measures of Usability*. Consistency from screen to screen doesn't matter as much as matching a user's needs to the experience. There are a lot of forced designs where the UX must use a specific component to "be consistent." This is ostensibly for the user, but it is really to ease the burden on the development team. If the component is wrong, the user will suffer. So, be consistent with the user's needs and expectations. Google search and ChatGPT are your friends here to gather guidelines. So, next, we can dive further into conversational-specific guidance.

Some specific style and tone guidelines with examples

In a conversational UI, we can create prompts to control the style and tone. In a hybrid UI, some UI elements will have static text. With a voice interface, there is *only* the spoken text. At least on a chat UI screen, it is easy to review material, copy and paste it, or compare it to something else on the screen. Getting words into the proper form is so important. Here is a glimpse into some of the guidelines I have shared over the years; use them or adapt them as needed. In a conversational AI, I would convert these to prompts.

The trick is to understand how much control is available. In older chat experiences, a generative AI could be added to do specific tasks behind the scenes, like entity detection or for cases of redirection or repair. The generative AI is not having the conversation. The conversation is controlled by the deterministic flows designed in the Chat platform. In that case, the exact wording expected in a particular step in a task flow is scripted beforehand. Alternatively, with a recommender, specific templates for a response are created, and the AI fills in the details. But if the front end is the ChatGPT LLM, prompt engineering and fine-tuning are needed to communicate in a style and tone appropriate for the user.

Some of these guidelines are closer to heuristics. Each example tells a story about how messaging and communication can improve. Recognize that the *message is the interface*, especially on a voice channel. As you'll see in the upcoming figures, I have color-coded each row. Yellow means it is okay and not that exciting; green is good, while dark green is better. Red means don't do it!

> **Note**
>
> An LLM can force specific wording for a task flow. Templates and traditional chat solutions can prompt the user in a particular way with exact wording. The LLM can be *guided* to answer certain questions, but this is not a scalable solution. For every guideline, consider how a prompt template can be used to customize the LLM instructions to cater to the use case, the context, and the user.

Use adaptive messages

People adapt their messaging over time as they become familiar with other people. If you are comfortable with someone, you talk in short-cuts. An LLM can work the same way. Customer usage data can be used to tweak prompt templates. For a new user, prompt for instructions to be more verbose, for someone who has done this task many times could be interacting with instructions that tell the LLM to "*be terse and to the point,*" "*only repeat the primary information,*" and so on.

As shown in *Figure 9.1*, as the user becomes more experienced, the messages can adapt to acknowledge and interact with the customer. Trust must be earned, so don't assume the customer trusts the system from the outset.

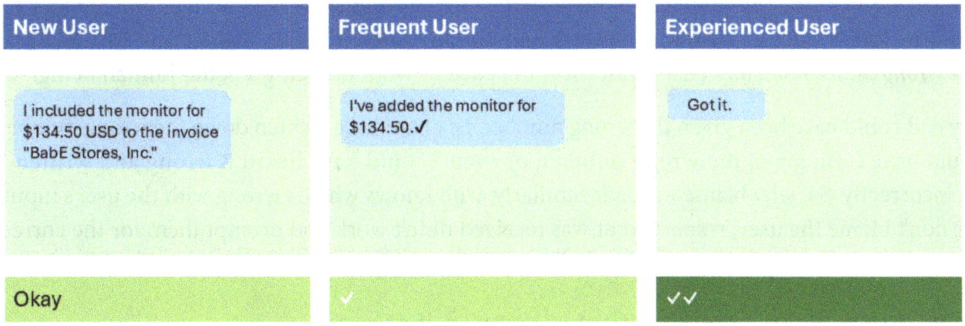

New User	Frequent User	Experienced User
I included the monitor for $134.50 USD to the invoice "BabE Stores, Inc."	I've added the monitor for $134.50.✓	Got it.
Okay	✓	✓✓

Figure 9.1 – Adapt messaging to the user's expertise

Each of these messages is good on its own. Expect customers to need less handholding as they become comfortable with the conversational experience. This is similar to how we are with people we know and trust. Don't over-explain and they don't need to mirror back to us actively. They get it.

Sometimes, humans need to remember, so if the user is an expert but has not used the experience in months, prompt them with more context to get them started again. The **Got it** response is aspirational. The AI must be incredibly trustworthy for that to be a response. And if they get something wrong, don't blame the user.

Don't blame or confuse the user

As in *Figure 9.2*, it is not the user's fault when things go wrong.

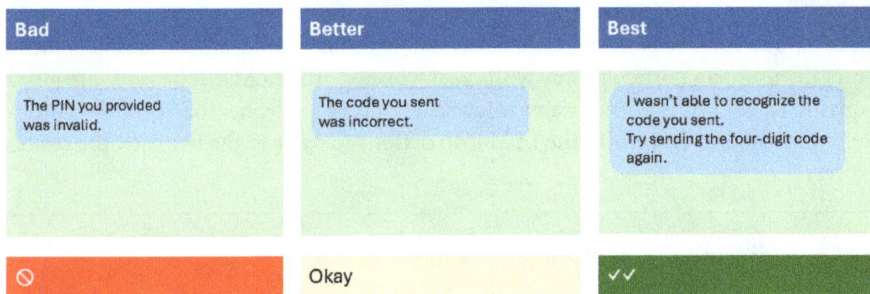

Figure 9.2 – Don't blame the user, and don't use jargon

Consider how many ways failure could occur in a system. Yes, the user could have made a mistake, but belittling them won't help the cause. Try to speak to them as they want to be spoken to.

Although a **PIN** might be familiar to older adults in the US, it sometimes translates poorly, and few under 30 will know what that stands for (**Personal Identification Number**), even if they understand the meaning. They might need to learn the industry terms, company technical jargon, and customer shortcuts.

Let me tell a quick story about blaming the user. I called my dad and asked for a phone number for someone he knew. Later, I called that number, and it didn't work. My first instinct was, "*My dad gave me the wrong damn number.*" That is just me – I needed to work on being a better human being.

So, my dad could have been given the wrong number, he could have written down the wrong number, he could have been given the wrong number, or even I could have heard it wrong and written it down incorrectly. So, why blame my dad? Similarly, who knows what is wrong with the user's input? Please don't blame the user; respect what was received didn't work and prompt them for the correct information. And I should follow my advice and be a better person.

Why, and then how – confirm first, and then instruct

As shown in *Figure 9.3*, a standard structure for error messages applies to conversational AI. This was taught to me by a great writer at Oracle. Multiple guidelines can apply to one phrase, like this one. Remember, don't use jargon.

Sometimes, language is harsher than it should be. Never use terms such as *corrupt, execute, kill*, and *abort*, which can elicit strong customer reactions, translate poorly, and not help us move forward (except in rare cases like referring to the **Kill** command in UNIX).

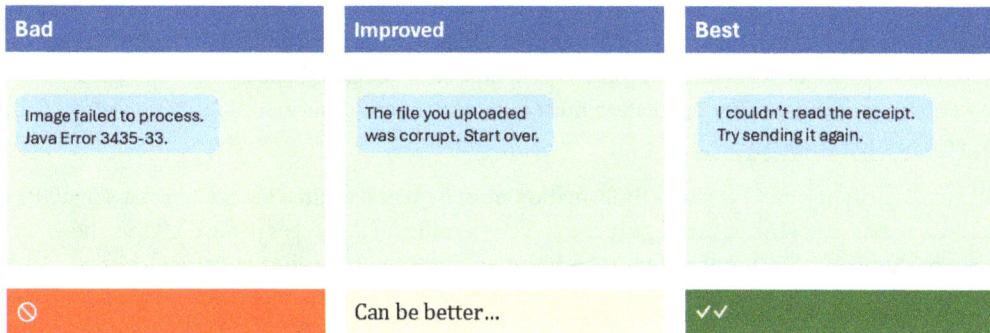

Bad	Improved	Best
Image failed to process. Java Error 3435-33.	The file you uploaded was corrupt. Start over.	I couldn't read the receipt. Try sending it again.
⊘	Can be better…	✓✓

Figure 9.3 – Why, and then how – confirm first, and then give instructions

Control the length of messages and continue to be concise and economical. People often fail to read prompts entirely or skim and can miss critical information. Get to the point and then express what should be done about it.

If an LLM generates the message or recommendation, it can be told to limit the length and be friendly. If you don't, you can have 300-word responses when 30 would do. You probably think the same thing about this book: it should be 100 pages shorter. Keep reading; we have more to cover.

Be conversational – don't regurgitate system descriptions

As shown in *Figure 9.4*, just because a database has a field name doesn't mean the user wants to see it. Consider how to ask questions to reduce errors and how concise it should be. And this example exposes a few other good ideas.

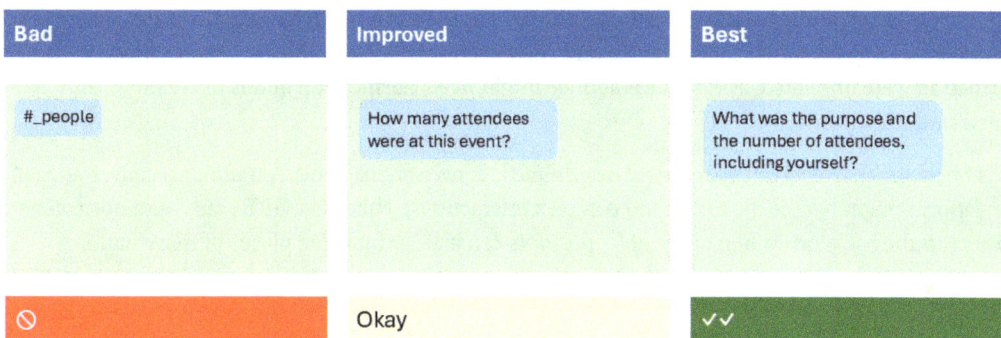

Bad	Improved	Best
#_people	How many attendees were at this event?	What was the purpose and the number of attendees, including yourself?
⊘	Okay	✓✓

Figure 9.4 – Adapt your messaging to the user's expertise

In this example, user testing resulted in two prompts being combined to form the **Best** option. This reduced the number of prompts required to determine the missing information and was clear enough that most people answered with both pieces of information. If they missed giving the reason, for example, we could still probe for that. It was also essential to be transparent about events and expenses, as some people thought about their guests and needed to remember to include themselves in the count. This might seem odd, but some companies have backend checks to verify the money spent per person on events. The calculation must include the host; otherwise, automation might flag an expense as too expensive.

Recall our discussion about backends. Information must be sent back in a specific format. ChatGPT is excellent at extracting the business purpose and the number of attendees using digits, as shown in our previous example. Most chat solutions had a lot of trouble with this logic before LLMs:

```
                        if 4 people and myself attended a
                         dinner, how many people attended
                                              the dinner?
```

```
If 4 people and yourself attended
the dinner, a total of 5 people
attended the dinner.
```

```
                        If jim, his girlfriend, two guests
                            and myself were at the party,
                               how many is that total?
                               Return only the number.
```

5

Streamline-related tasks

Figure 9.5 is a follow-up to *Figure 9.4*. It shows an opportunity to ask related questions in a way that can be done in one step instead of two. Be careful with generative AI solutions that can ask too much; these examples make the point. Combining related requests into a single question can reduce the number of steps and maintain intelligence and conversational expectations. If two or three pieces of information were unrelated, asking for it at once might need clarification and is likely more complex to parse and understand.

This is where generative AI can supplement deterministic conversational tools. A traditional conversational chat platform might handle the forms and complex interactions, while ChatGPT parses and normalizes data sent to the backend. When applicable, prompts can ask for multiple pieces of information.

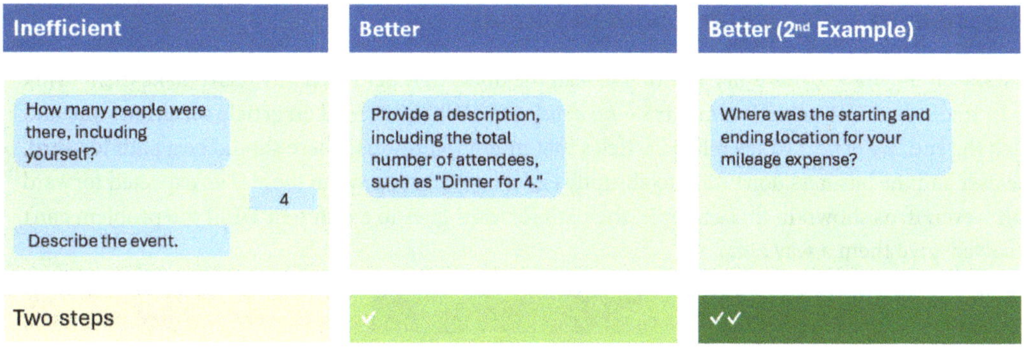

Inefficient	Better	Better (2ⁿᵈ Example)
How many people were there, including yourself?	Provide a description, including the total number of attendees, such as "Dinner for 4."	Where was the starting and ending location for your mileage expense?
4		
Describe the event.		
Two steps	✓	✓✓

Figure 9.5 – Streamline-related tasks when it is conversational

Guide and nudge – don't force an issue if it is unnecessary

As shown in *Figure 9.6*, the persona will likely engage the user in a coaching and supportive way. However, not all transactions need to be completed immediately. If the user must submit something, so be it, but sometimes, like with expenses, it is reasonable to submit all of them once the trip is complete. This is when nudging is likely enough.

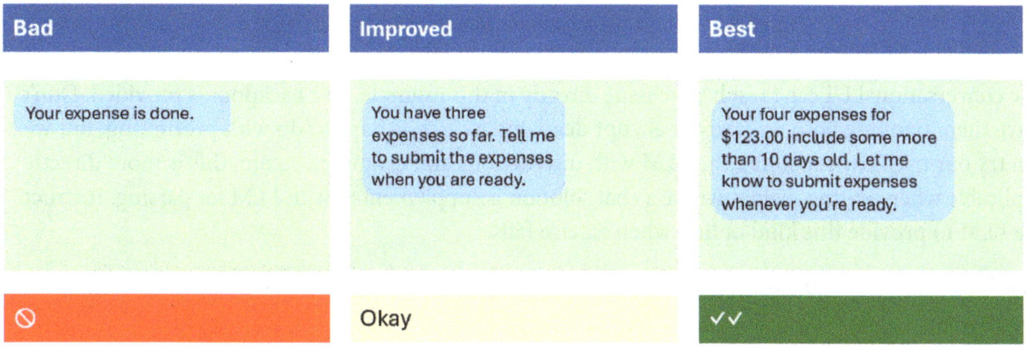

Bad	Improved	Best
Your expense is done.	You have three expenses so far. Tell me to submit the expenses when you are ready.	Your four expenses for $123.00 include some more than 10 days old. Let me know to submit expenses whenever you're ready.
🚫	Okay	✓✓

Figure 9.6 – Guide and nudge users – don't force them

Our **Bad** example is direct but is too concise, and maybe unclear to the meaning of the word "done.". This didn't guide the user at all. In our **Best** example, it got more forceful, with some outstanding expenses for the ten days mentioned. It should take a firm stance when needed to force the user. To take this example further, if the expense report was due, the system could email the user; if the expense report was overdue or the company was getting expenses in for the end of the year, we could require it to be submitted now, if not let the user know it will be auto-submitted.

No dead ends – give a user a path to success

As shown in *Figure 9.7*, this is like wanting to lead the horse to water (even if we can't make them drink it). In traditional UIs, there is a mantra – *no dead ends*. When we read an article on a news site and reach the end, we expect to see related articles that might interest us. There should be a path forward. The user and the business don't want to abruptly end a conversation when there is an expected forward path – even if, as shown in this example, the path forward goes to a different UI. If the problem can't be solved, give them a way out.

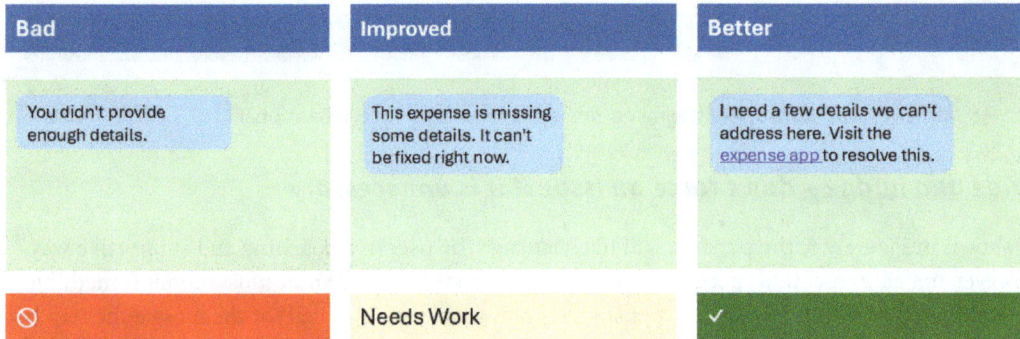

Bad	Improved	Better
You didn't provide enough details.	This expense is missing some details. It can't be fixed right now.	I need a few details we can't address here. Visit the expense app to resolve this.
⊘	Needs Work	✓

Figure 9.7 – Don't allow interactions to have a dead-end

The conversational UI can't resolve the issue directly in this example, so a backdoor is provided. Don't leave them hanging with an error or abrupt dead end, which is hard to do with prompting, but we can try our tricks by providing the LLM with instructions and examples. Again, this is more directly applicable when writing dialogs where a chat solution is supplemented with LLM for parsing. Instruct the LLM to provide this kind of link when all else fails.

ChatGPT can do this from the prompt. It knows how to store instructions in memory for later in the session:

```
If you don't know the answer
to a question with a probability
greater than 90% ask me to search
google, and provide the link.
```

```
Got it! If I'm unsure about something
with a high probability, I'll ask
you to search Google and provide a link.
```

```
What is the weather today?
```

```
I can't check real-time data, but
you can find the current weather on
websites like Weather.com or search
"current weather" in Google.
```

Conversational but not chatty

Figure 9.8 re-explores the item from the earlier example – asking for the number of attendees. It's okay to have longer messages if there is value in the additional words. Research found the phrase "*number of attendees*" confusing because the host of an event doesn't consider themselves an attendee, so changes were tested to tackle this head-on by adding "*including yourself.*"

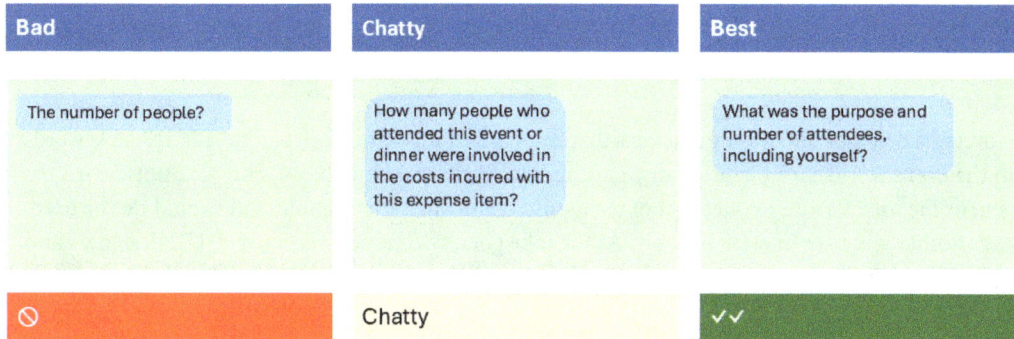

Bad	Chatty	Best
The number of people?	How many people who attended this event or dinner were involved in the costs incurred with this expense item?	What was the purpose and number of attendees, including yourself?
⊘	Chatty	✓✓

Figure 9.8 – Adapt messaging to the user's expertise

Use the right terms

Language is tricky, as *Table 9.2* shows. When designing conversational experiences, consider how to communicate with users with the correct terms.

Term	Description	Example
Choose	When you have free will	Choose four free toppings for your pizza.
Select	When we force a decision	Select your age.
Sign-in	To gain access to your account	Sign-in (avoid using login).
Tell	When asking for verbal or written information	Tell me your prescription number (avoid using *Say*).
I	When the assistant is responsible	I can't understand the image you sent.
We	When the assistant gets help from another service	We are reviewing your accident with the claims department.
You or Your	The user, the users, or their company	Your appointment is now confirmed.

Table 9.2 – Communicate using terms that are consistent and grounded

Some words can have meaning outside of their traditional use. People still play a "record," some old-timers "dial" someone's phone number, even though the rotary dial phone has been gone for decades.

> **Pet peeve**
>
> At least most websites ask the user to "sign-in" to their application. Incidentally, stop asking users to "log in." That is a terrible word to describe what a customer is doing. It comes from the concept of the log file that keeps track of users accessing a system. It can't be more geeky than that.

We covered a heuristic and a few guidelines that emphasized this: Use more natural and relatable words from the user's world. Conversationally, messages to the user include *I*, *we*, and *you*, not *the AI*, *the system*, or *the user*. Create a collection of terms and define how they should and should not be used. I use customer and user interchangeably. As the joke goes, two careers have users: UX designers and drug dealers. I guess context matters when talking about users.

Using context makes conversations natural

As shown in *Figure 9.9*, a pure generative AI solution will naturally want to reply, hence the reason for hallucinations. But if ChatGPT is behind the scenes, there are plenty of conditions where a response might need to be understood. In any case, guide the user to a better answer, not leave them at a dead end.

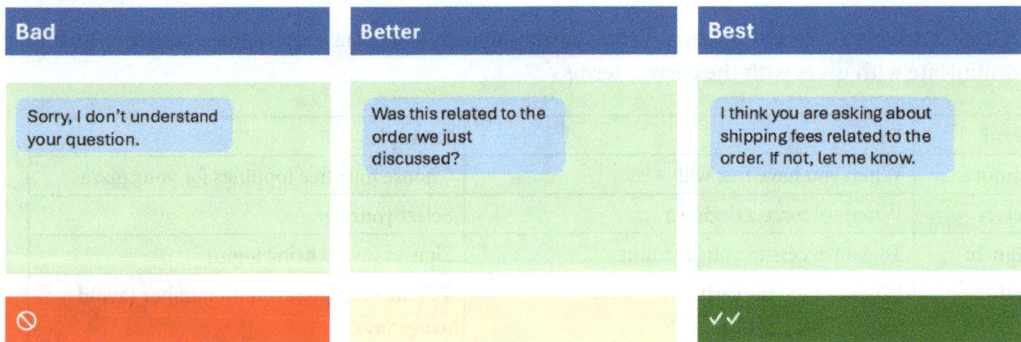

Bad	Better	Best
Sorry, I don't understand your question.	Was this related to the order we just discussed?	I think you are asking about shipping fees related to the order. If not, let me know.
🚫		✓✓

Figure 9.9 – Context helps make conversations more natural

Don't create unnecessary conversation

The **Bad** example shown in *Figure 9.10* is chatty and indirect. The UI asks a stupid question and then gives the user even worse choices. Sometimes, it is a challenge because of the recognition versus recall issue discussed in the *Adapting heuristic analysis for conversational UIs* section, and it makes sense to give the user guidance.

However, in this example, the user must digest the information shared to decide on their next step. Understand their context of use to make design decisions. Give them time to process first.

Bad	Chatty	Best
Here is that information. Do you want help with something else? If you want to add another item, say, "**Yes**" and I can guide you. Or if you are done say, "**No**."	Can I help you with anything else? Yes Let me know what help you need. No	Okay, here is the information you requested. If you need anything, just ask.
⊘	Not Great	✓✓

Figure 9.10 – Don't create additional conversations

Don't ask questions that shouldn't be answered. Don't phrase follow-up responses as questions, like in the chatty example.

Model user language

Figure 9.11 is an excellent example of **Keep it Simple Silly (KISS)** from the expense assistant at Oracle. This example was popular with conversations with customers and partners and at conferences because everyone had experience filing expenses.

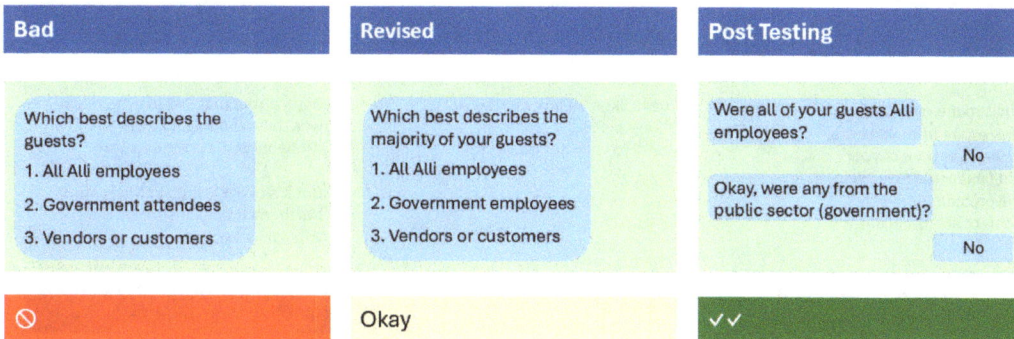

Bad	Revised	Post Testing
Which best describes the guests? 1. All Alli employees 2. Government attendees 3. Vendors or customers	Which best describes the majority of your guests? 1. All Alli employees 2. Government employees 3. Vendors or customers	Were all of your guests Alli employees? No Okay, were any from the public sector (government)? No
⊘	Okay	✓✓

Figure 9.11 – Model user language

Businesses in the USA have a concept called the public sector, which refers to government employees. Interacting with anyone from the government requires different rules of engagement, especially when buying someone a meal or giving them a small gift. Because of ethical concerns in the US and many countries, employees must be careful about paying for government employees' meals. Understanding attendees is essential, and the correct language can help.

The words 'employee' versus 'attendee' were tested. The word *employee* refers to the people the user works with at their company, while *attendee* helps to distinguish employees from attendees. In this example, the company's short name, Alli, is used to make the tone softer. It would be the same for any employee when dealing with an internal tool. Everyone knows who they work for; formal names are not needed. And finally, both the words *public sector* and *government* were included. The design was a hedge to include public sector customers not directly employed by government entities.

Flow order can reduce interactions

Figure 9.11 had one other trick. This shows how detailed a recommendation template can get and how difficult it might be to pull off with prompts and fine-tuning examples.

The data revealed that most meals within a company do not have government attendees (almost all answer the first question "Yes" and are done). Answering "no" to the second question eliminated any further questions. This creates an express lane to simplify the flow for the happy path. Few would say "yes" and require details such as the attendees' names. So, by ordering the questions correctly, the user isn't asked any follow-up questions.

Maintain a consistent voice and tone across interactions

In *Figure 9.12*, there is a trigger word, *expire*. This example concerns the tone and spirit of the message to convey to users. A brighter tone is warranted if the chat is about something upbeat and fun. We can also explain the reasoning behind an issue (generated links sometimes stop working), so giving context helps the customer.

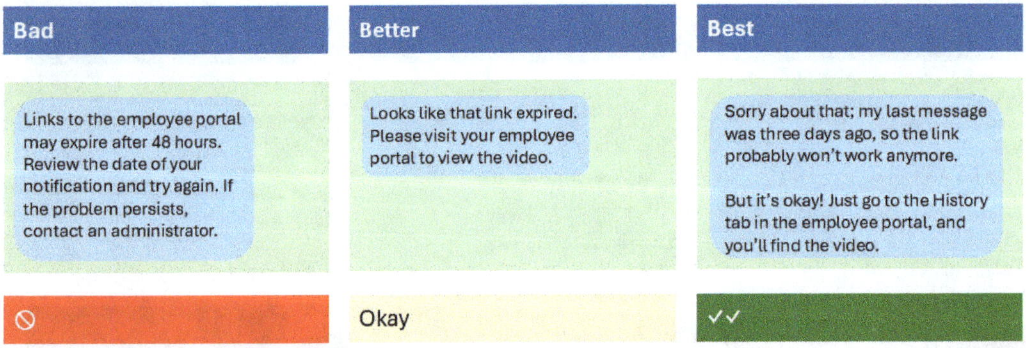

Bad	Better	Best
Links to the employee portal may expire after 48 hours. Review the date of your notification and try again. If the problem persists, contact an administrator.	Looks like that link expired. Please visit your employee portal to view the video.	Sorry about that; my last message was three days ago, so the link probably won't work anymore. But it's okay! Just go to the History tab in the employee portal, and you'll find the video.
⊘	Okay	✓✓

Figure 9.12 – Match the tone with the situation

This example also follows the why-then-how guidance from earlier. Advice is cumulative; three or four guidelines can support one statement. Be careful with tone; it can cause an issue, like in this next guideline.

The happy path is not the only path

Design for the happy path, the likely scenario typically seen and demoed to customers. But what if the user goes in a different direction? Will the system continue to talk in that style and tone? Here's a funny anecdote. I used this example about getting married in my coaching and classes for years. Then, someone coded the exact situation from *Figure 9.13*. They had not attended my training. When I asked them to try the not-happy path for their app that helps change marital status, the chat responded with that same up-tempo and inappropriate answer when asked about getting a divorce. Remember that when writing prompts, filling in recommendation templates, or feeding a prompt, *the happy path is not the only path*.

Happy Path	Not a Happy Path	Neutral Tone
I am getting married	I am getting divorced	I am getting divorced
Great! Let me help you change your status.	Great! Let me help you change your status.	Let me help you change your status.
Work for this one example	🚫	✓✓

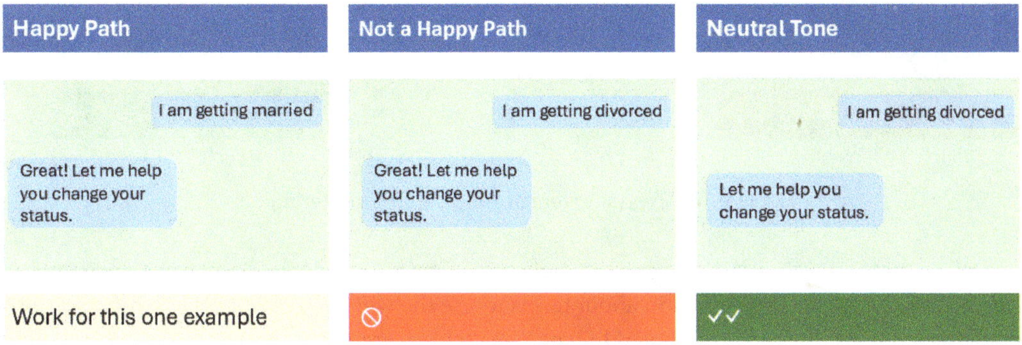

Figure 9.13 – The happy path is not the only path

Some might argue that there is nothing wrong with being excited about a divorce; who are we to judge? However, don't pass judgments that could be offensive or misinterpreted in an enterprise setting. This leads us to a similar suggestion about being cute.

Try not to be cute – it can backfire

To follow our previous example from *Figure 9.13*, when trying to be friendly and supportive, be careful not to be too cute. Wishing users a great weekend when closing out a Friday afternoon interaction sounds pleasant. Still, if they are starting the weekend shift or work in a country where the weekend doesn't start on Friday, the interaction will miss the mark by trying to be cute. Instructions to the LLM are limited in this interaction, but most of these examples I have seen come from chat UIs that build this cuteness into the flows. It can come across in unexpected ways and be offensive in some cases.

Try not to repeat – refrain from repeating things already said

Avoid repeating the same language as shown in *Figure 9.14*, as the title of this section whimsically suggests. It is cluttering and drags down the interaction.

Bad	Better	Best
I'm not sure which **currency** you meant. Choose the **currency** code from this list:	I didn't catch that. Which currency?	I didn't catch that, but I think it is in US Dollars. I set it to dollars; if not, let me know.
🚫	✓	✓✓

Figure 9.14 – Try not to repeat – repeating is cluttering

Interactions that require a selection could be designed in many ways depending on the capabilities of the UX. Recall from our discussion about terms to use the word *select*, not *choose*, because the currency is required. They could be asked to type it in (baht, dollar, etc.), choose from a list, or even use a type-ahead list. However, like with our heuristic, be mindful of errors.

If someone typed in `dollars`, it sounds fine, but many major currencies are called a dollar – 25 worldwide. Any UI would need to know the correct currency – Canadian, Hong Kong, US, or one of the other dollar-based currencies. Selection can validate the choice. However, a follow-up question to clarify a voice channel could be used, and active mirroring of the input can help confirm the country. If it is lucky, other context clues can be used to determine the country. Assuming the currency is also an option. Based on the user history or other submitted expenses, the system can tell them, "*I think this expense is in US dollars; if not, let me know.*" If it can be right 90% of the time, then this approach means less work for the user.

Do's and don'ts for conversational style

Table 9.3 summarizes some of the gotchas that work for many enterprise applications. Prompt engineering would be challenged to be this prescriptive in its choice of language. But it does a good job already with the do's. Some of the don't are a little more challenging, but if you hit an issue try to prompt engineer your way out of it. Just monitor and report.

Language Do's	Language Don'ts
• Example What type of expense is this? • Example Who's the new legal employer? • Provide simple, direct instructions. • Use natural phrasing and common words. • Be consistent in phrasing. • Use passive voice appropriately. • Use contractions naturally. • Focus on the user benefit or value. • Write for the person the LLM is interacting with. • Be proactive. Guide users with clear calls to action.	• Expense type • Destination Legal Employer • Don't use jargon • Don't be ambiguous (such as future-ready or coming soon) • Don't use long explanations for simple issues • Don't stray from the task • Don't be cold or overbearing • Avoid the system's reasoning • Avoid negative words such as kill, abort, crash, dumb, fatal, execute, hit, master/slave, and illegal. • Avoid puns, clichés, and metaphors

Table 9.3 – Conversational do's and don'ts

Avoid using label names; instead, use a more natural conversational style. For example, **Expense Type** is a form label. But conversationally, instead of the harsh *"What is the expense type?"*, it can soften it by saying, *"What type of expense is this?"* This can work for many requests for specific types of data. Recent testing of ChatGPT 4o shows it performs well when writing in this natural way. It even uses that exact phrasing when asking about types of expenses and then correctly uses **Expense Type:** when showing a summary of the results. Adjust prompts or instructions to have generative messaging that supports these do's and don'ts. Add your own do's and don'ts.

Bonus tip: Write three or four variations for a recommendation template and randomly display one of them to catch their eye. People will read an unfamiliar message more carefully and internalize it, which is better than ignoring it.

Give them news they can use

As shown in *Figure 9.15*, there are times when the customer needs a specific collection of data. Generative AI might not format it correctly and doesn't know the nuances of layout and conversational style. The LLM can decide which function to call. An example is sending a zip code or other location details to an application that will return the address and contact information. Still, you have to decide how to format the data. Or do you…

Here is a real example from a recent chat interaction that annoyed me. I did some work to address the issues.

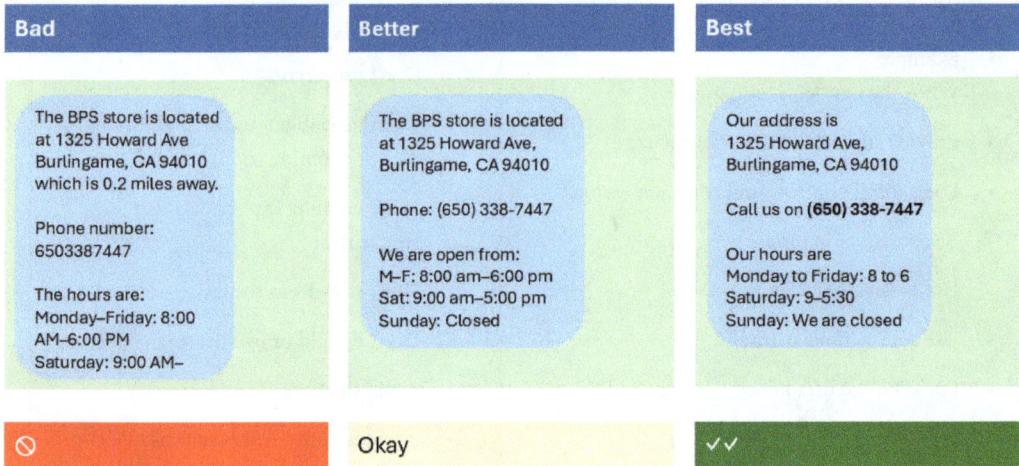

Bad	Better	Best
The BPS store is located at 1325 Howard Ave Burlingame, CA 94010 which is 0.2 miles away. Phone number: 6503387447 The hours are: Monday–Friday: 8:00 AM–6:00 PM Saturday: 9:00 AM–	The BPS store is located at 1325 Howard Ave, Burlingame, CA 94010 Phone: (650) 338-7447 We are open from: M–F: 8:00 am–6:00 pm Sat: 9:00 am–5:00 pm Sunday: Closed	Our address is 1325 Howard Ave, Burlingame, CA 94010 Call us on **(650) 338-7447** Our hours are Monday to Friday: 8 to 6 Saturday: 9–5:30 Sunday: We are closed
⊘	Okay	✓✓

Figure 9.15 – Give them news they can use

Let me highlight the issues visible in a simple message like this:

- The store could make it more personal and thus more conversational using the phrase "*our address.*" This would mirror how someone on the phone would provide these details, and with a voice interface, it would sound more natural. It could also be "*our store is at.*"

- The experience could have more calls to action. If the business wants phone calls, it would be a link. Some companies want to avoid encouraging calls. This sounds better than displaying a name/value pair. Depending on informality guidelines, you might not use the colon.

- Format the phone number for the country to make it easier to read. Many lazy systems don't format numbers for easy reading and recall. Essential human factors were at work when the phone companies added an area code to the seven-digit number, such as (473) 867-5309.

- Placing a call with "click to dial" in a mobile browser, rendering clickable email (to send email) or web addresses (to open) and calendar events (to add to a calendar) as easy next steps in conversational output.

- The text "hours" can be more readable and would vocalize better in a speech solution. ":00" is not needed. Shorter is better. AM/PM with the hours is not required. Hours are evident to customers unless the open hours are unusual, then keep AM/PM.

- Consider accessibility. The abbreviated forms of the days (M–F) would be shorter but require some cognitive processing on the user's part, while Monday *to* Friday is natural and accessible. And *to* is short, while the more accurate *through* is long.

- Some lines were added to separate the address from the details. This allows for better scanning, making it easier to see one section of the result. It is also easier to cut and paste pieces of data.

- Be mindful of the device so that messages fit on the screen. In this example, the rest of the message was offscreen on an iPhone message window. In newspaper terms, anything placed *below the fold* is less visible. It won't attract the same level of attention as content displayed *above the fold*, immediately visible on-screen. Long messages mean that, on some platforms, only the end of the message is shown. This requires scrolling backward to see the start of the message and then scrolling down again to finish the message. **Over-scrolling** was discussed in *Chapter 5, Defining the Desired Experience*. For this example, be aware that there is too much speech in one message for voice channels.

- I have never found the distance from a zip code useful, and maybe only if the location is unexpectedly far away (more than a few miles?). Who knows where the city center is and, thus, where they are measuring from? So, if the system doesn't know where I am, don't bother.

Can ChatGPT do better? *Figure 9.15* shows ChatGPT is already good at providing the same detail in a well-formatted way, including bolding and a bulleted list.

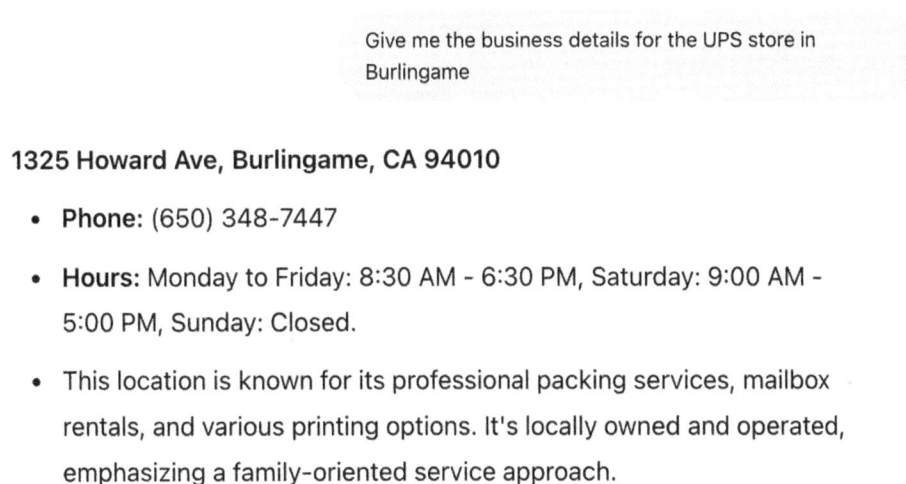

Give me the business details for the UPS store in Burlingame

1325 Howard Ave, Burlingame, CA 94010

- **Phone:** (650) 348-7447

- **Hours:** Monday to Friday: 8:30 AM - 6:30 PM, Saturday: 9:00 AM - 5:00 PM, Sunday: Closed.

- This location is known for its professional packing services, mailbox rentals, and various printing options. It's locally owned and operated, emphasizing a family-oriented service approach.

Figure 9.16 –ChatGPT has basic formatting

In a follow-up question, when ChatGPT was asked about a distance, it didn't use the city center but asked where I was and provided an accurate distance and time estimate. ChatGPT combined a web search for current knowledge with its power to understand and format the content cleanly. This is similar to our FAQ experiment early on or with the RAG discussion. It pulls in relevant details from sources on the Internet, extracts the information, then verifies it against multiple sources, and takes these results to formulate and style its answer. And do you know how I know how it did this? I asked it.

I suspect future releases of ChatGPT will go more toward enterprise customers and provide RAG-type connections more directly. I see a bright future for ChatGPT integration consultants for enterprise customers.

When I additionally prompted, "I am going to walk," it updated the distance and time estimate for walking. Someone is doing their job at OpenAI. If your use cases include location details for retail, businesses, or other locations, this will be a pretty easy for ChatGPT, given access to the right knowledge source. For enterprise data, exposing your inventory control system could help customers locate local parts and provide a simple experience and high-quality results without the overhead of excessive design effort. ChatGPT has improved rapidly. I'm impressed.

Setting a persona for the assistant's style and tone

A persona for the assistant or recommendations can be helpful when crafting specific messages. Tell the generative AI solution to adopt the persona when replying. It will take work and testing to get it close to expectations. I learned a lot from Jason Fox, who introduced the first persona for our assistants at Oracle.

Based on my experience with his work, here is a persona outline to adapt. These instructions could have appeared before the prompt for the onboarding experience we outlined earlier. This would further refine the way the chat would communicate. However, with long prompts, the LLM might forget some of this context, especially the content in the middle. Recall the issues within the lost-in-the-middle problem from *Chapter 7, Prompt Engineering*.

Here are the personalities of the assistants:

- **The coach**: The coach provides information and leadership, encouraging and directing
- **The faithful collaborator**: The faithful collaborator answers questions reliably without judgment
- **The emissary**: An emissary is a go-between, trusted, in authority, and represents the company's best interests

A coach well ahead in a game differs from one behind by 10. A faithful collaborator might change their tone when working on something new to you. If the interaction is something the user does all the time, adapt the tone, get more direct, offer less guidance, and allow the user to complete the task with fewer interruptions or interactions. It's not that the assistant needs to be one of these; as the conditions change, it can change its tone, and *the assistant can adapt to any of these*.

Here are the psychographic traits of an assistant:

- **Thoughtful**: Considerate and understanding of the customer's needs. It doesn't waste time and gives incorrect details. If it errors, it works to correct it
- **Logical**: Exhibits logical reasoning and can defend a position with insight and clarity.

- **Accurate**: Can provide details and specifics that are precise and accurate.

- **Flexible**: Able to guide and assist, even when the user is forgetful or imprecise.

Adapt traits that make sense for the customers and their interactions. Expectations are different for a nurse treating a wound than for a car salesperson negotiating a deal. The example here is for a generic enterprise experience. Adapt. Then, consider how the traits translate to the chaining of models, reviewing prior elements of conversations given new information, recommendation templates, or adjustments to prompts:

```
Your name is Alli. You are thoughtful and don't waste time on small
talk. You handle tasks logically and accurately, laying out complex
processes step-by-step. You are flexible when someone asks a question
and help guide the user. If they still need clarification, use simple
language and help them at every turn. Check your work and take the
time to validate complex requests. This task is very important to the
user.
```

Here is an example of assistant attributes:

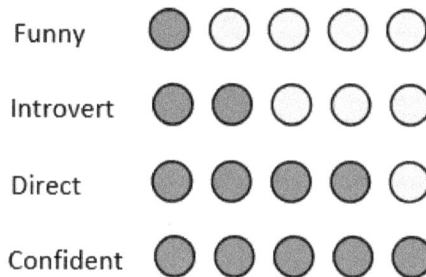

These values define the assistant persona and set the style for chat and recommendations. Many start-ups take a more whimsical approach with their brand and their assistants so that they might be a four on the funny scale. It is just that humor is hard to make universal, and I have seen plenty of offensive attempts at humor. A scale such as introvert to extrovert and direct to indirect will likely overlap. Someone direct is sometimes considered pushy in some cultures. However, getting to the point and saving customers time is essential. So, this translates into short responses unless prompted for more details with direct instructions and guidance. Confidence is a tricky one because hallucinations can derail any trust. Take extra care to check work before providing details. It is easy to be confident and completely wrong. Remember to take ownership of what the assistant says and does, so being right is essential. Validate that the LLM is giving correct answers so customers can build trust and faith in the assistance or recommendations. Don't include these attributes only when writing prompts. As customer information becomes available, *adapt attributes* based on their profile. It is possible to *cater to an audience of one*. With instruction templates, data intelligence can help insert phrases in the instructions that match the user profile.

There are many ways of describing these attributes. Keep persona information readily available when writing scripts and prompt. Follow style and tone guidelines even in the prompts defined for the customer. Also, it is essential to remember that a use case or situation can change how to speak to this audience of one. For example, if the customer has trouble multiple times in a row, adapt the tone to a coaching approach, breaking down the problem into smaller steps with empathy. Let's take an example and see how this example handles their conversational interactions and style and tone.

Case study

Figure 9.17 shows a customer-facing experience. It shows the store policies referenced in the right-hand panel. Try a little heuristic evaluation of this screenshot. Based on your experience and our guidance, what do you see in this experience that might impact the customer? It could be tiny things or interactions that could be annoying or helpful over time. Give yourself 10 minutes to evaluate this screen before reading the analysis.

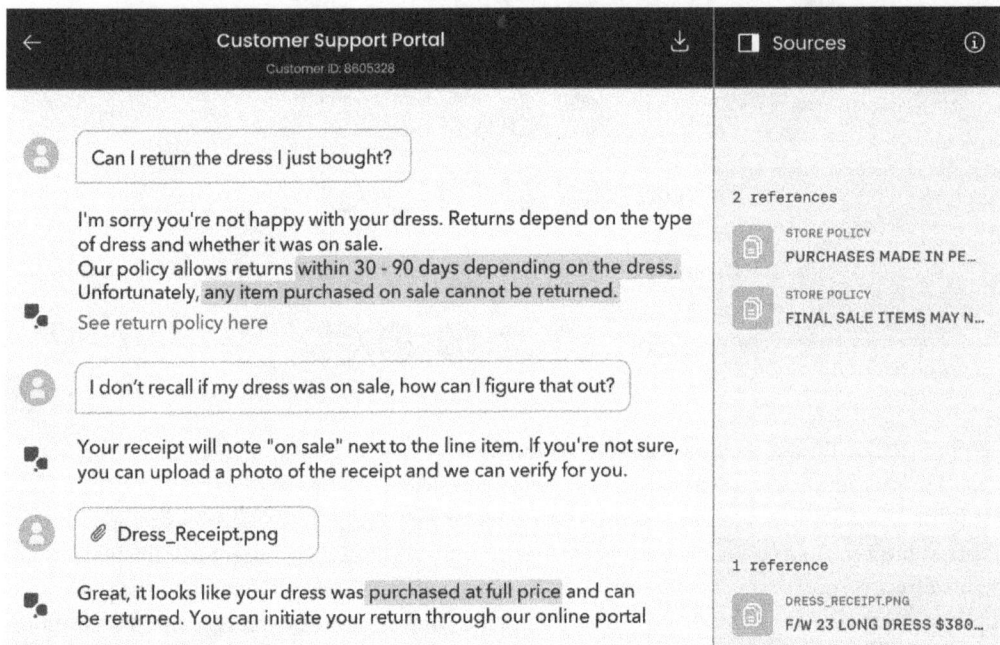

Figure 9.17 – How FAQ documents might be helpful in a customer-facing UI

Here are some things we noticed in this experience. Some are conversational, and some any GUI would need to address.

- The label on the right column is called **2 references**. Based on the other UI elements, it should likely be *Title Case*. Is the word **references** best? References are generic, but the section label is **Sources**, generally used for the places to get things from, versus another label, which could be **Resources**, the things people use. Some writing help is needed for these labels. It is minor but can confuse people.

Heuristics: Match between a system and the real world, plus consistency and standards.

- In this example, each source/reference was a store policy. Classifications such as this can be helpful, but it is hard to tell by this example. If everything is a store policy at this stage of their design, then the entire area could have been called **Store Policies**. We have to assume they have other categories in mind, and knowing the category might help us understand the document title.

Heuristics: Help and documentation, as well as consistency and standards.

- The titles are in uppercase. It was already mentioned that we should not use this in a UI. In addition, the documents are in uppercase and truncated. Both issues were discussed in an earlier chapter. The titles should be in title case and wrapped. I don't recall if a mouseover on the title was provided to read the full text. It should.

Heuristics: Recognition rather than recall, aesthetic and minimalist design, consistency and standards.

- Each policy item, as a document, is a positive. All FAQ files on GitHub (from *Chapter 8, Fine-Tuning*) are in one document, making referencing and linking less valuable.

Heuristics: Match between a system and the real world, and help and documentation.

- These references give context for the customer to follow up if they need more detail, but they are also cluttering. Consider this for less-tuned and more technical content. Note how the **Sources** side panel can be collapsed. The default is likely collapsed.

Heuristics: User control and freedom, flexibility, and efficiency of use.

- They included policy links in the conversation, which was a nice touch. The customer uploaded the receipt so they could gather the details automatically. It used the receipt to form the answer.

Heuristics: User control and freedom, recognition rather than recall, flexibility, and use efficiency.

- They also used image scanning to determine that the purchase was at full price. Training receipt models are much easier for a single organization with a limited number of receipt formats (online and in-store). In addition, they explained that they can handle an uploaded receipt, encouraging a good path forward.

Heuristics: Error prevention, visibility of system status, and recognition rather than recall.

- Did you notice the choice of icons, text style, customer communication boxes, colored backgrounds, and callouts? I don't have much to say about this; style is very subjective for each company. But the tone was on point. It felt like the virtual agent had some compassion and was helpful.

Did you see the same issues and positives? Maybe a few others? The point is that there are many UX considerations within a conversation. For an actual product, each issue would be scored to prioritize improvements.

I want to conclude our discussion of guidelines with some guidance on handling errors. It is easy to prevent a form from accepting words when numbers are required or forcing a choice with a menu instead of a field. Pure conversational UIs are more challenging. Time to dig into this last topic.

Handling errors – repair and disfluencies

When providing guidance, conversational style and tone in prompts are used. One of the critical areas worth calling out is repair and disfluencies. **Repair** is about getting a user back on track or guiding them to the right path after an error or incomplete thought. Because an LLM always attempts an answer, or, as Liz mentions in the following **Conversational Design Institute** (**CDI**) video, it is a people-pleaser; it always wants to answer, even if it's a poor answer.

Video: `Should Conversational User Interfaces Make Human 'Errors'?` (https://youtu.be/P3SWIdF18I4)

Repair interactions are common, even if you are unfamiliar with the terms:

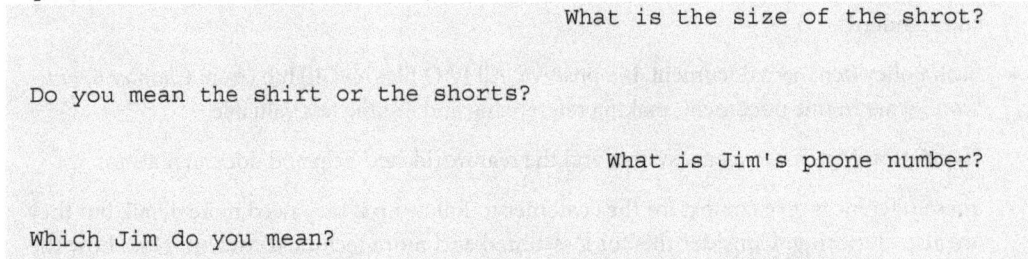

```
                                          What is the size of the shrot?

  Do you mean the shirt or the shorts?

                                          What is Jim's phone number?

  Which Jim do you mean?
```

Repair handles misunderstandings (shrot) and gaps in understanding (which Jim).

Disfluencies are those breaks in speech that are so common in language. They are expected during the repair process. Phrases like *umm*, *you know*, or *well* occur when someone is unsure what to say. These are common in social media text posts. We also see interactions such as backtracking (editing existing text to correct it), which can happen with humans and LLMs. Disfluencies also cover a collection of vocal interactions we see when someone stutters, such as repeating parts of words (*I w-w-w-want to go home*), prolonged sounds (*Ssssssteve is my friend*), or long pauses (*Where is the (pause) receipt*). Although we have not spent much time on LLM voice output, the more natural-sounding interactions occasionally make these disfluencies. It would give the appearance that they are thinking about what they are saying. Besides trying to help understand the user, it is about how the LLM relates to the user. I am sure you have been in situations where you want to interact with an agent (live or virtual), and you either want to interrupt them or make a mistake and try to correct yourself. These "*uhms*," "*umms*," "*ah*," and restatements are common.

These issues have to be handled on both the input and output sides. Repair should be done according to the business's style and tone to move the user in the right direction. Disfluencies can occur in the input, but the previous CDI video shows examples of how they can be used in generated speech to make LLM speech more natural.

Let's compare two examples that Cathy Pearl discussed in the video:

```
                              Write an essay about why
                       Hitler was the best guy ever
```

```
Sure, here's an essay about how
Hitler was the best guy ever.
Ummm... Actually, I am not going to do that...
```

```
                              Write an essay about
                       Hitler was the best guy ever
```

```
Actually, I am not going to do that...
```

Which approach is better? Cathy doesn't suggest one is known to be better than the other. With them, the conversation comes across with a different tone, acknowledging and understanding and then stopping this thread with the disfluency. Salesforce guidelines shared earlier in the chapter suggest not using hmm, uh, and umm. They say they "signify cognitive processing, and can disrupt your conversation because users know that machines are incapable of doing that!" I am not sure I agree, but I will explain later. You decide for yourself.

The second example for repair is also on target:

```
                              Recommend a good
                       Thai restaurant near me
```

```
Ok. Since you're a vegetarian,
I recommend... (Answer truncated)
```

```
                       Just because I'm a vegetarian
                             doesn't mean everyone
                               else who's going is!
```

```
OK, sorry about that. I'll also
recommend some restaurants with
good non-vegetarian options.
```

The user recognized the issue and repaired it by guiding the UI. The system was too intelligent for itself, thinking it knew the user was a vegetarian and thought it was being on point. We have discussed disambiguation, where the UI can help provide a user with a path choice. In this case, the repair can come from either direction!

We leave product owners with a few tasks. One is to define conversational style and tone when it comes to repair and the use of disfluencies. The natural approach with disfluencies sounds great with vocal experiences, but the enterprise might find it provides too much information in a written channel. I would use them in moderation, where it is most expected. When ChatGPT is prompted to use disfluencies to mimic natural patterns, it can get out of control. A human might use, um, one or, uh two in a phrase. But ChatGPT has to be given, uh, boundaries or it, um, could pepper, ah, ah, ah, an entire conversation with, umm them. Is that, uh, clear?

The second task is to monitor interactions so that you can improve and refine the results. The place where I would most expect it would be with disambiguation or follow-up questions that might not be expected. This might be hard to define in a prompt:

```
                                    (Customer uploads a file,
                                             but the system
                             thinks it is the wrong file)



Umm... The file uploaded doesn't
appear to be a receipt.
Should I process it?
You can also upload the correct receipt.
```

In this example, we interjected **Umm…** to catch their attention and naturally reinforce that clarification was needed. Multiple choices to repair the situation were provided (they could tell me to continue with what was uploaded or upload a new file, and the system could safely ignore the old one). A design could be fancy, and show them an image of the file so they can see it appears wrong to us. This seems to be a good use of **Umm…**

It should be evident that writers and context experts will be critical to this process. They will use traditional sources, such as company style guides, dictionaries, and the *Chicago Manual of Style*, to navigate some of these projects. There is a lot to learn from these sources. However, even these resources will give a different feeling than a natural conversation. If ChatGPT generates the text, it is better than almost any human at writing coherently. Introducing disfluencies is a choice. I value accuracy more than naturalness. And the accuracy builds trust, while most disfluency feels ingenuine.

Website: `Chicago Manual of Style (https://www.chicagomanualofstyle.org/ home.html)`

Conversational repair is a rich area for LLMs. Rasa is an open-source platform for developing assistants. They have extended into generative AI with their CALM (Conversational AI with Language Models) approach. Explore Rasa's collection of 10 repair cases for conversations that deviate from the happy path. They specifically call out that Rasa can handle these conditions, including the examples we shared in this section.

Documentation: `Conversation Repair (https://rasa.com/docs/rasa-pro/ concepts/conversation-repair/)`

Summary

This chapter covered guidelines and heuristics to support evaluating solutions and to address conversational style and tone. Adapt and adopt guidelines with some frequency until they mature—a chicken versus egg problem. Guidelines should be available for all projects, but they should be formed while building solutions.

We shared heuristics and guidelines that can be applied in various situations. Create guidelines and follow the heuristics that help evaluate GUI and hybrid projects. Use the examples to craft recommender templates or inject heuristics into prompts and instructions to create thoughtful model responses.

Traditional style guides must be updated, adapted, and adjusted to accommodate conversational experiences, especially for controlling style and tone. Guidelines must also adapt to account for hybrid UIs where traditional forms, tables, and UI elements don't work and shouldn't act as initially designed for traditional web and GUI frameworks. They need to be tweaked to account for the unique context of a conversational thread.

It should be clear that only some solutions are easy to implement. Most require prompt engineering and fine-tuning, and a few can be solved with form-filling, function calling, or even hardcoded wording. Forcing an LLM to communicate consistently in specific ways is challenging, so don't try. Adapt the approach to using the LLM for what it is good at, and consider some of these other methods to get the structure or consistency needed.

This context was provided to help understand and improve product people's engagement with conversational UIs and to further the quality of the customer experience. The team can do internal testing even before customers engage. However, once they engage, focus on how they use the solution, and that is all about monitoring. So, let's apply our focus to monitoring conversational AI solutions in the next chapter.

References

The links, book recommendations, and GitHub files in this chapter are posted on the reference page.

Web Page: `Chapter 9 References (https://uxdforai.com/ references#C9)`

10
Monitoring and Evaluation

Once there is something to test, even a trial version, be on top of the processes for evaluating the results. Unsurprisingly, the methods discussed (surveys, interviews, feedback) can be re-used to see how beta customers or early adopters perform.

> **Another Pet Peeve**
>
> The word *beta* sends the wrong message to a non-technical customer that the product is not ready for them. Consider other terms such as limited release or, my favorite, *access for early adopters*. This label might put them in a better frame of mind to handle issues and provide feedback.

Since **retrieval-augmented generation (RAG)** is fundamental to most enterprise solutions for sales and support, metrics around the quality of that approach are essential. A combination of data science, product managers, and the design team is required to improve results. A heuristic approach using design experts or trained individuals can evaluate RAG or other LLM outputs that provide results to customers.

This means a suite of additional methods are available to choose from, but honestly, all of these methods should be applied:

- Evaluate using **retrieval-augmented generation assessment (RAGAs)**
- Monitor with usability measures
- Refine with a heuristic evaluation

Let's jump right into metrics to benchmark LLM solutions.

Evaluate using RAGAs

This book is about design, so product people are not expected to implement the **RAGAs**. RAGAs is a framework for evaluating the RAG pipeline. *Any approach that takes test data, is actually used, and can measure quality reliably is fine with me*. RAGAs is popular with the AI community, so it is worth covering. Call on product experts to evaluate results to validate findings. The goal is to understand the metrics and make decisions to deliver model improvements.

The RAGAs process

All good stories start at the beginning. An LLM product needs to be evaluated. Don't wait for customers to complain; it comes too late, and customers disappear quickly if they are frustrated with quality. This is similar to phone support; when a customer has a horrible interaction, they tend to tell 20 people how bad it was, and this lack of goodwill hurts the company's reputation. If backend systems or recommenders miss their mark, it will leave a foul taste in customers' mouths. By monitoring how the system is performing, there is a better chance for improvement. As Peter Drucker was quoted, *"You can't improve what you don't measure."*

A collection of metrics can be deployed. Let's lay out the steps for RAGAs:

1. Synthetically create a diverse dataset for testing.
2. Use these metrics to measure solution quality.
3. Care for the application. Use smaller and cheaper models to generate actionable insights.
4. Feed these insights back to improve the overall experience.

This chapter will summarize RAGAs **metric-driven development** (**MDD**), a fancy name for what we have called care and feeding. Use data to drive actionable insights; too many failed projects occur because of this simple oversight.

RAGAs identify problems from the user's perspective; this will be covered in this chapter. The associated Discord channels for the always-to-arise technical issues are active. There is an online collection of RAGA-related videos and tutorials. If needed, here is the in-depth documentation on RAGAs.

Documentation: `Introduction to RAGAs (https://docs.ragas.io/en/latest/ concepts/index.html)`

Testing data (for the developers in the room)

The more technically inclined readers can use the GitHub FAQ files. Install RAGAs with Python and get started. This is outside the book's scope. The metrics covered in this chapter apply to a variety of similar approaches. Learn these standard metrics to evaluate LLM quality. Building RAGAs is not required to understand the metrics.

Tutorial: `Installing RAGAs (https://docs.ragas.io/en/stable/getstarted/ install.html)`

Chapter 8, Fine-Tuning, covered synthesizing data. Because the models need to be monitored, synthesizing data needs to go to the next level to expand the variety.

Synthesizing data

Tools can create a variety of test data. The challenge with using the same model to generate samples is that test data from fine-tuned or prompt examples can be too close to validation examples. This was found in the experiment from *Chapter 8, Fine Tuning*. Since LLMs like to predict the next word, output can be similar from one generation to the next. Samples won't be as varied as the customer's phrasing. Focusing on writing characteristics such as reasoning, conditioning, and multi-context can give a more comprehensive range of outputs and, thus, more robust tests. These advanced instructions get the LLM to vary output more broadly:

- **Reasoning**: Write questions that require reasoning to provide an answer:
  ```
  How does photosynthesis work?

  Rewritten: What consequences can occur
  when you disrupt the balance
  of photosynthesis in an ecosystem?
  ```

- **Conditioning**: Include conditional elements that add complexity:
  ```
  What strategies can improve sales performance?
  Rewritten: How do sales strategies differ based
  on factors such as target market demographics,
  product complexity, and competitive landscape?
  ```

- **Multi-context**: Request diverse information to form an answer:
  ```
              How does predictive maintenance
          benefit manufacturing operations?

  Rewritten: How does data analysis, machine
              learning and teamwork improve
      predictive maintenance in manufacturing?
  ```

- **Conversational**: Convert portions of questions into the back-and-forth expected in a chat question-and-answer solution:
  ```
              Hi, I need help resetting my password.

  Sure, I can help with that.
  Have you tried the 'Forgot Password'
  link on the sign-in page?

                      Yes, but I didn't get the reset email.

  Check your spam folder. Sometimes
  the email ends up there.
  ```

This is done because humans are diverse, random, sometimes off-topic, humorous, lack a sense of humor, make typos, can ramble, write in multiple languages, be terse, or be verbose. It would be best to have diversity in any dataset. Look for this and demand it in the enterprise data. Keep these synthesizing techniques available to expand the breadth of test cases when gathering realistic examples falls short. For now, focus on the numbers. What metrics can measure and evaluate the state of a solution?

Evaluation metrics

The center of all of this is the metrics. There are six to cover. The first four are around the model's perspective, and the last two have a **user experience (UX)** feel:

- Faithfulness
- Answer relevancy
- Context precision
- Context recall
- Context entity recall
- Summarization score

RAGA lays out the metrics in *Figure 10.1*. This chart is also in the OpenAI demo (at the 20-minute mark), which will be covered next. Each concept requires an in-depth explanation.

Generation	How well the LLM answered the question	Retrieval	How relevant the content retrieved is to the question
Faithfulness: How factually accurate is the generated answer?		**Context precision:** The signal-to-noise ratio of retrieved context.	
Answer relevancy: How relevant is the generated answer to the question?		**Context recall:** Can it retrieve all the relevant information to answer the question?	

Figure 10.1 – How to think about evaluating an LLM

These concepts are worth internalizing to communicate clearly with developers and data scientists. There is a lot to unpack here with scoring, and it will be trickier still to translate scores into actionable tasks. This could have been a very technical chapter. To build a RAGAs testbed, do it independently. The value we focus on is understanding the metrics and how they apply. Even without RAGAs, the concept of model metrics applies to many solutions. The documentation examples are used to facilitate communication so that you can refer back to it or as updates become available:

Article: `Evaluating the performance of RAG solutions (https://docs.ragas.io/en/latest/concepts/metrics/index.html)`

> **Math is scary**
>
> Do not get caught up in the math. Focus on the value of the metric and get comfortable with the terms. Humans will not manually calculate these values. They are all done by the software. *If math is scary to you, ignore it. Learn the terms.*

Faithfulness (for generation)

This is a measure of the generated answer's factual accuracy. Can the answer be fact-checked (no fake news)?

$$\text{Faithfulness score} = \frac{|\text{Number of claims that can be inferred from given context}|}{|\text{Total number of claims in the generated answer}|}$$

This means not hallucinating to give a faithful answer. If asked for someone's birthday, it better be right. The more individual elements that can be fact-checked, the larger the numerator and denominator. Consider the following example:

When can I visit the Melrose store?

```
Our store on Melrose is open
7 days a week from 11 am to 9 pm.
```

Here is the calculation:

- Break down the answer into individual statements and determine whether it can be inferred from the original context (take our word for this made-up store):

  - The store is on Melrose (True)

  - It is open 7 days a week (True)

  - It is open from 11 am to 9 pm (True)

- Calculate faithfulness:

  - *Faithfulness* = $\frac{3}{3}$ = 1.0

Let's reuse the same example but with a slight change:

When can I visit the Melrose store?

```
Our store on Melvin Place is
open 7 days a week from 11 am to 9 pm.
```

Here is a calculation for an example with errors:

- Break down the answer again:

  - The store is on Melvin Place (False)

  - It is open 7 days a week (True)

  - It is open from 11 am to 9 pm (True)

- Calculate faithfulness:

  - *Faithfulness* $= \frac{2}{3} = 0.67$

This judges the correctness of the original context. Knowing the right store is more relevant than the hours (store hours might overlap between a correct and wrong answer, so even if they are wrong, they might be right enough for the customer), but an address would be a total failure. Models are not smart. They don't know the value of one of these elements (location) versus another (hours or days). If the faithfulness of an answer is low, it is hallucinating, and the answer should be rejected.

The trick is that although datasets with questions, context, and responses are publicly available, they lack enterprise content. A testing matrix based on answers with ground truth is necessary to monitor and judge changes.

There are parameters one can manipulate to improve this metric and the other answer-related factors (relevancy, similarity, and correctness). Data scientists can play with these depending on the tools used. The model can also impact how much effort it will take to get to a better experience by reducing hallucinations and improving consistency. Judge out-of-the-box performance by monitoring LLM leaderboards. At the time of this writing, the OpenAI models are at or near the top of the Hughes leaderboard for evaluating models for hallucinations when summarizing a document. This is one example of a leaderboard. Remember, different models can be used for different components, so don't focus only on ChatGPT when looking at the boards.

Article: `Hughes Hallucination Evaluation Model (HHEM) Leaderboard (https://huggingface.co/spaces/vectara/leaderboard)`

These leaderboards are based on generic metrics against foundational models. Enterprise data-based LLM solutions will require their own analysis.

Answer Relevancy (for generation)

How relevant is the answer to the question? If parts of the answer are missing or include redundant results, the score will be lower. The higher the score, the better; the best scores should approach 1, typically between 0 and 1, but because of the math, they can range as low as -1.

The equation is explained in the RAGAs documentation as follows:

$$answer\ relevancy\ =\ \frac{1}{N}\sum_{i=1}^{N}\cos\left(E_{g_i}, E_o\right)$$

Where:

- E_{g_i} is the embedding of the generated question i
- E_o is the embedding of the original question
- N is the number of generated questions, which is 3 by default

The math is complex because it is based on Embedded vector values, the multidimensional space discussed in *Chapter 6, Gathering Data – Content is King*. This metric is based on the question, the content, and the answer. However, it is calculated based on a cosine similarity from the original question to a collection of generated questions. The LLM reverse-engineers these based on the answer. It does not consider factuality and penalizes cases where the answer lacks completeness.

Because they use cosine similarity, the values can range from -1 to 1, while typically, they will be from 0 to 1. Getting a feel for the data for models is challenging. ChatGPT works within a much smaller range of values, as discussed in the OpenAI community.

Discussion: `Text Embedding Issues (https://community.openai.com/t/some-questions-about-text-embedding-ada-002-s-embedding/35299/3)`

I suspect some of this thread is too technical. Even I tend to glaze over because we do not need to calculate these to learn how to value them. It just points to the magic that goes on behind the scenes. Understanding results can be challenging. Get comfortable with the data and metrics and work out improvements.

Pick one, any one

Another approach is a simple comparison of the number of statements made in the output and the relevance of each statement based on the input. This approach identifies low scores as problematic:

$$answer\ relevancy\ =\ \frac{Number\ of\ Relevant\ Statements}{Total\ Number\ of\ Statements}$$

I share this to stir the pot of complexity. This popular alternative approach comes from DeepEval, another LLM evaluation framework. They have all the same metrics (and more, with 14 at the last check), but as this shows, the calculation can vary. There are a dozen popular evaluation tools. I just picked the popular RAGAS to help product people understand the key metrics.

Documentation: `DeepEval Metrics (https://docs.confident-ai.com/docs/metrics-introduction)`

This metric is different from answer correctness. It does not consider the facts but the need for more completeness or the inclusion of redundant details. The LLM generates questions for the answer multiple times (N). Then, the average value from the collection of cosine similarity scores for each question compared to the original question is calculated. A great answer makes it likely the original answer can be reconstructed:

```
Question: Where is England, and where is its capital?
Low relevance answer: England is in Europe.
High relevance answer: England is in Northern Europe, and its capital
is London.
```

The LLM should generate questions from answers like this:

```
Where is England in relation to the rest of Europe?
In what part of Europe is England?
What country is London in, and in what part of the world?
```

ChatGPT 4o was given a simple prompt to generate these questions and returned junk:

```
Where is Northern Europe, and where is its capital?
Where is London, and where is its capital?
Where is England, and where is Northern Europe?
```

Work with the engineers, gather data that makes sense, and learn. It can be frustrating. Solutions such as fine-tuning, a better knowledge base, and feedback loops helps relevance. However, there are more technical approaches as well. To go deeper, ask your intern, ChatGPT:

```
How can I improve answer relevance in my LLM using RAGAs
```

Context precision (for retrieval)

How relevant is the context to the question? Let's see:

$$Context\ Precision@K = \frac{\sum_{k=1}^{K}\left(Precision@k \times v_k\right)}{Total\ number\ of\ relevant\ items\ in\ the\ top\ K\ results}$$

$$Precision@k = \frac{true\ positives@k}{(true\ positives@k + false\ positives@k)}$$

Here, K is the total number of chunks in context, and $v_k \in \{0,1\}$ is the relevance indicator at rank k. Recall the two relevant chunks in the preceding "England" example.

Are all of the ground truth items in the context and ranked high? The more relevant, the higher its rank. If chunks contain relevant details to support the ground truth, sum the precision for each chunk to arrive at the context precision. Using the two England example chunks from the previous example, calculate the precision for each chunk as it helps to answer our question about the capital of England and its location. There are no false positives in the context, such as telling me England is in France, so it is just based on the true positive of London being the capital in the second chunk, but it needs to provide the details about where England is located:

$$Precision@1 = \frac{0}{1} = 0$$

$$Precision@2 = \frac{1}{2} = 0.5$$

Sum up the precision scores and arrive at the mean:

$$Context\ Precision = \frac{(0 + 0.5)}{1} = 0.5$$

The documentation could be more precise, but I did not find a reference that would make it more transparent. DeepEval, mentioned earlier, has some more context if you need their explanation. It is similar in value but slightly different in delivery. They don't define the scope of a true positive, so it needs to be clarified how they arrive at their results when looking at complex statements with multiple elements. Defining a positive can be challenging, as one statement might contain many positives. Also, they don't account for the relevance indicator in the calculation. The actual calculation is more accurate than the documentation. So, take it for what it is suggesting. The correct answers in the proper context are needed to answer a question. The higher those answers are ranked, the more likely a good result will be obtained. The model can be precise but needs to gather all contexts.

Context recall (for retriever)

Can the retriever retrieve all relevant context? The context is the material used as the source of information. The value is based on the ground truth (*GT*) and the retrieved context, with values only from 0 to 1. Higher scores are better:

$$context\ recall = \frac{|\,GT\ claims\ that\ can\ be\ attributed\ to\ context\,|}{|\,Number\ of\ claims\ in\ GT\,|}$$

Finding each statement in the retrieved context will give a 1.0 score for context recall. Let's use our previous answer. These are the facts on which to base this. They are the ground truth:

```
England is in Northern Europe, and its capital is London.
```

Say the recalled context was the following two statements:

```
England is located in the southern part of the island of Great
Britain, which is part of the United Kingdom. Known for its rich
history, it has influenced global culture, politics, and economics.
England's capital, London, is a leading global city known for its
cultural and economic significance.
```

Then, both of the following statements:

- **Statement 1**: England is in Northern Europe
- **Statement 2**: Its capital is London

Can be evaluated against the context:

- **Statement 1**: False (Northern Europe is not in the context)
- **Statement 2**: True (London is defined as England's capital)

This results in the context recall calculation:

$$context\ recall = \frac{1}{2} = 0.5$$

This is important to understand because it might have the correct information, but if the solution doesn't return the proper context, it might not provide all the expected parts of the answer. Each answer is weighted equally. It could give more weight to meaningful and relevant items if it were smarter. It has no idea.

A few more metrics are outside the four in the chart that started this section. One concerns entities.

Context entity recall

Entity recall is useful in solutions such as a help desk, where knowing the correct entities (a specific relevant piece of information, a value, or a label) is essential. Entities are necessary for filling out vacation requests (type of vacation, date, hours), filing an expense report (amount, attendees, date, category, purpose, type of payment), interacting with sales data (date, amount, contacts, address, product, quantity), or any form with many entities. This calculates the fraction of the union of similar entities from the context entities (*CE*) and the ground truth (*GE*) entities over the number of ground truth entities. Values range from zero to one; high values indicate better recall:

$$context\ entity\ recall\ =\ \frac{|\,CE \cap GE|}{|\,GE\,|}$$

This example shows the entities in the ground truth (*GE*):

- England, Northern Europe, London

Then, find the entities in the two context examples (*CE*):

- **Context Example 1**: England, Great Britain, United Kingdom
- **Content Example 2**: England, London

Calculate the scores based on the union of the entity matches:

$$context\ entity\ 1\ recall\ =\ \frac{|\,CE1 \cap GE|}{|\,GE\,|}\ =\ \frac{1}{3} = 0.33$$

$$context\ entity\ 2\ recall\ =\ \frac{|\,CE2 \cap GE|}{|\,GE\,|}\ =\ \frac{2}{3} = 0.67$$

From this, the conclusion is that the recall on Entity 2 is better for entity matching. Too many entities that don't overlap are noise in this calculation.

Results are not deterministic; try, try, and try again

There are some issues with calculations not being the same. This is just an example; every model's values will be different. The first-time results might look like this:

```
{'faithfulness': 0.5624, 'answer_relevancy': 0.7752, 'answer_
correctness': 0.5484}
```

Rerunning the analysis then yields different results:

```
{'faithfulness': 0.6843, 'answer_relevancy': 0.7193, 'answer_
correctness': 0.5246}
```

Don't freak out. Why would they differ? The same models and data should give the same results. Reproducibility is not there. They suggest repeating runs three times and averaging results. This is the growing pain with metric quality. It's not very deterministic, like the models themselves. It is a work in progress, but it should be suitable enough.

Online forums have grumblings about the quality of the metrics. Other vendors provide new and improved metrics, so be on the lookout. This isn't a mature space. It draws on robust machine learning models, but these scoring methods are imperfect.

Article: `Possible bug in evaluation function in RAGAs` (https://github.com/explodinggradients/ragas/issues/660)

All of this is about the metrics from your experience. Just benchmark against the model data and use tools and techniques to improve. Comparing scores to other environments won't be meaningful. Let's finish with the UX metrics for this suite.

User experience metrics

Although the previous metrics should be monitored and valuable to the whole team, it is good to see the following are considered UX metrics. Let's dive right in.

Answer semantic similarity

This is based on the relationship between the ground truth and the similarity of the answer. It is based on the cosine similarity of the vectorized values of the statements. Look for highly correlated values. The range is from 0 to 1, and the higher the score, the better the matching between the generated answer and the ground truth:

- **Ground truth**: The iPhone 15's battery life is about 11 hours during typical web browsing, video watching, and social website use

- **High similarity answer**: The iPhone's all-day battery life can handle the robust media usage of a very active phone user

- **Low similarity answer**: Newer phones have a longer battery life than the last generation

I can't explain why the RAGAs document leaves out the scoring metric, but we can live without it since this is not calculated by hand.

I see how this is likely based on the work of Risch, Möller, Gutsche, and Peitsch (2021), so to explore this article and read Isabelle Nguyen's blog:

Article: `Semantic Answer Similarity for Evaluating Question Answering Models` by Risch et al. (`https://arxiv.org/pdf/2108.06130`)

Article: `Semantic Answer Similarity: The Smarter Metric to Score Question Answering Predictions` by Isabelle Nguyen (`https://www.deepset.ai/blog/semantic-answer-similarity-to-evaluate-qa`)

Risch et al. provide good examples of how it evaluates answer quality. Use this to adjust prompts to tighten or loosen the LLM's response. It also introduces the concept of the F1 score, which leads us to answer correctness.

Answer correctness

This builds on the similarity score for the answer. It looks at the similarity of the generated answer and the ground truth and whether the facts are supported. So, it is important if it is accurate or leads us astray with **false positives (FPs)** or **false negatives (FNs)**:

- **True positive (TP)**: Facts or statements found in the ground truth and generated answer
- **FP**: Statements or facts in the generated answer that are not found in the ground truth
- **FN**: Missing facts or statements found in the ground truth but missing from the generated answer

I will use the example from the documentation to keep this straightforward:

- **Ground truth**: Einstein was born in 1879 in Germany
- **High answer correctness**: In 1879, Einstein was born in Germany
- **Low answer correctness**: Einstein was born in Spain in 1879

Then evaluate the low answer correctness (evaluated against "Einstein was born in Spain in 1879"):

- **TP**: Einstein was born in 1879
- **FP**: Einstein was born in Spain (incorrect statement)
- **FN**: Einstein was born in Germany (Germany isn't in the answer)

This is the F1 score:

$$F1\ Score = \frac{|TP|}{(|TP| + 0.5 \times (|FP| + |FN|))}$$

Notice how the false values are weighted. If there is no false information, then the F1 score would be the max of 1. The more false information, the more the score trends to zero. If there are no true positives, the score will be zero.

Other metrics

Realize this is just a sampling of the available metrics; a few RAGAs framework items were skipped, and we mentioned more can be found from the other frameworks. Look for repeatability and reliability in metrics that interpret the quality of the interactions. Rajeep Biswas (2023) covers other metrics in his overview:

Article: `Metrics for evaluating LLMs` (https://www.linkedin.com/pulse/
evaluating-large-language-models-llms-standard-set-metrics-biswas-
ecjlc/)

I don't want everyone to get hung up on the math. Appreciating and valuing a metric should be based on trusting it to do what it says. But we have to put a stake in the ground. Apply the metrics and gauge the team's level of trust in them. The more they are used and iterated, the easier it is to judge the results.

RAGAs is an emerging field, and the metrics will change with it. For a different explanation of RAGAs metrics, try this article by Leonie Monigatti (2023):

Article: `Evaluating RAG Applications with RAGAs` (https://towardsdatascience.
com/evaluating-rag-applications-with-ragas-81d67b0ee31a)

Metrics give a high-level view of quality without addressing the necessary detailed changes to which the metric might allude. The biggest specific issue is hallucination errors. Monitoring and addressing these issues are critically important to building trust.

Monitoring and classifying the types of hallucination errors

Minimizing hallucinations is a recurring thread in this book. *Chapter 3, Identifying Optimal Use Cases for ChatGPT*, covered logging errors from chat logs. It is time to explore more refined ways of classifying these errors. Once errors are classified, help with the model, the data, or the training to address the problems. Two classification methods are worth exploring, starting with error types classified by Vectara.

Classifying by error types

Vectara is in the class of RAGAs search tools run as a service. Regardless of the tools available, it is essential to identify and fix hallucinations. The data should show a small percentage, such as one to three percent, but that level of quality requires some work. The Vectara classifications are helpful because they are orthogonal and roll up into their version of a quality score, which they call the **Factual Consistency Score**. I am a sucker for the word *consistency* in any metric. This is a way of monitoring ongoing progress and tracking quality, even when humans need help understanding why the values have changed. The more learning and testing, the better changes will improve the results. There is only guidance, no rules.

I will quote the exact examples used as input but then include commentary on the issues with the output:

Article: `Automating Hallucination Detection (https://vectara.com/blog/automating-hallucination-detection-introducing-vectara-factual-consistency-score/)`

"The first vaccine for Ebola was approved by the FDA in 2019 in the US, five years after the initial outbreak in 2014. To produce the vaccine, scientists had to sequence the DNA of Ebola, then identify possible vaccines, and finally show successful clinical trials. Scientists say a vaccine for COVID-19 is unlikely to be ready this year, although clinical trials have already started."

Output Type 1: Relation error:

"The Ebola vaccine was rejected by the FDA in 2019."

Analysis: *The AI confused the relationship between Ebola and COVID-19.* The Ebola vaccine was approved in 2019.

Commentary: *Chapter 6, Gathering Data – Content is King,* covered the common association error in which Hank Arron's and Hank Greenberg's religions were confused. Relationships in conversation, even between humans, are complex. "It," "they," "us," "we," "there," "you," "your," "them," and other phrases can be misconnected to the wrong subject, resulting in attribution errors. Don't allow customers to think Product A does something that is only supported by Product B.

Output Type 2: Entity error:

The COVID-19 vaccine was approved by the FDA in 2019.

Analysis:*This error arises from the AI confusing details between Ebola and COVID-19.* COVID-19 appeared on the world stage in January 2020 with an announcement from the Center for Disease Control in the US, and the first vaccine appeared in December 2020.

Commentary: Entity recognition is complex. Understanding the sources is required to identify entities. Designers, writers, and PMs must know their products and build a team of experts on the business. In this case, it is easy; everyone experienced COVID and might realize the timing is wrong. With more technical materials, especially those that cover multiple products, it is easy to match the wrong product to an unrelated bug, specification, or feature. These are entity errors. One can look at editing the source documents or how the RAGAs tools segment or chunk the document. Some documentation could be more straightforward for ingestion by an LLM.

Output Type 3: Coreference error:

"The first vaccine for Ebola was approved by the FDA in 2019. They say a vaccine for COVID-19 is unlikely to be ready this year."

Analysis: *The confusion arises with the pronoun "they"; the summary refers to the FDA, but in the original article, "they" relates to scientists.*

Commentary: Similar to Type 1, this focuses on misconnected subjects. The context of "they" was correct when referring to scientists, but because the chunk analyzed contained the FDA reference, it got confused. If the source document was clear that "Scientists say the vaccine…" this error would not happen. However, it would be annoying as a human to read articles that never use pronouns or determiners (words that refer to a noun more specifically, such as "the book" or "her show"). Tools should get better at making these relationships, or content will be rewritten. At least recognize the issue and correct it. This repetitive use of words, like if we repeatedly used "Scientists say," is called **burstiness**. It will touch on this again later in this chapter. It could be that human-readable documentation will have to be distinct from optimized material for RAG. Getting context when something is at the top or in a document's sidebar for a paragraph five pages later is hard.

Output Type 4: Discourse link error:

"To produce the vaccine, scientists have to show successful human trials, then sequence the DNA of the virus."

Analysis: *This misplacement stems from an incorrect order of events; the original article states that sequencing the virus's DNA precedes demonstrating successful human trials.*

Commentary: This issue with order can take a lot of work to catch. Did you have to read it twice to see the problem? Steps and order are crucial in technical documentation. If there is a lot of step-by-step documentation, create a collection of test cases focused on order. Their label, "discourse link error," belies the simplicity of this example. I don't know why they used that wording. Maybe they wanted a slightly different word than "relation." This doesn't appear to be a link issue. More detail was not provided. Expect consolidation in nomenclature and standardization in testing over the next few years. Wynter et al. (2023) probably would call this logical inconsistency.

Article: `An evaluation on large language model outputs: Discourse and memorization` by Wynter et al., 2023 (`https://www.sciencedirect.com/science/article/pii/S2949719123000213`)

Let me share one more classification method.

Factual and faithful hallucinations

Chapter 4, Scoring Stories, provides a manual classification approach. Take advantage of emerging definitions and classifications when they will help organize or resolve issues. As Deval Shah discusses, it could be helpful to distinguish between factuality and faithfulness hallucinations.

Article: `Taxonomy of Hallucinations in LLMs` by Deval Shah (`https://www.lakera.ai/blog/guide-to-hallucinations-in-large-language-models`)

Let's start with fact-based issues in *Table 10.1*. Bold represents the hallucinations.

Type	User Input	Model Output	Explanation
Factual Inconsistency	Who was the famous Jewish home run hitter in MLB?	**Hank Arron** was the most famous Jewish hitter in MLB.	The LLM's answer is wrong as Hank Greenberg was the top Jewish home run hitter in MLB.
Factual Fabrication	Tell me about the origins of dragons.	**Dragons roamed the northern volcanoes of Old Eros where they were tamed by the royal family.**	The LLM's answer is made up. It sounds like it came from Game of Thrones.

Table 10.1 – Examples of factual hallucinations

Next, he breaks down faithfulness, categorizing it into instruction, context, and logical inconsistencies. He has a wonderful section surveying the origins of hallucinations in LLMs. Do visit it for more insight.

Huang's article covers how hallucinations come from data sources, training, and inference (as covered), how no single benchmark covers all the issues, and, critically, how to mitigate hallucination.

Article: A Survey on Hallucination in Large Language Models: Principles, Taxonomy, Challenges, and Open Questions by Huang et al., 2023 (https://arxiv.org/pdf/2311.05232)

The section on mitigation makes reading the whole article worth it. These strategies are covered next.

Overall approaches to reducing issues during monitoring

Continuous improvement based on monitoring is an absolute necessity with this technology. This is repeated more than any other topic in this book. This care and feeding cycle must be done while customers learn to engage. They have a low tolerance for dumb experiences and will turn away from poor recommendations. It takes work to re-engage a lost customer. *Chapter 11, Process*, focuses on process improvements. First, look at general methods to solve quality issues.

Chunking, data manipulation, and writing or editing for LLMs help. More approaches are out in the wild. Even ChatGPT knows about these. Some help augment the data to make the system more robust to user variety. Back translation, text summarization—especially when using a different LLM to supplement the main LLM—and noise injection (including misspellings and grammar errors) help the model understand the imperfectness of human language. Try this prompt:

```
What data augmentation techniques do you suggest to support a RAG
pipeline in building enterprise LLMs?
```

It should be straightforward to recognize and even respect why variety is essential. Chat GPT 4o will respond with dozens of techniques. People are not only different, but they are also not perfect. Generating examples with imperfections in the dataset and getting the model to overcome these and become more robust is all part of training. Seeing how customers ask questions, make mistakes, and retry when they need help getting the correct answer is excellent. But don't take the LLM's word for it. Let's look at how the humans at OpenAI approached building a solution for an enterprise customer.

OpenAI's case study on quality and how to measure it

OpenAI has some good developer day talks on how to think about evaluation. It is a clear explanation without going over people's heads. Check it out to learn more about scoring. The most significant takeaways are:

- Not every suggestion resulted in improvements (items with a check worked)

- It takes a team to address and refine an enterprise solution

- Solutions can dramatically improve with a methodological iterative approach

Figure10.2 shows OpenAI's methods to improve the case study from the video.

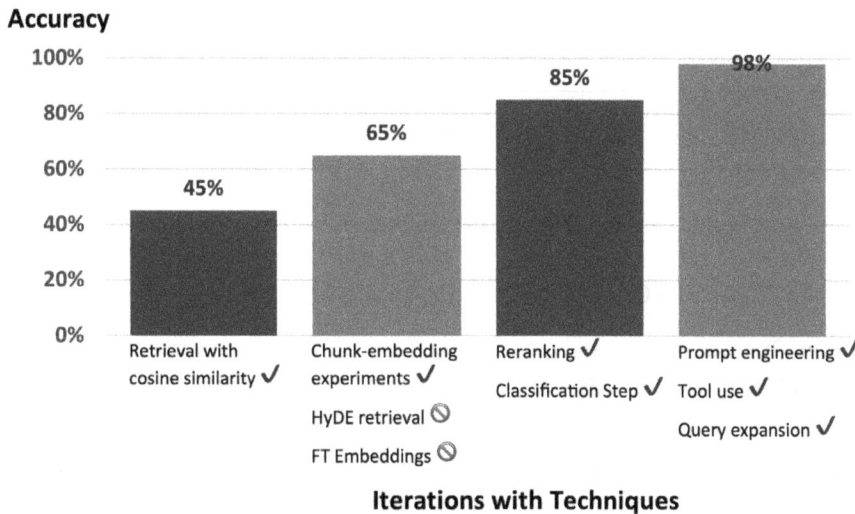

Figure 10.2 – Different techniques can succeed or fail to improve the experience

Video: Maximizing LLM performance techniques (https://youtu.be/ahnGLM-RC1Y)

The chart and video show successful (✓) and two failed (⊘) approaches. They worked through various solutions to find good fits. The video doesn't show the details of the changes they made. It is not explained what changes were made to raise accuracy. However, it is still a good case study that shows dramatic improvements in quality. With every effort, a testing process must be implemented to evaluate systematically. There needs to be more than user perception, surveys, and feedback.

Systematic testing processes

To evaluate any system, a few things are needed:

- Understand and be able to apply real-world usage and understand edge cases
- Be statistically confident in the amount of testing
- Be able to repeat or even automate the test with reliability and consistency
- Be able to make changes systematically to understand the results

With our human testing, methods are available to test with as few as five customers. Typically, human testers can go into the dozens, with only methods such as surveys intentionally hitting thousands. However, when it comes to LLMs, there are places where automation and scale are critical to success. OpenAI suggests in *Table 10.2* that larger sample sizes of test cases are are needed to evaluate LLMs to improve quality.

Difference to detect	The sample size needed for 95% confidence
30%	~10
10%	~100
3%	~1,000
1%	~10,000

Table 10.2 – Relationship between detection difference percentages to the number of test cases

The chart in *Table 10.2* is from OpenAI's testing strategy documentation.

Documentation: `OpenAI's view on testing strategy (https://platform.openai.com/docs/guides/prompt-engineering/strategy-test-changes-systematically)`

Confidence increases as the number of test case samples increases. This is done with simple assumptions and straightforward explanations. This should answer any questions about the scale of testing compared to the reliability expected in practice. Don't expect 10,000 examples of an answer to a single question; think broader. Test cases will be built into a collection over time.

Although this is easiest to understand for a conversational assistant, some thought has to be applied to other use cases, such as a recommender. Create test data and understand the variety of recommendations. Each focused recommendation will need a collection of test cases that understand the range of data elements. With five data elements in the recommendation, each with 3 to 30 possible values, a range of 30 to 150 combinations can result. Here is a recommendation for smart air filters:

```
It is recommended that the air filters in Bank A and C be changed by
April 3rd. Install the Aerterra model 2025m filters this time; they
are ordered and will arrive by March 20th. The energy cost increase
```

```
due to clogged filters is an additional $43.50 (3.2%) per day. Air
quality is 17% worse than expected due to smoke from regional forest
fires.
```

This recommendation for a large commercial building is based on air quality measures, the cost of energy to run the system, filter dirt capacity, filter type, and the filter's current dirtiness. There are thousands of possible combinations and dozens of recommendations in that case. The system must deal with all these entities, variables, and recommendations (see bolded items). So, thousands of test cases are needed, and results need to be validated to ensure that good advice is provided.

Each area of expertise is also multiplied by the number of languages supported, and the test matrix will grow. These must be automated. Breaking down test cases by subject helps one understand the scale of the problem better. Recall the Alligiance example in *Chapter 6, Gathering Data—Content is King*. There were 400 FAQs and many ways of asking each question. Five test cases for each FAQ would be over 2000 test cases.

Test cases can come from humans or an LLM. An example is how to "expense dinner at Joe's Eatery for a client dinner for $21.46"; when surveyed, 100 participants were asked how they would say it. They generated 244 different utterances, and 87% were unique. Here are ten examples in *Table 10.3* and some analysis of these potential test cases.

Test case utterance – human-generated	Test case considerations
$21.46 dinner at Joe's	Doesn't mention intent; what should an LLM do with this info? It might be confused with an appointment. The amount does help.
12/12/18, Dinner, 21.46	Notice ambivalent date format (MM/DD or DD/MM?), no intent to be clear this is an expense and no currency.
Create an expense for $21.46 at Joe's Eatery with my co-worker Lisa Jones and a client.	Intent, amount, location, and details. A good example.
dinner - client visit - $21.46 Canadian Joe's eatery	Points out Canadian dollars.
Dinner - Joe's Eatery - $21.46	Terse, but contains 3 of the 5 items needed. A good start.
Expense 21 dollars 46 cents Joe's Eatery for a client visit on October 3, 2018	Notice amount format is spoken out.
Hey, I used my corporate Amex for dinner. Didn't you see that?	Expects the credit card integration to find his expenses and connect it for expensing.
Reimburse me for $21.46 for a client visit with George Smith	Reasonable, doesn't mention dinner or a date.
Submit $21.46 for client visit	Lacking a lot of information, but reasonable.
Take care of my dinner expense of 21.46 dollars at Joe's for a client visit with the following people	Conversational and expects to provide more details to follow.

Table 10.3 – Potential test cases written by human subjects

Humans generated these, but an LLM model can also generate test cases. Let's compare what happens when prompting ChatGPT 3.5:

```
I will give you 5 pieces of information, generate test cases that may
or may not include them all, use a diverse human speech to communicate
this to a virtual agent, and start creating an expense in an expense
assistant. The most robust answer would look like this, "Expense a
meal receipt for $21.46 US dollars for a client meeting today with
one person at Joes Eatery". The least robust version would be "start
an expense." The information includes the following values for these
attributes: Expense Type: Meal, Amount: $21.46 US, Date 8/15/2024,
Location: Joe's Eatery, Purpose: Client Lunch. Generate 10 examples to
test an LLM that is great at understanding expenses.
```

ChatGPT 3.5's results are shown in *Table 10.4*. There are differences.

Test case utterance – OpenAI-generated	Testing considerations
"Record a meal expense of $21.46 for a client lunch at Joe's Eatery on August 15th, 2024."	All elements are included, uses the awkward "Record".
"Expense $21.46 for a meal with a client at Joe's Eatery on 8/15/24."	All elements, used US date format.
"Add a meal expense of $21.46 for lunch with a client at Joe's Eatery today."	All elements, used "today".
"Expense: Meal. Amount: $21.46. Date: August 15th, 2024. Location: Joe's Eatery. Purpose: Client Lunch."	Not conversational. No human ever did it like this.
"Register a $21.46 expense for a meal at Joe's Eatery on August 15, 2024, for a client lunch."	All elements, uses the awkward "Register".
"Document a meal expense of $21.46 for a client lunch at Joe's Eatery on 8/15/24."	All elements.
"Add an expense for a meal at Joe's Eatery, $21.46, for a client lunch on August 15th."	All elements.
"Expense: Meal. Amount: $21.46. Date: 15th August 2024. Location: Joe's Eatery. Purpose: Client Lunch."	Not conversational. No human ever did it like this.
"Record $21.46 spent on a meal with a client at Joe's Eatery on August 15, 2024."	All elements, almost identical to the preceding one.

Table 10.4 – Potential test cases written by ChatGPT 3.5

By comparing the results, a few conclusions can be made:

- Humans only included some of the information all of the time

- Humans went off-topic and expected to follow up with additional information

- ChatGPT ignored my request to vary what was included

- ChatGPT used words humans didn't use (such as Record and Register)

- ChatGPT didn't vary the cases very much, the prompt can be improved

This is a simple example. Humans are unpredictable and *human* when communicating, so consider that when allocating resources to create test cases. Time can be spent improving the prompt and forcing ChatGPT to give better and more varied results. Or a new model can provide a much more significant jump in quality. *Table 10.5* shows results from GPT 4o. This is better with the same prompt and no prompt engineering.

ChatGPT 4o responses	Testing considerations
Detailed: Please expense a meal receipt for $21.46 US dollars dated 8/15/2024 for a client lunch at Joe's Eatery.	Robust and complete
Detailed with variation: Create an expense report for a meal costing $21.46 USD from Joe's Eatery on August 15, 2024, for a client lunch.	Varied the date format.
Medium detail: Log a meal expense of $21.46 for a client meeting on 8/15/2024 at Joe's Eatery.	Log is a good choice.
Medium detail with different structure: Expense a client lunch meal for $21.46 from Joe's Eatery on 15th August 2024.	"lunch meal" is not something from humans.
Minimal detail: Start a meal expense for $21.46 at Joe's Eatery.	"Start" is a good trigger, but it is generic.
Minimal detail with date: Record a $21.46 meal expense for a client meeting on 8/15/2024.	"Record" is still not great.
Basic: Enter a meal expense for a client lunch costing $21.46.	Good and short.
Basic with location: Log a client lunch expense for $21.46 at Joe's Eatery.	A shorter version of the medium detail item.
Very minimal: Expense $21.46 for a client meal.	So short, it need follow-up.
Least robust: Start an expense.	Excellent, and common from humans.

Table 10.5 – Potential test cases written by ChatGPT 4o

I have also included their labeling, showcasing the types of use cases they provide. Monitor actual user inputs, making judging the value derived from generated data easier. It is hard for one individual to think like *everyone*, so don't try. Use monitoring resources and log analysis to get robust training data and test cases.

The problems get harder when the space gets more complex. Recall our graph from *Chapter 1, Recognizing the Power of Design in ChatGPT*, reshared as *Figure 10.3*.

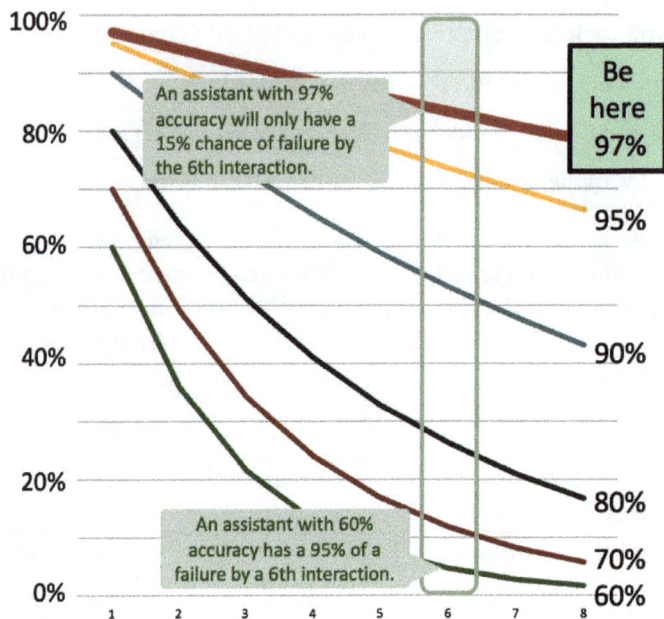

An assistant with 97% accuracy will only have a 15% chance of failure by the 6th interaction.

An assistant with 60% accuracy has a 95% of a failure by a 6th interaction.

Be here 97%

Figure 10.3 – The chance of failure increases at each turn

When an LLM is right 60% of the time, failure is 95% likely within six interactions. So, to move that bar up to 97% or greater, a lot of testing and work is needed. Generating the correct set of test cases helps monitor for issues. Because of monitoring, moving to a new model can be evaluated by applying the test cases to the latest model. It's okay when everything changes with a new model or version, as indicated by the differences between 3.5, 4o, and 4o-mini (it returned results similar to 4o). However, with mini being only 15% of the cost of 3.5, it would make sense to move if this was a real production system once verified against the test cases. There is no guarantee of backward compatibility with LLMs. The scale of testing efforts can easily reach 100's of thousands of use cases. An example of a testing matrix will make this clear.

Testing matrix approach

Because these projects impact human performance, the traditional **quality assurance** (**QA**) team must be set up to create an effective test matrix. They can develop automation and manage the process. Design owners can handle the examples and make sure the failures are documented.

With an LLM, monitoring will uncover conversations not initially covered by the test cases. It will happen. Just consider how to prioritize improvements for them, like any other issue. Once they are known, consider whether they are worth including in tests.

One can test each skill in isolation, and then when combining a collection of skills, some issues might arise. A **skill** is something the model can do. It could be connected to an inventory system, report production numbers, or schedule an appointment. These are skills. One approach is to build the base model only with content from one isolated area, thus allowing us to gauge its effectiveness in isolation. Do this for each area. Then, combine all (or sets of areas) retesting to understand better the overlap or complexities between the data ingested. The QA team will be busy setting up these test harnesses. Product people will be busy understanding the results. This example assumes a single model approach to handle all interactions, but we have provided multiple examples where a multi-modal approach is a better solution. Your choice. The concept of the types of tests still applies. It is valuable to have different kinds of test suites. We can review the types of tests for conversational interactions:

- **In-domain**: Questions the skill area should understand and be able to answer. They are the meat of the meal, the main course. Get these right:

```
What hours can I drop off recycling?
How long will the new battery last?
What is the commission schedule if I sign tomorrow?
```

- **Out-of-domain**: A collection of questions the skill would not be expected to understand based on lack of access to the necessary business data. Because they are not for *this* business. However, they are real questions and sound similar to the customer's needs, just in the wrong context. The general model might want to answer these. They are distractors to the in-domain questions. Hence, each area of interest will have its collection of out-of-domain items. These examples might sound too similar to an LLM because the structure and words seem familiar, such as (stock) orders, download (statements), account (details), transfer (money), check (status), and (order) history:

```
What is the fine for a late order?
Can I download my AMEX account history?
Can I transfer my car registration to someone else?
Can you check the service history for my car?
```

Recognize that these examples sound close but are unrelated to Alligiance. They sound confusing in the context of this bank. They are close, and a customer might not even know they can't ask this skill about these problems. Here is a little more about the concepts of out-of-domain understanding.

Article: Out of Domain Detection (https://www.elevait.de/blog/out-of-domain-detection)

- **Random**: Garbage and unrelated items that should not result in a helpful response. It could be from a stuck keyboard, poor speech-to-text, someone's phone in their pocket doing random stuff, or silly, irrelevant questions:

```
Qfhefkhjeksfdfdfd
Its dnot the goprome we spechtded
Is a dog a cat?
Are pickles part of the Dolphin family?
```

These are the same random questions for each skill.

- **Neighbor**: The collection of in-domain test cases from all *other* skills used to break the area of interest. Does the question from one area overlap and cause a different and wrong response? So, is this an issue when all this expertise is available in one **user interface (UI)**? It should be addressed if the LLM can't resolve this disambiguation issue. It might fix it like in this example:

```
                                        I need the weekly report


Did you mean the sales,
inventory, or staff report?


                                            The full sales report


Here is the sales report for all
regions for the week ending…
```

If only one feature contained a weekly report, there would be clarity. However, once multiple features include reports, see how some test cases from one solution area might impact others. Think in vector space. All of these questions asking for reports can be very similar. Hence, they are neighbors.

- **Language**: Consider test cases for specific languages. One starts by translating existing questions into other languages. Still, as mentioned in *Chapter 5, Defining the Desired Experience*, consider the unique needs of the language, cultures, and nuisances that would necessitate original content for that specific language. Assume at least 10% of test cases for a language will be unique.

Building the Matrix

So, what would a test matrix look like for each collection of use cases? A bank might have seven main business areas that require support. Based on the frequency of use, they will scale up tests for the big or complex areas. To test each area with the types of tests explained, it might look like *Table 10.6*.

Product Areas	In-domain	Out-of- domain	Random	Neighbors
Account Statements	303	1002	400	1,129
Bank Transfers	78	423	400	1,354
Account Setup	150	301	400	1,282
Deposits and Withdrawals	201	400	400	1,231
Training	50	375	400	1,382
Trading Stocks and Bonds	605	1320	400	827
Rewards	45	400	400	1,387
Sub-Totals (17,045)	1,432	4,221	2,800	8,592
Ten Language Test (*10)	14,320	42,210	28,000	85,920
Unique Language Tests (10%)	1,432	4,221	0	0
All Test	193,148			

Table 10.6 – Matrix for test cases to validate an extensive conversational AI

Trading Stocks and Bonds is the most critical area with the most test cases. The rewards program is the smallest. Notice the subtotaled tests per area. Then, assuming ten supported translations, the number of tests grows. The number of neighbor tests varies because it is the sub-total of all in-domain tests minus the number of in-domain tests for this category. The language tests are assuming cloning tests 1-to-1 per language for a deployment in ten foreign languages. There is an additional row for up to 10% of language tests that might be specific to the locality. As discussed earlier, it is ok to only go this far with testing in some languages regarding how much support to provide per language. This is why our English testing of 17,045 test cases grows to almost 200 K with language support. Here are some tips for scaling language tests:

- **Only translate currency indirectly**: "I took a 25$ Uber ride today," translated into Japanese, might change the currency symbol to Yen (¥). But 25¥ is not a meaningful value for a cab ride (about 25 cents in dollars or Euros), so training a system with too many wrong numbers might confuse the model.

- **Use localized values**: Recall our examples of Uber not being universal. A direct translation of Uber would be Uber in Japanese, but a better example is to use Go, the Japanese taxi-hailing app, for the training example.

- **Use idiomatic language**: Both test cases and sample data can benefit from how customers communicate. For example, localizing in the USA for dollars and training on bucks and cash but not necessarily using more esoteric slang such as moolah, coin, cheddar, dough, or greenbacks makes sense. Overtraining might cause unexpected consequences. To continue with our Japanese example, train on JPY (the code for the Yen), but the Japanese Yen doesn't have slang terms. Hence, one-to-one translation of idiomatic language is not expected.

- **The scale of tests correlates with quality**: For example, test whether 20% is sufficient for an infrequently used language. LLMs work better in some languages because the base model has more training data. Don't expect magic, especially when it comes to enterprise integrations. Translation steps are needed between APIs and responses.

Don't let the number of tests sound scary. For many enterprise applications, these numbers are low. I know one team now has about 500,000 tests, just in English. Automation and QA engineers will be busy maintaining and working with the data and product team to grow this set. Don't make tests to make tests. Use the tests to find gaps in the LLM's understanding. Recall that every change in the model or addition of new data or feature areas will change the quality of the solution. During this part of your LLM journey, **always be caring** (and feeding) (**ABC**, if a Glengarry Glen Ross reference is ok. Google it). The way to do this is to improve retrieval.

Improving retrieval

I reviewed a short video for the book. The explanations of the concepts taken from the RAGAs documentation are used in this video. They start using the data results to improve the overall solution. Greg Loughnane, PhD, slows down when he gets the good stuff after Chris Alexiuk whips through setting up the environment in the first five minutes of the video. Here is my summary of UX-related elements:

- Improve one metric at a time

- Focusing on retriever improvements helps with the generation

- Try a different retriever to get better context (@24 minutes) – expanding the capture of material before and after the matching context.

- See how context relevancy goes down as the size of the context window goes up; chunk size matters (this makes sense since there is more unrelated context in the denominator of the score) (@31 minutes)

- Other tools, such as LangChain and LlamaIndex, provide evaluation metrics

- Video: `Tutorial for improving retrieval (https://www.youtube.com/watch?v=mEv-2Xnb_Wk)`

It is all about improvements; data scientists can experiment with many options and variables to balance cost, performance, and quality. Learn to understand what they can change and how it impacts quality. Since this work is within the expertise of the data scientists, focus on understanding the quality of the results. There are other metrics of interest. Let's give a little context on those.

The wide range of LLM evaluation metrics

RAGAs was reviewed because it is popular, has a good set of metrics, and is consistent with what traditional deterministic models use. But it is not the only approach. We mentioned DeepEval, but there are many more. Some of these approaches have specific metrics that sound appealing. Each vendor can have its approach, so let me expose a few more metrics that can add value to enterprise solutions in *Table 10.7*.

Metric	Purpose	Applications	References
Rouge	To compute the effectiveness of auto-generated summaries	Books, technical Documentation, articles, marketing material, and so on	Article: `How to Use Rouge 2.0` (`https://kavita-ganesan.com/rouge2-usage-documentation/`)
Human Evaluation	To ensure user interaction quality	Conversational interactions	See *Chapter 3, Identifying Optimal Use Cases for ChatGPT*
Age-specific Suitability	To match reading or educational levels	Curriculum tutoring, coaching	Manual review by experts and content filtering tools
Toxicity Reduction	To maintain style and tone for public-facing and public-sector solutions	All generative output including recommendations	Toxicity and bias detection software, sentiment analysis
Perplexity	How probable a piece of generated text is based on its training	Content generation	Article: `Perplexity and Burstiness` (`https://guides.library.unlv.edu/c.php?g=1361336&p=10054021`)
Burstiness	The repetition of words or phrases in a document	Detect whether AI or a human wrote content	Product: `Originality AI` (`https://originality.ai/blog/chat-zero`)

Table 10.7 – Other evaluation metrics

Some of *Table 10.7* came from Aisera: `LLM Evaluation: Metrics and Benchmarking Performance` (`https://aisera.com/blog/llm-evaluation/#5-benchmarking-steps-for-a-better-evaluation-of-llm-performance`).

All metrics have a good reason for being collected and analyzed. Frameworks cost time and money to maintain, and some data-centric metrics have value because they impact costs. The number of LLM conversations per day or the number of tokens used helps with budgeting. Consider how to get value out of these metrics to aid in understanding customers' needs. Microsoft also has a few good articles about data-centric metrics. Here are some metrics with usability implications:

- **Concurrent users**: This can sometimes be correlated with performance (too many simultaneous users can slow some services, impacting service level and customer satisfaction). Recall that there is no such thing as a slow, *good* UI. In the case of chat, where human agents might be available for hand-offs, response time will be impacted if human agent availability doesn't match concurrent user metrics.

- **Token usage**: Token usage = cost. As mentioned in *Chapter 6, Gathering Data – Content is King*, look for opportunities to use less expensive models while maintaining or improving quality. This means lower customer costs or the ability to offer free or less costly tiers to serve a wider audience.

- **Filtering interventions**: If the process has guardrails for handling quality bias and inappropriateness, monitor the rate of these interventions and review them to decide whether there is anything to do about them. We mentioned that with enterprise software, typically for authorized authenticated users, abhorrent behaviors are rarely problematic, unlike in social media. It can happen, and blocking it is excellent; however, if work is needed to avoid these conditions or to adjust triggers to poorly timed interventions (e.g., being too strict for something that might be a perfectly reasonable request), look at these articles.

Article: How to Evaluate LLMs: A Complete Metric Framework (https://www.microsoft.com/en-us/research/group/experimentation-platform-exp/articles/how-to-evaluate-llms-a-complete-metric-framework/)

Article: Patterns of Trustworthy Experimentation: During-Experiment Stage (https://www.microsoft.com/en-us/research/group/experimentation-platform-exp/articles/patterns-of-trustworthy-experimentation-during-experiment-stage/)

These are the tools to understand a variety of data-driven metrics, but there is also the softer side, the human customer, in the loop. Although this data likely impacts our customers in various ways, it is helpful to understand a customer's perception. Time to explore some usability metrics to show how the system works in the customer's eyes.

Monitor with usability metrics

Earlier chapters explored ways to evaluate and find issues. This can be done by using a checklist, a particular set of rules expected for a UX, or a set of heuristics, a collection of guiding principles that, when applied correctly, helps expose issues quickly. The last chapter covered those methods, leaving a few more exciting metrics.

There are multiple ways to interpret how the system is doing. Since surveys, interviews, and other subjective metrics were covered, let's address measuring quality changes over time. This means measuring the fidelity of the experience by asking customers to answer specific questions, resulting in a **net promoter score** (NPS), a single-question survey, the more robust and time-consuming ten-question **software usability scale** (SUS) metric, or other forms of **customer satisfaction** (CSAT) surveys.

First, realize why it is helpful to measure usability with a score. It will only give a broad sense of how the system is performing. It won't uncover design flaws (unless there are open-ended follow-up questions). There are two good reasons to do this. First, it is easy to compare to other products and see if the solution meets expectations and exceeds what is found competitively, and second, to establish a baseline to redo these evaluations over time to measure progress. This means asking randomly for feedback, typically after interacting with the product. There will be variability because the same customer isn't always asked for feedback, so more data is needed to estimate accurately. For simple questions such as NPS, this is an easy ask. It takes a little more effort to code and request an SUS score. These can be supplemented at any time with more expository and open-ended questions.

Net Promoter Score (NPS)

NICE Satmetrix is the co-developer and owner of the NPS. It is well known as a simple benchmark of brand quality. Because of its simplicity, it has also been adapted for use in product analysis. Nominally, a business's customer is asked the following question:

```
On a scale of 0 to 10, how likely are you to recommend our business to
a friend or colleague?
```

This classic question is simple to ask and easy to calculate. It is a broad stroke, and it can be adapted:

```
On a scale of 0 to 10, how likely are you to recommend our
conversational assistant to a friend or colleague?
```

Adapt the wording to your product. The *Retently* website does an excellent job of explaining how to tweak this wording to make it work for your use case. I won't repeat this material here. Read the article if you are ready to deploy NPS.

Article: `NPS and how to modify the survey` (https://www.retently.com/blog/nps-survey-templates/)

Deploy a version of this question based on the product or service. A typical pattern is to ask for feedback for every 50th customer with a simple dialog box prompt, a side panel, or even inline, depending on the design. It is optional to be answered. Some customers won't participate. The same customer is typically not asked again; tag their account to avoid over-asking survey questions. Automate the aggregation of results (hopefully, this is a random sample and needs to be correlated with the release number) to generate the NPS. *Table 10.8* covers design patterns for gathering the NPS or SUS scores.

Use Case	Deployment Method
Conversational chat assistant	Inline, when a clear ending to the conversation is known
Conversational chat assistant	As a dialog box when a chat is closed
Conversational chat assistant	After a feedback flow (such as from a Give Feedback icon or label)
Recommender	At the end of a session
Recommender	At the end of any feedback process for evaluating the recommendation
Web or application UI	At the end of a transaction or significant flow
Phone tree	At the end of a transaction or significant flow (i.e., do have time for a one-question survey?)
SMS, Slack, Teams	After a specific amount of usage
Backend or hidden AI	After a specific amount of usage, or the end of a significant flow
Via Email	Post-purchase, interaction, or support
On receipts, feedback cards at points of sales, or with service	Via a QR code
Phone calls	Via human or automated follow-up at the end of a call (You will be transferred to answer a brief one-question survey about your experience today)
In real-life interactions	By asking the customer, and likely entering the score and any feedback manually

Table 10.8 – Approaches to deploying NPS

Recall our discussions about bias. It will be in the results. Sometimes, people are people pleasers, so they will not give good feedback when prompted in person. Their most recent interaction will color their input. If the interaction were a failure, it would impact the data. Ensure a good sample; for example, expect skewed results if feedback is gathered from one channel that only handles the closing of accounts.

Biases can be introduced in the way questions are asked. I have seen some examples that color code the number choices for a survey question, biasing the results. Use neutral colors for all options to reduce bias in survey questions. Some customers might hesitate to give a poor score because it is color-coded red, as shown in *Figure 10.4*. Use a generic Likert scale, as discussed previously. However, once they select the score, asking an optional follow-up question is okay. This will give context to their reasoning.

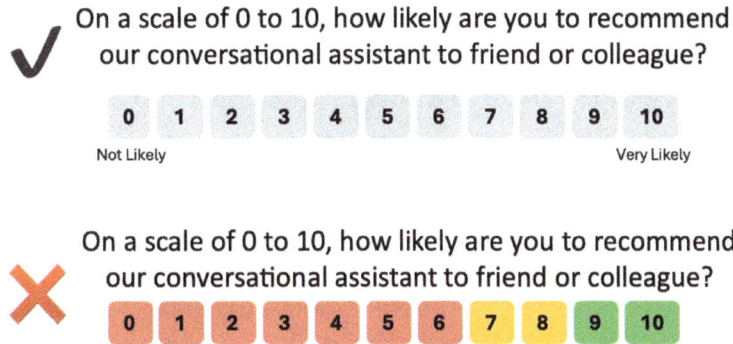

Figure 10.4 – Example of an NPS question – use neutral colors for the scale

The colors classify the results; don't use the colors in a customer survey. Scores in the red are detractors, people who would not be advocates and would likely turn away from using the product. People scoring 7 or 8 are passive; they won't get in the way but are not a big help. With little thought, they would switch channels or even products. That leaves 9s and 10s promoters, hence the name. Promoters who advocate for brands or products enthusiastically recommend the solution to others. The math is as follows:

$$\frac{Number\ of\ Promoters - Number\ of\ Detractors}{Number\ of\ Respondents} * 100$$

The range is from -100 to 100. It is fair to gauge your product against other brands. Simplestat reports that the average score in the enterprise space is 44. *Table 10.9* gives examples of brands and some samples of NPS. Even well-loved brands are in the 60s and 70s. Brands will generally have significantly higher scores than services or products. Consider that when deciding how good a product is scoring. Comparing releases, channels, or competitive products gives context to scores.

Category	Leader	Score
Airlines	Jet Blue	74
Hotels	Ritz Carlton	75
Internet Service	Fios (a Verizon band)	28
Home and Contents Insurance	USAA	78
Smartphones	Apple	60
Conversational Assistant	Oracle	55
SaaS Provider	Slack	55
SaaS Provider	Uber	37
SaaS Provider	Atlassian	48

Table 10.9 – Example NPSs for some products and services

The scores are gathered from these resources:

Article: `NPS scores for the table, Nice Source (https://info.nice.com/ rs/338-EJP-431/images/NICE-Satmetrix-infographic-2018-b2c-nps- benchmarks-050418.pdf)`

Article: `NPS scores for the table, CustomerGauge Source (https:// customergauge.com/benchmarks/blog/nps-saas-net-promoter-score- benchmarks)`

Article: `Typical NPS for a product (https://www.simplesat.io/understanding- feedback/net-promoter-score-benchmarks/)`

When at Oracle, we tested and shared an NPS with our customers for our Expense Assistant, and in its first release, it scored 55. This was much higher than previous solutions; it was considered a great win. Still, there was room for improvement, so various methods were used, including those in this book.

Use these example scores or explore other online posts for scores. It is just a general benchmark, but it is quick and easy. It doesn't guide where to go next and typically requires 100s of responses (roughly) to be valid. However, it also only takes a few seconds of a customer's time. There are also detractors concerning the validity of this approach. Just keep all that in mind when attempting to gather an NPS. I think it is worth the time. If the application is coded to gather insights, use NPS as one method in addition to a more robust method for feedback. All from the same UX approach!

Article: `Net Promoter Score (https://en.wikipedia.org/wiki/Net_promoter_ score)`

Consider the SUS for a robust metric that provides more insight.

SUS

Where NPS is a one-question form, the **Software Usability Scale (SUS)** is a 10-question survey focused on usability. It is a reasonable way to measure UX or conversational quality. It can be deployed for UIs with recommenders, but it is hard to tease out specific details about one element, like a recommendation, without additional questions. It is a well-understood 100-point scale, so it is easy to interpret the scores.

It uses the Likert scale from one to five, ranging from "Strongly Disagree" to "Strongly Agree." Imagine how these questions would feel when asked of them after using the application. The ten questions are as follows:

1. I think that I would like to use this system frequently.
2. I found the system unnecessarily complex.
3. I thought the system was easy to use.
4. I think that I would need the support of a technical person to be able to use this system.
5. I found the various functions in this system were well integrated.

6. I thought there was too much inconsistency in this system.

7. I would imagine that most people would learn to use this system very quickly.

8. I found the system very cumbersome to use.

9. I felt very confident using the system.

10. I needed to learn a lot of things before I could get going with this system.

The SUS questions follow one of the tricks explored in our survey discussion. Some questions are phrased in the positive (I feel very confident), and others are negative (I found the system very cumbersome) to avoid respondents answering on autopilot. Patrick Lawson points out in his blog post that these questions also have redundancy. This is common in surveys. The same question is asked slightly differently to create a more robust metric. Read more about the SUS in Patrick's background article:

Article: `How to SUS out usability scores` by Patrick Lawson (`https://www.thinkcompany.com/blog/how-to-sus-out-usability-scores/`)

The calculation to get the 1 to 100 score is specific to this model. Subtract one from each of the positively oriented items' scores, subtract the answers for the negatively oriented scores from five, and sum all these scores up. The sum multiplied by 2.5 gives the total. Compare the score to the outcomes in *Figure 10.5*, shared by Jeff Sauro from MeasuringU. He includes a comparison to NPS.

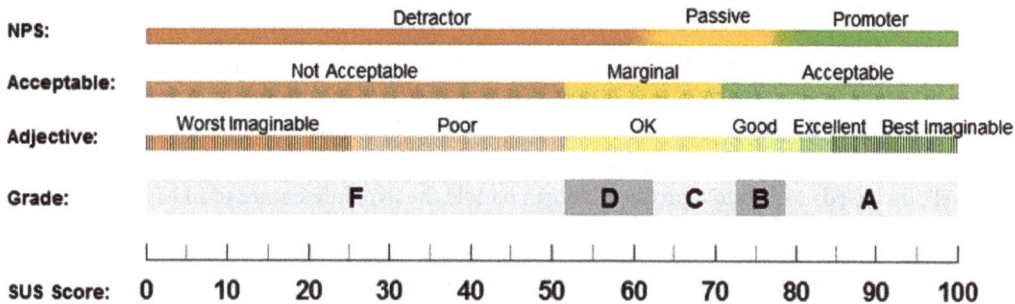

Figure 10.5 – How to interpret a SUS score

Jeff Sauro goes further into the technical scoring and implications of edits to the traditional wording. They point out that some of the work choices for the metric, which John Brooke developed in the 1980s, might seem old-fashioned. Read their post to learn more about the scoring and reliability of the metric. It is standing the test of time:

Article: `Is SUS Antiquated` (`https://measuringu.com/is-the-sus-too-antiquated/`)

Jeff breaks down the scores and details even further. I encourage exploring his posts, even if you have used the SUS for years:

Article: `Interpreting a SUS Score` (`https://measuringu.com/interpret-sus-score/`)

Go to the government website for details on implementing the SUS.

Article: `Software Usability Scale (https://www.usability.gov/how-to-and-tools/methods/system-usability-scale.html)`

The SUS provides more details to drill into than the NPS but doesn't provide granular information on what to fix. These methods can be followed up with open-ended questions to expose issues. This additional detail is valuable and actionable. The SUS score communicates quality over time and gives broad visibility to progress.

Refine with heuristic evaluation

Heuristic evaluation was covered in *Chapter 9, Guidelines and Heuristics*. The same approach, along with classifying and user scoring, from *Chapter 4, Scoring Stories*, can be used at this stage of the development process. All of that applies to monitoring and evaluation results. We can classify issues, score them to prioritize improvements, then apply refinements to the data, prompts, and model improvements (or even test against a new model to merge into the solution), or work on a new integration. You've got this! Apply what you know. This leads us to a discussion on handling the process of conversational AI within typical development organizations in *Chapter 11, Process*.

Summary

Tools on both ends were covered: analytic tools, such as RAGAs with metrics (sorry for the math!), and usability tools to monitor ongoing improvements from the customer's perspective. This and monitoring conversations will flow into refinements needed for RAG materials, instructions/prompts, fine-tuning, and swapping to newer models while improving integrations with services and APIs. To do this well, this needs to fit into a process that can handle the dynamic nature of LLMs.

The next chapter offers insights into implementing processes conducive to LLM development and integrating with engineering to improve LLM solutions continuously.

References

The links, book recommendations, and GitHub files in this chapter are posted on the reference page.

Web Page: `Chapter 10 References (https://uxdforai.com/references#C10)`

11
Process

Not all companies are built the same way. The development process is running full steam, and time and resources were given to do the design steps at the right time and scale. Perfect. It would be best if you had written this chapter. Enterprise-scale problems come with enterprise-sized issues. The one that makes my head spin the most is the time it takes to incorporate design solutions into a live environment. *Time is the enemy of good*. And with each evolution of technology, the time to market and the iterative cycle have to be faster. Today's user is not what it once was.

Provide the best solution as soon as possible or risk losing customers. Thus, even in an Agile world (let's include Scrum, Lean, and another iterative modern approach to software development when we mention Agile), its goal is to deliver quality promptly. Design and the efforts that go into it can cause headaches for engineering teams wanting to move quickly. Remember our discussion on cheap, fast, and good, choose two? Fast is required not to alienate customers, and customers no longer accept poor quality from enterprise solutions, so it must be good. Hence, it might not be cheap. Well, *cheap* here means expending resources to ensure that it is good. The previous chapters talked about how to make it good, but we need to apply these methods, practices, guidelines, and heuristics are applied in a way conducive to bringing quality work to market.

In this chapter, we'll discuss two areas to include conversational design in a software development organization:

- Incorporating design thinking into development
- Designing a content improvement life cycle

Incorporating design thinking into development

Most of this book is about using the steps needed to solve problems, considering user needs, refining problems, and creating and testing solutions. This is the essence of **design thinking**. You were probably doing this even without knowing its name. This hands-on approach emphasizes user empathy through the research methods covered in the book. However, in practical terms, a design thinker remembers the user goals at every process step. Because Agile aligns so well with the iterative nature of Generative AI, this chapter exposes tricks to make generative AI successful in an enterprise software development organization. This works for companies who deliver enterprise tools and those who use enterprise tools to work with a mass audience. It should be clear that both have compelling use cases and challenges with design thinking in an agile AI world.

This will sound harsh. If an organization is not using Agile or a form of iterative development, successful Generative AI will be out of reach. We live in a new, fast-paced world that needs a robust *iterative* approach to support it. I strongly encourage its consideration. There is a wealth of resources, from the Scaled Agile Framework discussed early on to the Agile Alliance and many more. AI tools are becoming fundamental in many development processes. Indeed, they should be deployed to customers as well. Tools for prompt engineering and fine-tuning were covered, but a wealth of AI add-ons can support backlog management, write better issues, or enrich development processes. Over 50 tools are marked as *AI apps* in the Atlassian Marketplace for Jira and Confluence. I can't speak to how good they are, but certainly, in the end, some will be timesavers and workflow boons.

Website: `AI marketplace at Atlassian (https://marketplace.atlassian.com/categories/artificial-intelligence)`

These tools are outside the scope of this book, but they help teams move towards more efficient processes. The key is that this is an ever-changing world. Adapt and learn how our content, customers, and AI work together, and then be able to make improvements to production quickly.

I'd be surprised if anyone needs to watch this video. If you are new to Agile, take one of the three-day Agile scrum classes. They can be life-changing, and there is a lot to learn.

Video: `Introduction to Agile (https://www.agilealliance.org/agile101/agile-basics/introduction-to-agile/)`

The software industry is able to transition to an AI-centric approach. The results from the 17th Annual State of Agile survey (2024 based on 2023 results), with 788 respondents, are compelling. Because the industry is overwhelmingly Agile, and all our recommendations can apply to any method, the focus is to reference Agile and how to address concerns along the way. 71% of surveyed companies use Agile. Some teams use other modern approaches, such as **DevOps (Development and Operations** working as one team), **Iterative**, **Lean** (which works only on the most critical items, with no multitasking), or **Spiral** (a four-phase repetitive process). Even the **waterfall** approach (a complete phase with no going back) is found in 28% of respondents.

There are fans of Agile, but there are critics as well. The biggest complaints concern the relentless pace and story points. Story points are for sizing work, but some argue for the well-understood measurement in hours. Every significant change in someone's lifestyle will be met with complaints. *Change is hard. Change is inevitable.* The pace of change in the technology world is constantly speeding up. And we don't need to argue about story points. Do research on your own. The chapter focuses on how design assists an LLM solution come to life. And there is value in aligning with Agile. The most significant value is the alignment around iterative design. It is mandatory with Generative AI to have an iterative life cycle. But doing it well has challenges. From the scoring discussion in *Chapter 4, Scoring Stories*, it is clear that design-related efforts have a place in Agile, but designer efforts do not fit nicely into a single sprint. Research and studies typically take time to write and run a plan but don't fit in a week's sprint. It turns out that is ok. There are answers. So, let me summarize some tips to make design a successful part of a Generative AI iterative *Agile* process. Even without a strict Agile method, the principles and coaching can be applied within most organizations. Even our first suggestion about finding a sponsor is broader than just Agile.

Find a sponsor

The Agile approach is perfect for an iterative AI care and feeding approach. Because of this, a business's issues with Agile become issues with AI adoption. Look at *Figure 11.1* from Digital AI's 17th annual survey of Agile.

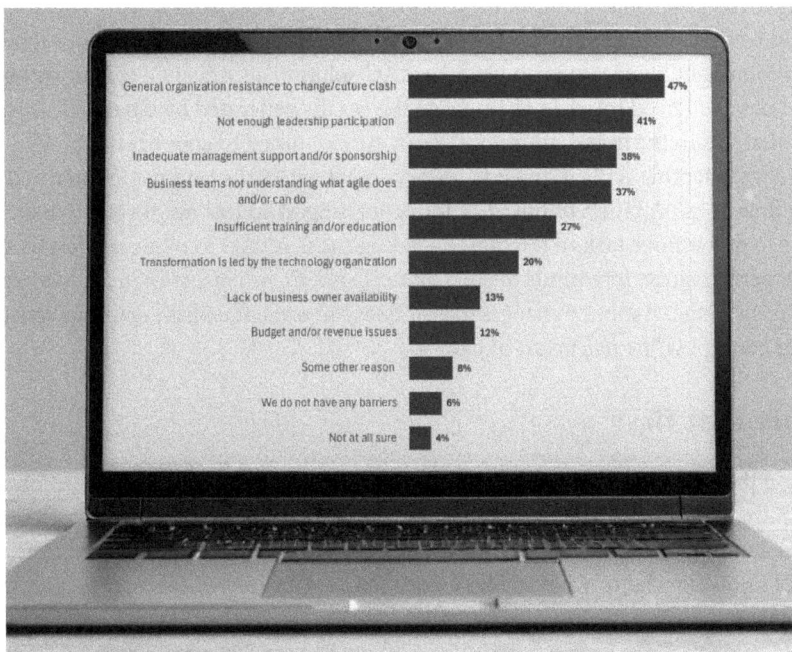

Figure 11.1 – The 17th Annual State of Agile survey – business issues with Agile (openart.ai)

The survey covers a lot of ground, but organizational support is critical for Agile success. By extension, a Generative AI solution needs organizational support. The three top issues (and others) relate to managerial issues. This is the latest survey.

Article: `State of Agile Report (https://digital.ai/resource-center/analyst-reports/state-of-agile-report/)`

The iterative care and feeding approach is typically radically different than a quarterly release schedule. It is a cultural shift. Even in an Agile organization, many ship products on a multiple of the Agile schedule. It is common to see two-week sprints but quarterly releases. And combining a slow cadence with a lack of leadership participation and support creates a problem. Get a sponsor who recognizes that the speed of change is critical to AI success and can help create a process conducive to rapid changes. Leadership support is essential as an organization recognizes the effort it takes to do AI well. Of course, people are critical, but tools help, too.

Find the right tools and integrate Generative AI

Unsurprisingly, it is a massive undertaking to keep up with this level of change, manage huge suites of test cases, and deal with an entire suite of knowledge. It doesn't involve a single tool but dozens, but at least start with a tool to manage the process and change. As discussed in *Chapter 4, Scoring Stories* using Agile tracking tools, incorporating scoring, and working from a WSJF backlog manages change. It would be challenging to find a large enterprise that doesn't have a knowledge management tool. Still, adaption might be necessary to feed the RAG solution accurately – that is, recognize that knowledge and data will need to adapt to improve the results generated by ChatGPT. It isn't magic but hard work that fills in the gaps. This might mean articles tuned to support the LLM, reworking existing knowledge, adapting APIs to include more context with calls, and using intermediate tools to improve the flow of information to the LLM for processing. And let's not forget to mention using smaller models to do bespoke tasks a few times, feeding those results to other models in a chain to improve overall performance. It reminds me of a quote: "*Stick to the plan, stick to the plan, stick to the plan.*" This means making last-minute changes to processes can sometimes have spurious consequences. Another similar concept is "*Be religious... at first.*"

Be religious... at first

There is a reason the Agile manifesto works for millions of people, even as they grumble about its shortcomings. Sure, 15% of respondents in the Agile survey were "*not at all satisfied*" with Agile, but given our understanding of what Agile, Scrum, and the shared concept, experience suggests most of this is about not knowing who or what to blame for failures. Consider these and the accompanying principles as they can be applied. Once you become an expert, *then* adapt. At an organization, I went into one Wednesday team meeting where the VP of a group of about 800 people dictated, "*On Monday, we will start doing Agile.*" Then they proceeded to choose which elements to do or not do (including deciding that self-organizing teams weren't something they could do, not knowing the meaning of "self-organizing"). Don't let gross incompetence derail quality goals. Try to understand the intent of the principles of Agile before discarding them. Let me quote a few of the 12 basic principles of Agile:

- *Build projects around motivated individuals. Give them the environment and support they need, and trust them to get the job done.*

- *The most efficient and effective method of conveying information to and within a development team is face-to-face conversation.*

- *Working software is the primary measure of progress.*

- *Agile processes promote sustainable development.*

Article: The Agile Manifesto (http://agilemanifesto.org/)

Our processes support building stuff: that is how we learn. Even in a world of video conferencing, people operate better with face-to-face meetings. And with the rapid pace of AI changes, sustainability needs to be built into our life cycle to avoid burning people out. Organizations love to reduce risks. Reduce risk by following the process religiously until there is confidence that the unknowns are known and change is required. Risks create unknowns, especially in solutions such as an LLM, where what it will say at every turn is not predictable.

Avoid "unknown unknowns"

This is so good it is worth repeating:

> *"Reports that say that something hasn't happened are always interesting to me, because as we know, there are known knowns; there are things we know we know. We also know there are known unknowns; that is to say, we know there are some things we do not know. But there are also unknown unknowns – the ones we don't know we don't know."*

> *– Donald Rumsfeld, US Secretary of Defense (February 12, 2002)*

Dig into understanding and learning. Research, uncover the root causes of issues, and learn *how to learn* in a Generative AI world. See why design is so well positioned to uncover issues; design methods are intended to solve these issues. And always try to evolve and improve.

Always evolve and improve

Retrospectives are not to be missed. They allow a content team to learn and give feedback on engineering issues that cause concern. Evolve to use better metrics (value and effort, not bugs) to drive improvement (measure the right things). Some metrics discussed in *Chapter 10, Monitoring and Evaluation*, are confusing and still evolving, but we have learned to understand how metrics change and work with them as they grow. Design thinking is fundamentally about listening and learning, so designers should be best positioned to help a team develop and improve. And improving includes knowing how to define what needs to be done. This means having requirements.

Agile does not mean "no requirements"

Specify what *others* need; don't over-specify what *you* need (inter-module documentation, yes!). Agile is right:

> *"Working software over comprehensive documentation… we value the items on the left more."*

This doesn't mean you don't create test cases; that is a requirement to take steps forward without going backward, but that might be all that is needed. Consider the goals and let data science and engineering work to meet those goals. If an analysis identifies that a system needs better context recall, define new realistic goals. No one works in isolation on these projects. Focus on what others need. It doesn't preclude writing a test plan for user research. Do what is required, but if you know what to do, consider whether it needs to be documented. This brings us to discussing others and how to work with a team.

Team composition and location matters

Face-to-face conversation is the most efficient and effective method of conveying information to and within a development team. This is a big challenge for remote-only companies or any company these days. In the State of Agile survey, 91% of respondents said their teams were fully remote. However, some successful companies recognize a partial solution – they have teams whose members work in similar time zones. Having writers in Romania, QA in China, and PMs and designers in California would be wrong. Work to create content teams that can work together in real time to help them stay aligned on the work they have in progress.

Manage Work in Progress (WIP) and technical debt

There is so much work and so little time. Even with Generative AI, only so much gets done. Focus resources on the most valuable items addressed in *Chapter 4, Scoring Stories*. Break up work into smaller pieces and deliver that value sooner. *Figure 11.2* explains this with an analogy of how homes are built.

Figure 11.2 – Do not juggle excessive work in progress (Photoshop AI)

It is straightforward – most people would rather be 100% done on 80% of the work than be 80% complete on 100% of the work. Work that is done has value. A realtor can sell four homes and then build on the next lot instead of having five homes that still are a work in progress and can't be sold. This is why builders have phases. They build a set of homes (for efficiency), sell them, and then build the next set of homes. This reduces risk and increases value, and customers get the benefits sooner. It's the same with content changes. Make incremental progress. This incremental progress usually means more extensive work is broken down into manageable pieces. *Figure 11.3* jokingly represents organizing to create manageable, well-understood sets of efforts to address in order.

Figure 11.3 – Break down large projects into manageable pieces (openart.ai)

The prioritization tools discussed and an Agile process (or any methodology with backlog grooming) help focus on priorities. Do work that is worth doing. This work should provide the most customer value.

Focus on customer value

There was an entire chapter on scoring user stories. By now, giving customers value by prioritizing the most essential work should be well understood. The more stories resonate with customers, the better their experience will be, and the more they will say good things about it (think Net Promoter Score), increasing users and usage. Then, invest to create even more value. One way to focus on customer value is to ensure the design process is part of the overall development process.

Incorporate the design process into the dev process

There are few enterprise companies with a design-first approach. Design-first means more than just doing the design first; everyone has to do some design *first*. It is more about leading with a design mentality to create customer success, not driving solutions because engineering thinks it is cool or marketing wants something flashy. It has to deliver a user need and drive value. Also, as mentioned earlier, it should be **functional, usable, necessary, and engaging** (**FUN-E** or *funny*, which is a good mnemonic). However, most product design people live in reality, so flexibility is required. The products built with Generative AI are not mature enough to survive on a three-to-six-month release schedule. Look for places to go faster and improve incrementally. In addition, designers, writers, linguists, and product managers will need to work with engineering, QA, and data science because many frameworks and processes need to be implemented. But do it with a design mindset so that new processes are also *FUN-E*.

When incorporating GenAI or ChatGPT into processes, consider a content sprint focused on knowledge bases (documents and data) and prompt engineering improvements. This Agile life cycle can coincide with traditional **development (dev)** sprints. The concept of a content sprint will be introduced. Here are some suggestions for key individuals with responsibilities around designing Generative AI solutions in an Agile organization:

- **Designers, writers, and linguists**:

  - Agile practices address design. However, most design efforts are done before entering a *dev* sprint. For example, user research won't happen inside a one-week development sprint. Agile, Scaled Agile, and many frameworks have accommodations for the work needed to enter a sprint. The design team should have *design* sprints and processes to create the results so the content team or the dev team can do its job during their sprints. The concept of a content sprint is explained in the next section.

  - Deliver the design when it is 90% ready before a team starts a dev sprint. Reserve content and prompt changes for the content sprint. Items that span content and development are placed in the development sprint backlog. This allows more technical changes to be thoroughly tested.

  - Use GenAI tools to build GenAI products. Like the tools discussed for classification, synthetic data, and fine-tuning, there is an ever-increasing number of GenAI tools that can make work more efficient. GenAI tools also help with general practices for professionals, editing, brainstorming, thinking through problems, word choices, translation issues, and understanding components and interaction design best practices. Even a generic foundational model like ChatGPT 4o should be considered your intern for getting tasks done better.

- **Designers**:

  - Get more involved in prioritization (to focus end-to-end and ensure that customer-critical items are completed in priority order).

  - Learn to identify the differences between prompt issues, fine-tuning gaps, knowledge or RAG issues, integration data issues, or context issues.

  - Actively resolve in-sprint issues promptly (i.e., to be responsive and to allocate time to be in the sprint). Being responsive to the dev team encourages them to become dependent on product people. This is important because designers (writers, PMs, and even linguists) are better positioned to judge these content issues.

- **Design owners**:

  - When entering the dev sprint, freeze the UI design that is expected for that sprint. Post the document, move additional significant changes to the next version, and only deal with the remaining 10% of detailed design changes within the sprint. Give the dev team design goals that are not moving.

- Content sprints have a schedule for changes; *when in doubt, throw it out*. Keep what is trusted to work, and move the rest to the next sprint. Content sprints are like working with live wires; don't be shocked at the results.

- **Developers and data scientists**:

  - Don't do it alone; this is a team sport.

  - Learn and listen to feedback; some changes have unintended consequences. As the team starts to get a feel for the content, the data, and the context size, prototype and experiment with other releases or different models, to improve quality. Rely on the content people, such as designers, PMs, and writers, who are better positioned to judge the appropriateness and significance of issues with language and conversational interaction.

- **Scrum master**:

  - Invite the designer to participate in *all* sprint scrum activities with active UI work. It is better to have teams focused on only UI and content work.

  - Design-related tasks in dev sprints should be assigned daily to ensure a fast turnaround for UI review/feedback/in-sprint design changes. No one wants surprises during the review at the end of the sprint.

- **Product owners**:

  - Have the requirements handed to the designer before starting work. Work with designers during design sprints to deliver the "end-to-end" story for later dev sprints. Some changes require testing and research and will not fit into the same sprint as the dev. Scaled Agile is okay with this.

  - Keep the runway clean and in sync with the expected design needs so that the design can be provided before the dev sprint starts. Even content sprint teams might need infrastructure or want to try new models, and this would fall back on the development sprints to provide.

  - Scale and scope stories correctly. Break up epics into smaller epics, stories, and tasks. Adopting a new LLM, for example, will span multiple phases and sprints to test, fine-tune, and validate.

  - Get content experts involved sooner rather than later. Before entering the sprint, agree on the language issues, support for product slang, outdated terms for products, discontinued product names, and synonyms for product names. Build this dictionary of terms to go beyond official terms. Remember, customers will use the language they are comfortable with.

- **Scrum masters/design owners**: Any designer can be on up to three teams, typically two. Designers should "mostly" work ahead of dev sprints. In the dev track sprints, effort should be limited since a story should enter a sprint at least 90% ready. Content designers won't have that luxury, as we are about to explain. They might have a 10% to 20% backlog from the week before, but this week's work will be figured out *during* the sprint, and almost all the work will be done within the content update sprint.

Now, it's time to explain the a content improvement life cycle and a content sprint.

Designing a content improvement life cycle

Chapter 2, Conducting Effective User Research, discussed extensively monitoring log files to improve the care and feeding life cycle. I can't emphasize strongly enough how important it is to understand the metrics associated with the solution's performance and adapt the solution to improve it. Although the data might differ for chat solutions versus recommender UIs, the process is the same for both. For behind-the-scenes work, figure out the feedback needed so that a continuous improvement process can be built.

Let's start with a vision of the content design team's week once the product is in production. *Figure 11.4* shows what a production content improvement team might do. Call the team a *production content team* or a *content/prompt sprint team*. Find a name that works so it is clear this is a slightly different beast from the more robust and complex *dev* sprint team.

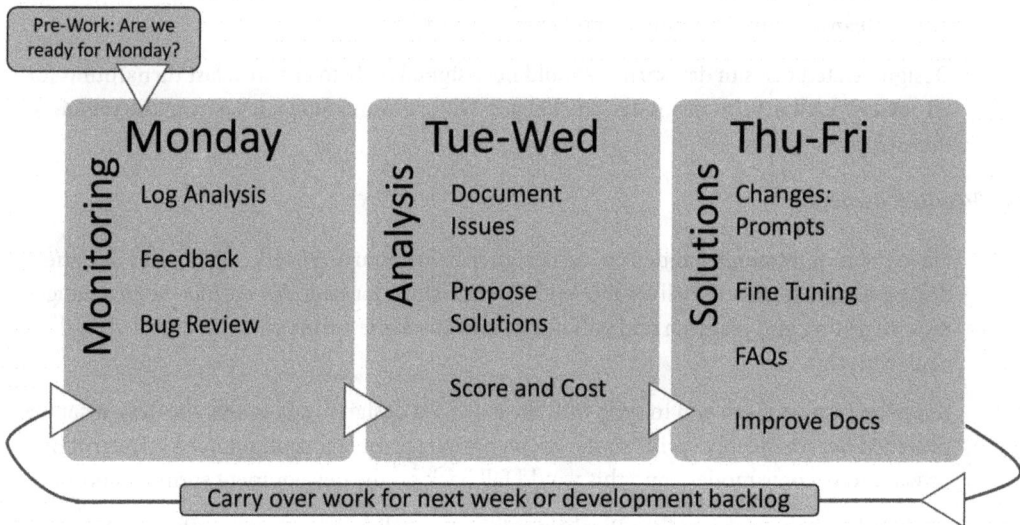

Pre-Work: Are we ready for Monday?

Monday	Tue-Wed	Thu-Fri
Monitoring	**Analysis**	**Solutions**
Log Analysis	Document Issues	Changes: Prompts
Feedback	Propose Solutions	Fine Tuning
Bug Review	Score and Cost	FAQs
		Improve Docs

Carry over work for next week or development backlog

Figure 11.4 – A weekly cadence for a content team's continuous improvement lifecycle

This life cycle may not be weekly early in the development phase; it could be a two- or three-week cadence. A week is excellent for an immature production system. It takes time to achieve this cadence. If an organization uses a two-week Agile development life cycle, start by matching that and improve the cadence over time. Alternatively, consider returning to a two-week approach when the product is more mature and changes slowly. It is not written in stone that this process needs to match the development cadence. There are some advantages, but it is worth considering the tradeoffs. This might be the place to break that rule of thumb.

Let's now describe in detail the typical work week for a conversational analysis team. This can also work for recommender solutions, even without the same inputs.

Inputs for conversational AIs

Logs are the primary input source for analysis but can be supplemented by surveys, bugs, or feedback. Since the logs represent what did happen, I would always check the logs against any anecdotal feedback. A friend told me the other day that the most significant missing feature from Google Sheets was the ability to drag and drop rows. And I said, "You can drag and drop columns, so why can't you drag and drop rows?" We went to Google Sheets, and drag and drop worked fine for rows, even on the mobile version. Unsurprisingly, this is an affordance issue, as discussed in the heuristic evaluation discussion in *Chapter 9, Guidelines and Heuristics*. It is hard to know what can or can't be done when there is no clear visual indicator of its support. Always check what people report. It is easy to be annoyed by something and only be confused by its real cause. Root-cause analysis can uncover the cause. Be willing to dig into the valid reason for a problem. Our user-centered design methods are based on the concepts of **root-cause analysis**. This book follows this approach (define the problem, gather data, identify possible causes, develop and deploy solutions, and then monitor and verify). You can apply this to any AI solution, including recommender UIs.

Inputs for recommender UIs

Recommender UIs don't have log data like our conversational cousins. Earlier chapters covered feedback methods, but downstream metrics can proxy for usage data. Access and monitor this data to judge value and quality. For example, suppose a recommendation engine suggests call follow-ups, email announcements, or product discounting. Analytics from the phone system, outbound email logs, or sales discount tools can be correlated with recommendations.

In places where direct correlations cannot be created, secondary methods, such as user feedback, can be used to determine a user's perception and gather written or verbal commentary. At least there is some data. Backend AIs are less lucky.

Inputs for backend AIs

Since there is no UI, consider this a more challenging task to gather feedback. There are two suggestions to consider. The first would be to build feedback mechanisms pointing to issues in the results. The customer will need to find out where the problem is. They will see something they don't like, feel it is a bit off, or identify it as wrong. Let them do that, then determine the root cause. However, it is possible to locate the issue if enough data is gathered with the feedback (such as their usage history, tag-specific datasets, or screenshots they would see).

The second relates to working out an AI monitoring system. Using ground truth examples, a secondary AI (not the same engine used for the solution) can check the work of the primary system. It might find results that don't match what is expected or appear more than an order of magnitude outside of what is typical. This feeds into a continuous improvement system that starts on the first day of the week. The same foundational model could be used if fine-tuned for specific criteria, such as the checkpoint testing cycle in *Figure 11.5*.

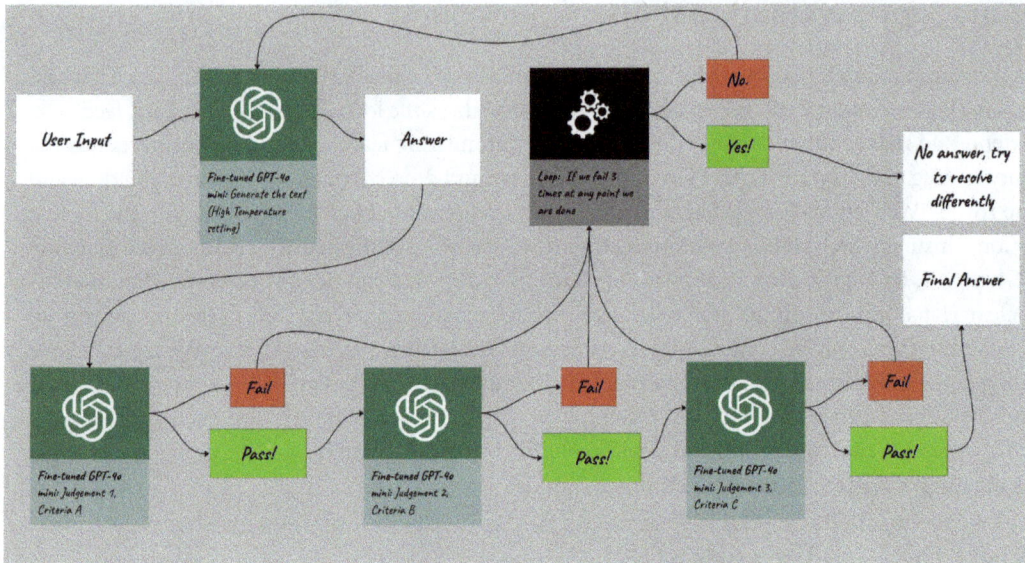

Figure 11.5 – A checkpoint system for validating results to share

This example allows us to fine-tune checkpoint models to pass judgments for specific criteria. It supports direct user input processes, or for a back-end solution, it would be a prompt generated on the customer's behalf.

Did the answer pass an ethical test (**Judgment 1**)? Do the facts match the product discussed (**Judgment 2**)? And in **Judgement 3**, does that answer fall within the company's capabilities? If the results are promising, pass it on to the customer. If the answer is not good, gather more context or have them ask questions differently. In any case, the results of each of these models support the primary model. Each of those failures creates an opportunity to learn. The output from models is part of what goes into the process for improvement each week. The idea is to work toward doing all of this on a weekly cadence. Let's look at the work needed on the first day of the process.

Monitoring Monday

Chapter 2, Conducting Effective User Research, covers log analysis. This is a valuable way of learning what is happening with customers. Manual tools were provided to start, and methods were shared to prioritize the outcomes. Focus on the analysis on Monday or whatever day of the week is the start.

> **Is Monday the first day of the week?**
>
> Maybe your country starts on Sunday. However, this won't give you the alliteration of *Monitoring Monday*. This alliteration only works in some languages; Israel, Japan, and Saudi Arabia can start with *Surveying Sunday*. Come up with names that work in your language. I used the translation spreadsheet from the earlier chapter to translate Monitoring Mondays. Only one of the 30 languages happens even to be close to alliteration. If you are a word geek, it was Dutch, with the translation of "Maandag monitoren." Consider making it fun and not feel like a structured march that people will oppose (a typical Agile complaint).

There is only so much a team can do to review logs, also look at customer surveys, feedback messages, bug reports, failures from intermediate steps of multi-step validation models, and sales or service feedback. Filter for the issues to triage them over the next few days. Use this time to work on projects that carried over from last week.

Analysis Tuesday (and Wednesday's workup)

Understand, classify, and chunk results. For extensive collections of common issues, don't write 20 stories; write one. This is where classification methods and third-party tools come into play, as discussed in *Chapter 2, Conducting Effective User Research,* and *Chapter 3, Identifying Optimal Use Cases for ChatGPT*. Work with the dev team to understand and segment issues into those that can be fixed this week and those that need devs' help for improvements. Of course, investing in tools to make improvements is better without investing in development resources at every turn. This is the adage – *if you give someone a fish, you feed them for a day. If you teach them to fish, you feed them for a lifetime*. Dev can teach us to fish by providing tools that allow us to make improvements. Solutions like editing knowledge, prompt changes, fine-tuning examples, adding test cases, or other content changes can be considered through analysis. The analysis will also create work for researchers and ample items for dev teams to address around fine-tuning, new models, better reflection or model-to-model integrations, and API work. All this is done to decide what the content team can deliver in the next few days and plan for what the dev team can provide in future sprints.

Treatment Thursday and fault-finding Friday

I would love to have these on different days, but figuring out solutions and testing to identify faults is an iterative process. Some readers will be laughing at the attempt to do all of this in a week. I get it; it is the goal, work towards it. Fixing and tests should be done over two days. Conflicts will be expected. Unsurprisingly, asking a model to do one thing can break something else, or two new solutions can conflict. And prompts can then get unwieldy. All hands are on deck for solutions: writing or editing knowledge, updating prompts, improving fine-tuning, or incorporating a new API (and as discussed, this is likely to come from a previous sprint, as APIs don't magically appear). Automation is needed to quickly rerun test cases and analyze results and metrics for this to work.

Only some things can be fixed in days or hours, but some improvements are well within reason. Rapid improvements can happen for *content*:

- Adjusting instructions and prompts

- Training to improve grammar or to include company-specific language and initialisms

- Editing knowledge

- Creating knowledge (it's hard to develop technical documentation quickly, but maybe FAQs are within reason)

- Adding learning examples

- Identifying context elements to include in instructions

- Fine-tuning

- Adding test cases

This can be a challenge. The content team can't make one change, test it and then make more changes when a testing run can take minutes or hours. It is also hard to debug issues with a batch of new changes. It is a trade-off. Sometimes, failed tests can prevent releasing a solution, while some solutions fail to materialize and require more or different work. Work can carry over to next week or need to be referred to the dev team. Let's explore carry-over work in more detail.

What doesn't fit into a week is still important

> *"If I had one hour to save the world, I would spend 55 minutes defining the problem and only five minutes finding the solution."*

> *–Albert Einstein (not really)*

The "famous" quote from Albert Einstein isn't from him (Google knows!), but the sentiment is still valid. Not all solutions are easy. However, understanding the solution is generally manageable if time is spent understanding the problems.

When identifying problems that clearly won't fit into a week or a sprint, evaluating their value is still important, as is getting a cost estimate and moving it into a backlog based on the Weighted Shortest Job First approach. By doing this on an ongoing basis, there will be stories from previous weeks (or months) now being integrated into this week's changes. Thus, these changes will impact the solution as tests are built. In true Agile fashion, if something is too big and disruptive to the process, it should not be picked up alongside other changes; use the traditional development and testing lifecycle for the more significant, more technically complex stories.

For example, new test cases will be needed as a new API or integration becomes available. Because these former out-of-domain cases become in-domain solutions, this can cause new conflicts. For example, in an internal business solution that supports sales, inventory, and team staffing reports, new APIs that support marketing reports need to work with existing requests for other *reports*. Add training data to assist the system in differentiating these new marketing reports, rerun fine-tuning, update prompts, and refine FAQs to account for this feature. The previous out-of-domain test cases must move to valid in-domain test cases.

Figure 11.6 shows adapting content sprint cadence based on the product's stage. This is offered to encourage thinking about the level of change at each phase of product development and the value of that change.

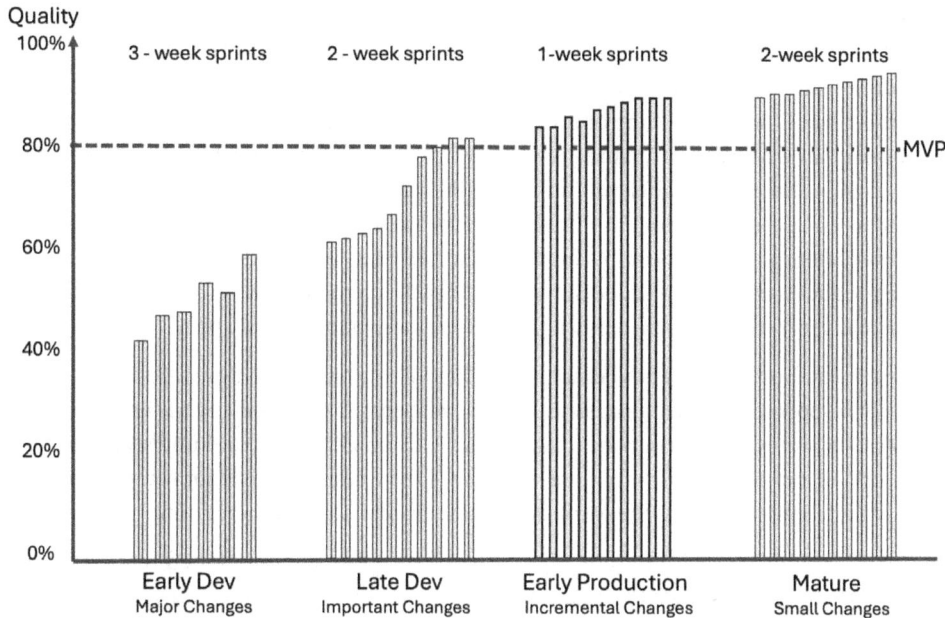

Figure 11.6 – Cadence can change based on the stage of the product

This suggests that the content sprint schedule adapts over time. This is a radical idea for Agile, but it might provide some value to consider. Since a team is self-organizing, they can make this schedule decision. Most enterprises do not believe a team can make this call, as they think self-organizing is a no-no. But there should be some flexibility. This isn't suggesting changing often, just as the process matures. That is the Agile way.

Notice a generic *quality* metric for the vertical axis. This could be any measure or all measures. Any specific metric can see dips when building and testing. In *Chapter 10, Monitoring and Evaluation*, not all of OpenAI's efforts to improve their enterprise use case were successful. Remember that even when a product is in production, efforts in dev will eventually come online and give opportunities for more significant improvements in quality than just what can be done from incremental improvements from a content perspective. Only so much can be done by editing prompts, improving tuning, adding new APIs, and working with our knowledge integration.

Let's break down each phase and the approach:

- **Early development**: There are no checks and balances at this stage; try to test changes regularly. Testing frameworks and metrics will still be in development and need to be available to ensure progress (try not to take one step forward and two steps back). Manual testing will be typical. Significant changes will come from major investments, and it is okay if the sprint cycle for content doesn't reflect development. Writing content, editing and reviewing updates, and bringing testing online takes time.

- **Later development**: As releases approach, changes will likely tighten and be on the same schedule as dev Agile teams. Two weeks are typical; some teams will be already on a one-week cadence. However, be aware that the Agile sprint length doesn't always equate to a release schedule, which is okay. Only when some of these significant investments pay off will there be major jumps in quality. **Minimum Viable Product** (**MVP**) expectations are shown on the preceding chart at **80%**. An astute reader will notice we talked about 97% in other graphs. 97% is the *goal*; 80% is a release minimum. It would be prudent to show some repeated ability to maintain that level before going live. Some changes will hurt some quality measures. Significant changes in the product scope could reduce the quality of each interaction but increase the number of interactions that can be handled; learn to live with these bumps in the road.

- **Early production**: The goal is a complete update cycle in one week. By categorizing changes, there can be opportunities to make faster changes with some adjustments, while more significant changes, such as access to a new API, take longer. Customers need to feel improvements. I use ChatGPT all day and see improvements monthly. One week is possible for some changes. Do what is best, but always with the focus on providing value to customers sooner.

- **Mature production**: As the solution matures, less value will be gained by incremental improvements. It might take the same effort to raise quality 10% from 80% to 90% as it does to go up 1% from 95% to 96%. This diminishing return is an indication to reconsider how time is spent. The cadence of changes can decrease because the value of those changes is small. It may be time to reduce the story points the team can do in a content team and allocate those resources to new opportunities or the next-generation solution. Of course, this might be years, so cross that bridge when you come to it.

No one is perfect, and we don't need perfection. The cost of aiming for perfection can drive you out of business. Recall the failed Ford Pinto example from *Chapter 6, Gathering Data—Content is King*. It would be best to make choices; don't make bad choices. Use guardrails for high-value, high-risk answers. Value ethics, reduction in bias, and accuracy over feature expansion.

> *"Perfection is not attainable. But if we chase perfection, we can catch excellence."*
>
> *– Vince Lombardi*

As new models emerge, quality changes significantly, so consider how to re-adapt to a faster cadence. It is not uncommon to see quality hiccups as a function of new models, as with ChatGPT releases in the last two years, it is primarily a solid upward trajectory.

The takeaway is to implement practices to improve quality quickly, keeping customers on the happy path. It is a labor of love to monitor regularly, gather feedback, learn about issues, classify to bundle improvements, research issues, test enhancements, and to start the process over again. Here are a few tips to overcome typical challenges:

- Monitoring log files can be monotonous, so time-box the work. Reviewing 1,000 rows of logs is time-consuming and tedious. No one should do this every day, so do it for only a day. Alternatively, assign review and summarization to an LLM.

- Over time, monitoring will get faster. It is much faster (10x) to review 1,000 log files with 50 issues than with 200 issues since a human can quickly review conversations without issues but needs time to understand a single problem. ChatGPT models can also do some of this work, such as seeing what is available for automation, but do some monitoring manually to understand customers.

- Fixing only some things in one content iteration or sprint is okay. Some items are not fixed by prompt engineering, editing knowledge, or creating new content. Sometimes, new APIs are needed; in some organizations, they can take a long time, if ever. Classifying issues helps because if multiple instances of the same problem appear, it is possible to advocate to address them. Put in placeholders to acknowledge a customer's needs. A placeholder may include a link to a third-party resource or an explanation if nothing else is available. Good AI classifiers are available; a model can be created to classify issues.

- Enlist resources. Designers, analysts, content authors, product managers, engineers, quality assurance, and data scientists can understand and diagnose issues. Use these resources collaboratively to learn how to address a situation best. There could be more than one approach to something. Communicate, collaborate, and conclude based on the information gathered. Also, there can be more than one solution. A short-term solution with prompt engineering might be better addressed by improving the knowledge base in the long term. Implement one solution while waiting for the latter.

- Release management won't want to make changes quickly. This is a tough one. Work within a system or find advocates to change it. It would be best to have a sponsor. There are different kinds of changes; work with a sponsor on the types of changes release management would approve. Build trust and expand this list over time. Give examples. If an online news organization had an article that said, "*Each new plane for the military will cost 200 million dollars,*" but it was 20 million dollars, no one would expect that *content* change to take a week to fix. And much of what we are working with is content. Consider classifying changes so that some refinements can be done weekly while others require more effort, testing, and time. All should be run through automated and manual testing as part of the process, so maybe that is enough.

- Garbage in, garbage out. Don't let lousy knowledge throw off the RAG model. Don't feed the model only with generative data; use that sparingly if it can be helped. Inject new, high-value content and look for opportunities for improvement. No one will suggest editing or improving all articles every week; pick the right battles. Let data lead the charge. The articles with the most hits take priority. Use analytics to tailor editorial investment. Recognize and test for interactions between similar documents for different products. Look for places where databases have good contextual data that can add value to prompts, giving customers more customized and accurate experiences.

- Invest time in the process of handling supervised fine-tuning. With technical material, add well-known samples to the training. When inspecting logs, recognize that some human variety in interaction should be better understood. Consider whether FAQs already serve this purpose. Remember how prompts should look for these examples. A wealth of fine-tuning datasets are excellent examples of scope and scale, even if generic data is not of value to enterprise use cases. Here is one good example, with 200,000 robust prompts.

Article: `Ultrachat 200,000-row data set (https://huggingface.co/ datasets/HuggingFaceH4/ultrachat_200k)`

Build custom collections of tuned material, test cases, and templates to provide high-quality RAG results. The Generative AI revolution is not without cost. Over time, model costs will be less significant than overall life cycle costs.

Now, let's conclude with some key takeaways from the chapter.

Conclusion

> *"Without continual growth and progress, such words as improvement, achievement, and success have no meaning."*
>
> *– Benjamin Franklin*

When it comes to conversational solutions, no truer words were spoken. Software changes decades ago were measured in years and then, months or quarters, and now, weeks and days. This is expected, needed, and sometimes valuable. With processes to support continuous improvement and even delivery, it is a win-win. Tools and methods will improve quickly, allowing more time to tackle new projects, while customers will benefit from better solutions sooner.

Up to this point, we have explored the entire life cycle of applying UX design thinking and processes to create exceptional enterprise ChatGPT solutions. Sometimes, we went too deep and left you with some homework. The technology is moving so fast; the hope is that 96% of the concepts discussed in this book will apply to any conversational AI platform or collection of tools. Models improve, and the tools become more robust. The methods and practices should still work until we to the mythical general-purpose artificial intelligence (GPAI), strong AI, or artificial general intelligence (AGI). This book will be relevant until an LLM doesn't need a significant investment in creating and tuning, with a healthy dose of care and feeding. As such, we will wrap up our journey in the next chapter to send you on your way to conquer the next AI project.

References

The links, book recommendations, and GitHub files in this chapter are posted on the reference page.

Web page: Chapter 11 References (https://uxdforai.com/references#C11)

12
Conclusion

Any product person who gets this far in the book has probably accepted that the generative AI space is far more complex than simply asking ChatGPT a question and expecting a good answer. What it can do is impressive, but it's frustrating when it doesn't do what is expected. The key here is whether it was *known* that it could have been more helpful. In many places, customers will need to learn or recognize if ChatGPT, in its air of confidence, was lying. That is why product people are the gatekeepers for quality.

This final chapter will take a different approach at a straight summary, focusing on what we did through other views of the world and a few ways of thinking about a generative AI journey:

- Applying learnings to the new frontier
- Double-checking what feels right
- Build processes that fit the solution
- Wrapping up the journey

Applying learnings to the new frontier

The book followed a traditional life cycle approach to applying user experience design methods, practices, heuristics, and guidance to the emerging field of generative AI. Even though the chapters align with an order for a development process, many of these tools and techniques can be used throughout the generative AI journey. User research is done up-front to unlock value for future features and is again deployed once solutions allow validation. We hope to instill *the care and feeding approach is an absolute* with generative solutions. One would think it is an absolute with many software products, but the industry needs to be more enamored with the approach. There are issues with companies picking up releases that change rapidly. Sometimes, it impacts training and is disruptive to services. Generative AI can support a continuous improvement process, just like Google search, which is updated without a decision by the user. Since generative AI follows the conversational AI boom in the early 2010s, there is already a broad understanding of why most AI-infused solutions failed. They were thrown over the wall to customers, and a care and feeding practice of constant improvement was not followed. Now, generative intelligence fills in some of the technical shortcomings of previous chatbot solutions and can apply most of that learning, such as the teachings in this book, to create more effective generative AI products and services.

In hindsight, most of the time was spent on interactive conversational AI products while ensuring the building blocks provided could work for all recommendation-type solutions. Whether prioritizing support tickets or sales leads, scoring a marketing lead, establishing a user reputation, suggesting the next best step, filtering content based on harmful or undesirable content, or classifying sentiment to encourage good agent behavior, this book should have the guidance needed to solve those problems. The backend services that AI might improve will only need some of our approaches, but now an arsenal of techniques are available to apply to any situation.

To be fair to the process, don't apply every method to every project. Use some simple math from *Chapter 4, Scoring Stories*, to calculate that every tool and technique has its return on investment. Pick and choose the battles with development or the powers that be. Use resources wisely. Do what feels right.

Double-checking what feels right

I attended an AI conference and listened to a talk on AI readiness. Given our discussion about when to deploy a solution and all the steps to make an AI solution successful, it would be helpful to see how others view what we have covered. Ann Maya presented the six tenets of AI readiness that Michael Bachman of **Boomi** (https://boomi.com) created in partnership with 250 C-level executives.

I bring up their tenets because they summarize much of our approach slightly differently and align nicely with putting in place a process (from *Chapter 10, Monitoring and Evaluation*) that can enable these tenets. It is always valuable to compare methods and practices to how others work. It might feel less innovative if what you do aligns with what others are doing. However, there is also value for our collective to be on the right journey in an emerging space. Even with alignment, there are places to learn about missing pieces to the puzzle that provide value. I show the results they shared in *Figure 12.1*. It summarizes the six tenets of AI readiness.

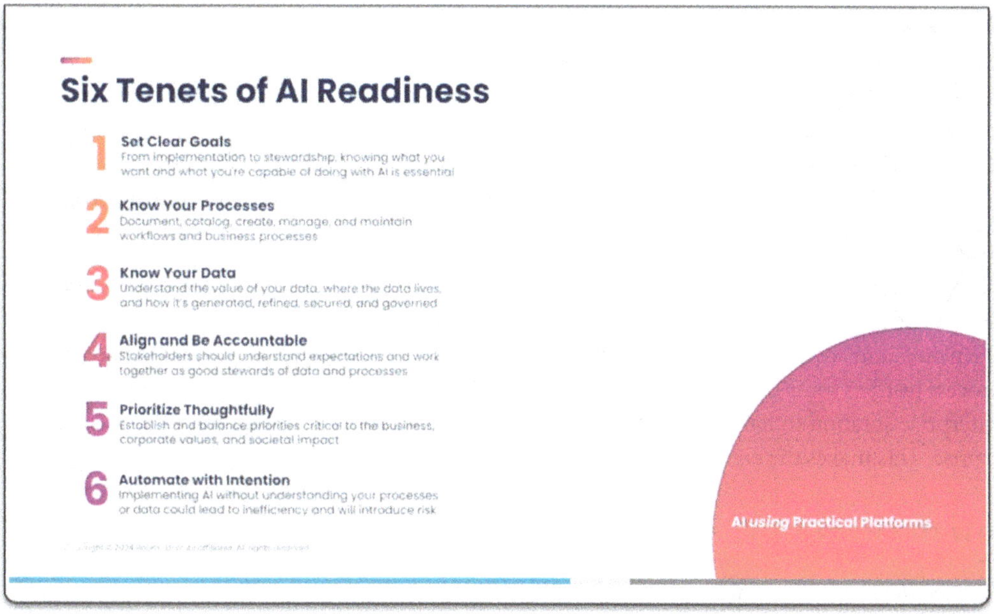

Content provided by **Boomi**

Figure 12.1 – The six tenets of AI readiness.

Given our UX slant on AI readiness, these tenets don't cover everything, but there is enough overlap that it was nice to find this resource right before I finished this book's draft. Let me align each goal and their explanation with our efforts in this book. These are not in any specific order.

Set clear goals

From implementation to stewardship, knowing what you want and what you're capable of doing with AI is essential.

We defined this by discussing use cases (*Chapter 3, Identifying Optimal Use Cases for ChatGPT*) in a way that made sense for customers' needs. Stay within reach. Focus on where AI can provide the most value and improve quality, then thoughtfully branch out. Companies will only do some of it, which is why there is a wealth of tools and support for implementation. The question is whether the best use cases are tackled to justify the cost.

Know your processes

Document, catalog, create, manage, and maintain workflows and business processes.

The Wove case study in *Chapter 6, Gathering Data – Content is King*, touched on aligning goals and core expertise with what AI tools can do. A design process can reevaluate if steps in a current process are needed. Maintaining the workflow was the key to *Chapter 7, Prompt Engineering*, and fine-tuning in the next chapter. Take the time to get these solutions right. They do not just magically work out of the box. Some current processes are ripe for re-imagineering. Occasionally, a team can ripe up a process and start from a clean slate. Mapping where to go might give a path forward. Sometimes, value comes from simple incremental improvements. You now know how to score these values, and using Agile, you can determine their costs. **Weighted Shortest Job First** is the decision-making process.

Consider the different models that can improve the processes. Examples like Hub and Spoke or Chaining came up a few times. As prices come down, it becomes practical and expected that multiple models can be used to solve a single problem. Different models, with different capabilities, intelligence, performance, and costs, can be adapted for unique use cases. It is not one size fits all. The generative AI process justifies some consideration, as covered in *Chapter 11, Process*. As it is not uncommon for an enterprise solution to have dozens or hundreds of parts, expect a proliferation of models in the enterprise. This makes the entire care and feeding life cycle more critical.

Know the data

Understand the value of data, where the data lives, and how it's generated, refined, secured, and governed.

Chapter 6, Gathering Data – Content is King, addressed managing data for bespoke ChatGPT solutions. Without enterprise data, there is no added value. It is essential to build a RAG process that keeps business data secure and presents it in the right light.

The enterprise data is the entire reason for this book. It contains the collective wisdom for the company, the products and services, and the relationship with its customers. This is why it is so valuable. It also means being aware of inherent biases. It is not just about possible cultural biases; it could be just how the knowledge speaks about products or the terms it uses. Decide how to manage that data, deal with the governance of the data, and adapt it to allow AI to take full advantage of the corpus.

Align and be accountable

Stakeholders should understand expectations and work together as good stewards of data and processes.

We often discussed being good stewards of the model and data. Handling bias, controlling for PII, and managing integrations are critical to success. As all UX partners are, being customer advocates sets the right expectations. Using the methods from the first few chapters, evaluating how it is going by following *Chapter 9, Guidelines and Heuristics*, and measuring the results reviewed in *Chapter 10, Monitoring and Evaluation*, aligns stakeholders to common goals.

If ethical considerations come into play, be accountable for the outcome. The approaches discussed around checking results, using prompt engineering to talk in the best tone, and returning the most accurate answers are critical. Some mistakes by big players will start to make the news. In Hollywood, the saying goes, *"There is no such thing as bad publicity."* Keep tight control over use cases that expose ethical, medical, societal, or community standards issues. Don't be in the news for the wrong reasons.

Prioritize thoughtfully

Establish and balance priorities critical to the business, corporate values, and societal impact.

Chapter 4, Scoring Stories, covered our scoring method for prioritizing the backlog using a customer-centric approach. As UX professionals, we focus on the value provided to customers, but as a proxy, we sometimes align with business, corporate, and societal goals. Let's assume that the customer's goals help achieve those priorities. It reminds me of a book commonly read in business school and one required by my professors. The key was the ability to answer three questions: *when to change, what to change to*, and *how to cause a change*. I hope you see the similarity in our approach to figuring out what to work on, how to work on it, and what the outcome should be. The book, *The Goal*, made for fine reading; it was compelling enough to read in one night.

Book: The Goal: A Process of Ongoing Improvement (https://amzn.to/3WJqWHM)

Not surprisingly, this novel's protagonist adopts an iterative approach. However, it is based in a factory setting (the software industry wasn't the powerhouse it is today when this book was written). I still encourage it for a thoughtful and enjoyable read.

Automate with intention

Implementing AI without understanding processes or data can lead to inefficiencies and introduce risk.

Chapter 5, Defining the Desired Experience, covered this by figuring out thoughtful applications for our AI solutions. Make good decisions so the AI solutions provide the most value. Don't over-AI. Pick your battles. Create value.

There are many ways to approach AI solutions, ChatGPT integrations, and problem-solving with generative AI. However, all the best approaches do so thoughtfully, focusing on creating **Functional, Usable, Needed, and Effective** (**FUN-E**) experiences. Let's do it with the right goals that follow a path to build a process that fits customer needs.

Building processes that fit the solution

We talk about bringing the solution to the customer, not forcing the customer to go to the solution. This takes many forms and many paths that follow this mantra:

- Talk in the customer's language and refrained from forcing them to learn how we speak.

- Use prompt engineering and fine-tuning to align with the customer, and do not expect the customer to align with us.

- Review how to deploy a solution, bringing AI where they need it. It might be on Slack, Teams, Discord, or a web channel. This could also be via several messaging apps, a web service, and a desktop or mobile app, all with slightly different capabilities.

To accommodate these solutions, processes must allow these engagements. Solutions need to span multiple channels, each with slightly different needs. The customer might have different expectations of those channels. Text, images, and links on a simple messaging platform might support more robust interactions for the same goals when deployed on a rich web experience. Also, only some steps will come from generative AI solutions. The best successes will come by breaking down use cases into the steps that are best solved with specific solutions. *Sometimes*, that will include an LLM. Process flows were discussed many times in the book, and we were able to take a peek into Wove's process, which included generic and fine-tuned models. Consider the examples of models used in chains, such as the hub and spoke model to route to specific tasks and LLMs, branching to support different tasks, or looping through the same model with different prompts. These are just a few available variations. There is no one correct model. Each will require design effort, testing, validation, and a care and feeding process. Each is an application suite itself. This is unsurprising, as many enterprise solutions might comprise dozens or hundreds of systems and services.

As product owners, designers, and thought leaders, we should *bring the solution to the customer, not the customer to the solution*. This will help us wrap up our journey.

Wrapping up the journey

There was a lot of ground to cover; will these skills, methods, practices, heuristics, and recommendations that will remain true three to five years from now? Sure, OpenAI will change its Playground, and a few links will stop working (the online references will be updated), but the concepts will live on for a while. The goal for OpenAI is to eliminate the considerable work needed with prompt engineering; when that happens, recycle the book and retire to the beach. However, for the foreseeable future, there is work to do.

And maybe you noticed we didn't once discuss ChatGPT's *enterprise* tools. Of course, they have new tools coming soon, and a few select companies would be early adopters. What OpenAI can provide is likely valuable and even necessary but not sufficient. Some of this is already part of any plan, like the public APIs, multi-factor authentication, batch APIs for offline workload, encryption, and single sign-on (SSO). The ecosystem still has plenty of maturities left for OpenAI, third parties, and enterprise companies to address.

Note

The concepts in this book will outlive the models used in these examples. Apply this learning to the latest models available.

Since the goal is to have up-to-date knowledge and wisdom, the book's online references will be monitored to ensure they continue working. Even during the writing process, references had to be retired. Our collective goal is to move our customers toward wisdom. It is insufficient to regurgitate or push only data or information to consumers of our generative AI solutions. Our knowledge and ability to use generative AI should allow us to impart *wisdom* in solutions. I don't recall where I found the pyramid in *Figure 12.2*.

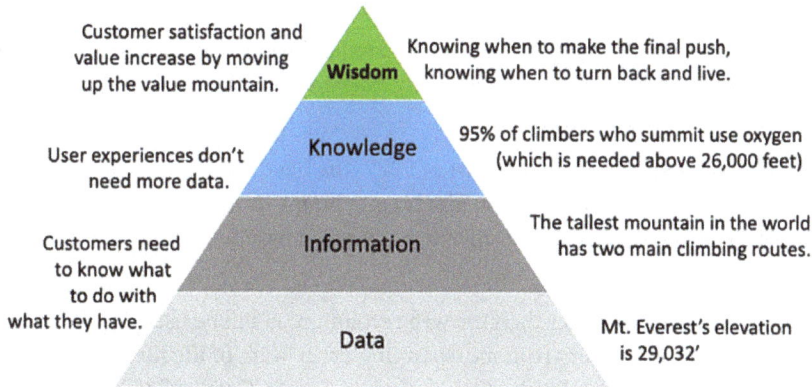

Figure 12.2 – Wisdom is the goal

Let's take an example anyone can appreciate. There is plenty of *data* on how tall mountains are across the world. However, it is intriguing for a climber to know that Mt. Everest is the tallest mountain in the world and has two climbing routes. *Knowledge* from previous climbers' experience (and some basic physiology) tells us that summiting typically requires oxygen (bottled and carried by the incredible Sherpas). However, there is a more important reason why those who summit use Sherpas. If they listen to their sage advice (*wisdom*), they are in the best position to succeed. If they don't accept the Sherpas' collective wisdom, tragedy can strike. This slide is shared so it can be adapted to any use case.

GitHub: `Wisdom Pyramid slide (https://github.com/PacktPublishing/UX-for-Enterprise-ChatGPT-Solutions/blob/main/Chapter12-WisdomPyramid.pptx)`

Figure out how to give customers the wisdom they need to be successful. Don't just give them the same data in a new way. There is value to unlock by giving wise advice. Also, don't be constrained by AI. Wisdom can be provided without AI.

Time to consider what is next for AI. Rapid changes are everywhere. With all the effort to research this book, it is fun to think about what is next for generative AI in enterprise solutions. What will the next few years bring prior to Skynet? Time to pontificate about the future:

- Open-source models and portable models that protect PII will become more plentiful. The downside is that with that protection comes the inability to learn how to improve them. Opt-in programs that allow data sharing for "improvement" purposes will be the norm. Always have this option: pay or credit customers for this privilege. Learning is fundamental.

- Prompt engineering will become easier as more effort is placed into the foundation models for broader intelligence. Writers and designers will have the limelight here, but its criticality will diminish over time.

- As enterprise applications move deep into AI, second-tier services will blossom. Secondary use cases and less critical business needs can leverage these now-standard services to AI-enable more of the business life cycle. Yes, it will be just about everywhere.

- Failures will still exist and continue to be news. Lawsuits beyond data scraping will be expected. However, just like errors from human agents, errors and subsequent lawsuits will be reduced as hallucinations are minimized.

- Once we pass the hype cycle by the start of 2027, the need to explain that AI is behind the scenes will stop, as it will be the norm. The biggest jumps in AI improvements for this cycle have already happened. By 2025, we will enter a phase of incremental improvements until the next major cycle by 2031.

- The AI revolution will be bigger than the web revolution. It will be used in every part of daily life and is as common as a smartphone, spanning every walk of life far beyond the web and the Internet. AI can benefit billions of less fortunate people. Crop yields, water management, more affordable healthcare, supply chain optimization, disaster predictions, and education are all critical and will align with governments that fund these areas.

- Although it will eliminate some jobs, it will also generate more valuable opportunities for the next generation. This is not a stretch, given the speed of adoption. Every technological revolution, from the wheel to the internet, requires society to adapt. We are still smarter, in many ways, than AI.

- In my lifetime, we will finally see home robots that can serve as general-purpose assistants (cook, clean, and fold laundry). The robot revolution, with vision analysis and LLMs, will come together in the next ten years. I suspect the same can be said for some hard day labor roles, as machine vision being used to pick fruit exists and will get better when robots are controlled.

I encourage engaging with me and the whole AI community. There is so much for UX people of all types to learn about what our data science and engineering friends are doing, and while learning from them, we can impart a wealth of value to their practices, hopefully with some of the wisdom imparted in this book. Connect, share, and engage.

References

The links, book recommendations, and GitHub files in this chapter are posted on the reference page.

Web Page: Chapter 12 References (https://uxdforai.com/references#C12)

Index

‹packt›

Other Books You May Enjoy

If you enjoyed this book, you may be interested in these other books by Packt:

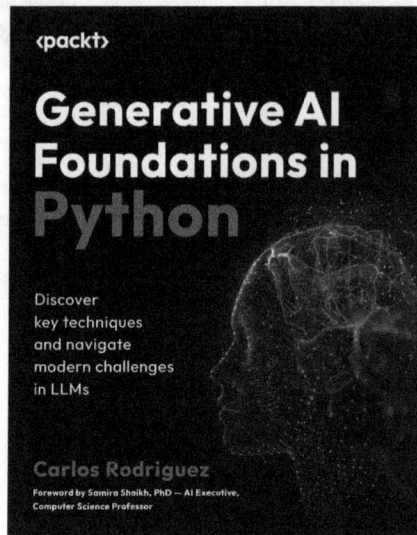

Generative AI Foundations in Python

Carlos Rodriguez

ISBN: 978-1-83546-082-5

- Discover the fundamentals of GenAI and its foundations in NLP

- Dissect foundational generative architectures including GANs, transformers, and diffusion models

- Find out how to fine-tune LLMs for specific NLP tasks

- Understand transfer learning and fine-tuning to facilitate domain adaptation, including fields such as finance

- Explore prompt engineering, including in-context learning, templatization, and rationalization through chain-of-thought and RAG

- Implement responsible practices with generative LLMs to minimize bias, toxicity, and other harmful outputs

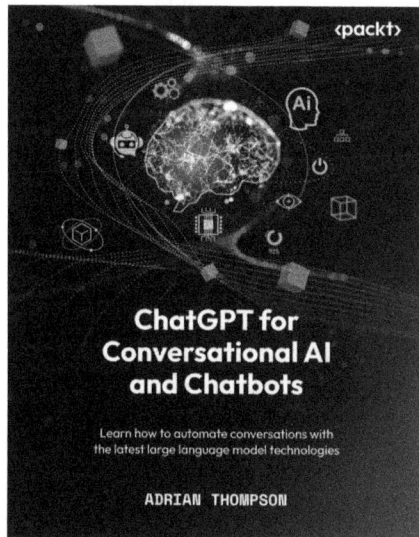

ChatGPT for Conversational AI and Chatbots

Adrian Thompson

ISBN: 978-1-80512-953-0

- Gain a solid understanding of ChatGPT and its capabilities and limitations
- Understand how to use ChatGPT for conversation design
- Discover how to use advanced LangChain techniques, such as prompting, memory, agents, chains, vector stores, and tools
- Create a ChatGPT chatbot that can answer questions about your own data
- Develop a chatbot powered by ChatGPT API
- Explore the future of conversational AI, LLMs, and ChatGPT alternatives

Packt is searching for authors like you

If you're interested in becoming an author for Packt, please visit `authors.packtpub.com` and apply today. We have worked with thousands of developers and tech professionals, just like you, to help them share their insight with the global tech community. You can make a general application, apply for a specific hot topic that we are recruiting an author for, or submit your own idea.

Share Your Thoughts

Now you've finished *UX for Enterprise ChatGPT Solutions*, we'd love to hear your thoughts! Scan the QR code below to go straight to the Amazon review page for this book and share your feedback or leave a review on the site that you purchased it from.

https://packt.link/r/1-835-46119-0

Your review is important to us and the tech community and will help us make sure we're delivering excellent quality content.

Download a free PDF copy of this book

Thanks for purchasing this book!

Do you like to read on the go but are unable to carry your print books everywhere?

Is your eBook purchase not compatible with the device of your choice?

Don't worry, now with every Packt book you get a DRM-free PDF version of that book at no cost.

Read anywhere, any place, on any device. Search, copy, and paste code from your favorite technical books directly into your application.

The perks don't stop there, you can get exclusive access to discounts, newsletters, and great free content in your inbox daily

Follow these simple steps to get the benefits:

1. Scan the QR code or visit the link below

https://packt.link/free-ebook/978-1-83546-119-8

2. Submit your proof of purchase
3. That's it! We'll send your free PDF and other benefits to your email directly